Published by NY Research Press,
23 West, 55th Street, Suite 816,
New York, NY 10019, USA
www.nyresearchpress.com

Encyclopedia of Thermodynamics: Selected Topics
Volume V
Edited by Barney Tyler

International Standard Book Number: 978-1-63238-176-7 (Hardback)

Printed in the United States of America.

Encyclopedia of Thermodynamics: Selected Topics

Volume V

Edited by **Barney Tyler**

NY RESEARCH PRESS

New York

Encyclopedia of Thermodynamics: Selected Topics

Volume V

Contents

Preface

This book was inspired by the evolution of our times; to answer the curiosity of inquisitive minds. Many developments have occurred across the globe in the recent past which has transformed the progress in the field.

An integrated analysis of the science of thermodynamics is provided in this book. Thermodynamics is one of the most interesting parts of physical chemistry which has significantly contributed to the modern science. Being focused on a broad spectrum of applications of thermodynamics, this book accumulates a series of contributions made by veteran scientists and researchers from across the planet. This book evaluates selected topics in thermodynamics with regard to reaction rates, Planck's law, statistical, chlorination, and theory of simple liquids. The book will appeal to students engaged in post-graduate courses. It will also serve as a good source of reference for those researchers who are interested in thermodynamics.

This book was developed from a mere concept to drafts to chapters and finally compiled together as a complete text to benefit the readers across all nations. To ensure the quality of the content we instilled two significant steps in our procedure. The first was to appoint an editorial team that would verify the data and statistics provided in the book and also select the most appropriate and valuable contributions from the plentiful contributions we received from authors worldwide. The next step was to appoint an expert of the topic as the Editor-in-Chief, who would head the project and finally make the necessary amendments and modifications to make the text reader-friendly. I was then commissioned to examine all the material to present the topics in the most comprehensible and productive format.

I would like to take this opportunity to thank all the contributing authors who were supportive enough to contribute their time and knowledge to this project. I also wish to convey my regards to my family who have been extremely supportive during the entire project.

Editor

1

Thermodynamics Approach in the Adsorption of Heavy Metals

Mohammed A. Al-Anber
Industrial Inorganic Chemistry, Department of Chemical Science, Faculty of Science Mu´tah University, P.O. Jordan

1. Introduction

Adsorption is the term that used to describe the metallic or organic materials attaching to an solid adsorbent in low, medium and high coverage as shown in Figure 1. Wherein, the solid is called adsorbent, the metal ions to being adsorbed called adsorptive, and while bounded to the solid surfaces called adsorbate. In principle adsorption can occur at any solid fluid interface, for examples: (i) gas-solid interface (as in the adsorption of a CO_2 on activated carbon); and (ii) liquid-solid interface (as in the adsorption of an organic or heavy metal ions pollutant on activated carbon).

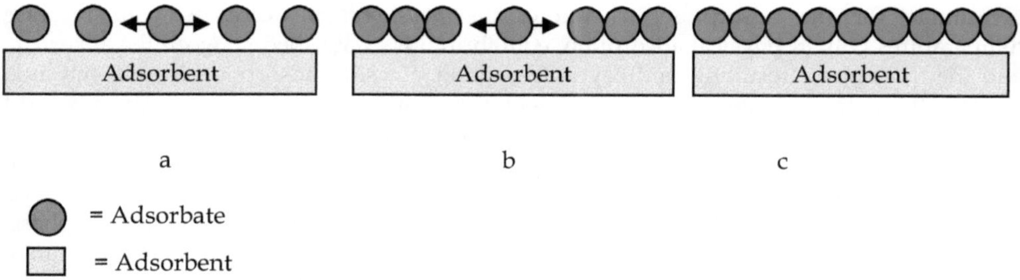

Fig. 1. a) Low coverage (no attraction between adsorbate metal ion/ molecules, high mobility, disordered). b) Medium coverage (attraction between adsorbate metal ion / molecules, reduced mobility, disordered). c) High coverage (strong attraction between adsorbate atoms/ molecules, no mobility, highly ordered).

we talk about *Chemisorption* and/ or *Physisorption* processes. However, *the chemisorption* is a *chemical adsorption* in which the adsorption caused by the formation of chemical bonds between the surface of solids (adsorbent) and heavy metals (adsorbate). Therefore, the energy of chemisorption is considered like chemical reactions. It may be exothermic or endothermic processes ranging from very small to very large energy magnitudes. The elementary step in chemisorption often involves large activation energy (*activated adsorption*). This means that the true equilibrium may be achieved slowly. In addition, high temperatures is favored for this type of adsorption, it increases with the increase of temperatures. For example, materials that contain silica aluminates or calcium oxide such as silica sand, kaolinite, bauxite, limestone, and aluminum oxide, were used as sorbents to capture heavy metals at high temperatures. The adsorption efficiency of the sorbents are influenced by operating temperature [2-7]. Usually, the removal of the chemisorbed species from the surface may be possible only under extreme conditions of temperature or high vacuum, or by some suitable chemical treatment of the surface. In deed, the chemisorption process depends on the surface area [8]. It too increases with an increase of surface area because the adsorbed molecules are linked to the surface by valence bonds. Normally, the chemi-adsorbed material forms a layer over the surface, which is only one chemisorbed molecule thick, *i.e.* they will usually occupy certain *adsorption sites* on the surface, and the molecules are not considered free to move from one surface site to another [9]. When the surface is covered by the monomolecular layer (monolayer adsorption), the capacity of the adsorbent is essentially exhausted. In addition, this type of adsorption is irreversible [10], wherein the chemical nature of the adsorbent(s) may be altered by the surface dissociation or reaction in which the original species cannot be recovered *via* desorption process [11]. In general, the adsorption isotherms indicated two distinct types of adsorption—reversible (composed of both physisorption and weak chemisorption) and irreversible (strongly chemisorbed) [10-11].

On the other hand *Physisorption* is a physical adsorption involving intermolecular forces (Van der Waals forces), which do not involve a significant change in the electronic orbital patterns of the species [12]. The energy of interaction between the adsorbate and adsorbent has the same order of magnitude as, but is usually greater than the energy of condensation of the adsorptive. Therefore, no activation energy is needed. In this case, low temperature is favourable for the adsorption. Therefore, the *physisorption* decreases with increase temperatures [13]. In physical adsorption, equilibrium is established between the adsorbate and the fluid phase resulting multilayer adsorption. Physical adsorption is relatively non specific due to the operation of weak forces of attraction between molecules. The adsorbed molecule is not affixed to a particular site on the solid surface, but is free to move about over the surface. Physical adsorption is generally is reversible in nature; i.e., with a decrease in concentration the material is desorbed to the same extent that it was originally adsorbed [14]. In this case, the adsorbed species are chemically identical with those in the fluid phase, so that the chemical nature of the fluid is not altered by adsorption and subsequent desorption; as result, it is not specific in nature. In addition, the adsorbed material may condense and form several superimposed layers on the surface of the adsorbent [15].

Some times, both physisorption and chemisorption may occur on the surface at the same time, a layer of molecules may be physically adsorbed on a top of an underlying chemisorbed layer [16].

In summary, based on the different reversibility and specific of physical and chemical adsorption processes, thermal desorption of the adsorbed sorbent could provide important information for the study of adsorption mechanism.

2. Factors affecting adsorption

In general, the adsorption reaction is known to proceed through the following three steps [16]:

1. Transfer of adsorbate from bulk solution to adsorbent surface, which is usuallymentioned as diffusion.
2. Migration of adsorbate (Fe^{3+} ion, where its ionic radius = 0.064 nm) into pores.
3. Interaction of Fe^{3+} ion with available sites on the interior surface of pores.

From the previous studies, it was shown that the rate-determining step for the adsorption of Fe^{3+} ion is step (3).

Normally, the driving force for the adsorption process is the concentration difference between the adsorbate in the solution at any time and the adsorbate in the solution at equilibrium ($C-C_e$) [17]. but, there are some important factors affecting adsorption, such as the factors affecting the adsorption of Fe^{3+} ions in the aqueous solution:

2.1 Surface area of adsorbent

Larger surface area imply a greater adsorption capacity, for example, carbon and activated carbon [18].

2.2 Particle size of adsorbent

Smaller particle sizes reduce internal diffusion and mass transfer limitation to penetrate of the adsorbate inside the adsorbent (i.e., equilibrium is more easily achieved and nearly full adsorption capability can be attained). Figure 2 represents the removal efficiency Fe^{3+} ions by natural zeolite through three different particle sizes (45, 125 and 250 μm). It can be observed that the maximum adsorption efficiency is achieved with particle size 45 μm. This is due to the most of the internal surface of such particles might be utilized for the adsorption. The smaller particle size gives higher adsorption rates, in which the Fe^{3+} ion has short path to transfer inside zeolite pores structure of the small particle size [19].

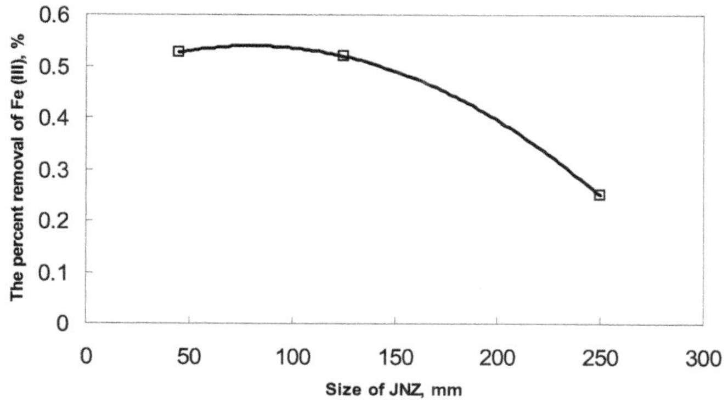

Fig. 2. Percent removal of Fe^{3+} ions (1000 ppm) *vs.* natural zeolite particle size: 1 g adsorbent/ 50 ml Fe^{3+} ion solution, initial pH of 1% HNO_3, and 300 rpm.

2.3 Contact time or residence time

The longer residence time means the more complete the adsorption will be. Therefore, the required contact time for sorption to be completed is important to give insight into a sorption process. This also provides an information on the minimum time required for considerable adsorption to take place, and also the possible diffusion control mechanism between the adsorbate, for example Fe^{3+} ions, as it moves from the bulk solution towards the adsorbent surface, for example natural zeolite [19].

For example, the effect of contact time on sorption of Fe^{3+} ions is shown in Figure 3. At the initial stage, the rate of removal of Fe^{3+} ions using natural quartz (NQ) and natural bentonite (NB) is higher with uncontrolled rate. The initial faster rate may be due to the availability of the uncovered surface area of the adsorbent such as NQ and NB initially [20]. This is because the adsorption kinetics depends on: (i) the surface area of the adsorbent, (ii) the nature and concentration of the surface groups (active sites), which are responsible for interaction with the Fe^{3+} ions. Therefore, the adsorption mechanism on both adsorbent has uncontrolled rate during the first 10 minutes, where the maximum adsorption is achieved. Afterward, at the later stages, there is no influence for increasing the contact time. This is due to the decreased or lesser number of active sites. Similar results have been shown in our results using zeolite and olive cake as well as other reported in literatures for the removal of dyes, organic acids and metal ions by various adsorbents [19, 21].

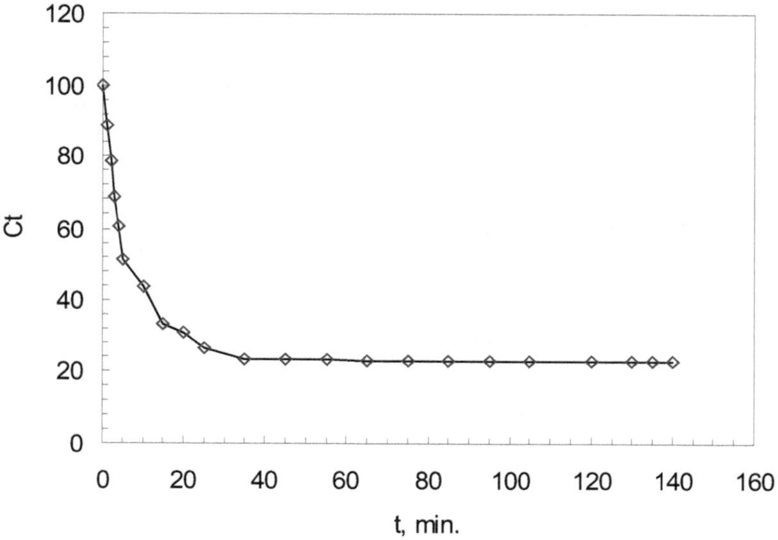

Fig. 3. Adsorption of Fe^{3+} ions onto olive cake. Variation of the Fe^{3+} ions concentration with time. (Initial concentration of Fe^{3+} ions: 100 ppm, Agitation speed: 100 rpm, pH: 4.5, temperature 28 °C).

2.4 Solubility of adsorbent/ heavy metals in wastewater/ water

The slightly soluble metal ions in water will be more easily removed from water(i.e., adsorbed) than substances with high solubility. Also, non-polar substances will be more easily removed than polar substances since the latter have a greater affinity for adsorption.

2.5 Affinity of the solute for the adsorbent

If the surface of adsorbent is slightly polar, the non-polar substances will be more easily picked up by the adsorbent than polar ones (the opposite is correct).

2.6 Size of the molecule with respect to size of the pores

Large molecules may be too large to enter small pores. This may reduce adsorption independently of other causes.

2.7 Degree of ionization of the adsorbate molecule

More highly ionized molecules are adsorbed to a smaller degree than neutral molecules.

2.8 pH

The degree of ionization of a species is affected by the pH (e.g., a weak acid or a weak basis). This, in turn, affects adsorption. For example, the precipitation of Fe^{3+} ions occurred at pH greater than 4.5 (see Figure 4). The decrease in the Fe^{3+} ions removal capacity at pH > 4.5 may have been caused by the complexing Fe^{3+} ions with hydroxide. Therefore, the removal efficiency increases with increasing initial pH. For example, at low pH, the concentration of proton is high. Therefore, the positively charged of the Fe^{3+} ions and the protons compete for binding on the adsorbent sites in Zeolite, Bentonite, Quartz, olive cake, Tripoli in which, this process decrease the uptake of iron ions. The concentration of proton in the solution decrease as pH gradually increases in the ranges from 2 to 4.5. In this case, little protons have the chance to compete with Fe^{3+} ions on the adsorption sites of the olive cake. Thus, higher pH in the acidic media is facilitated the greater uptake of Fe^{3+} ions. Above pH 4.5, the removal efficiency decreases as pH increases, this is inferred to be attributable to the hydrolysis [19-22].

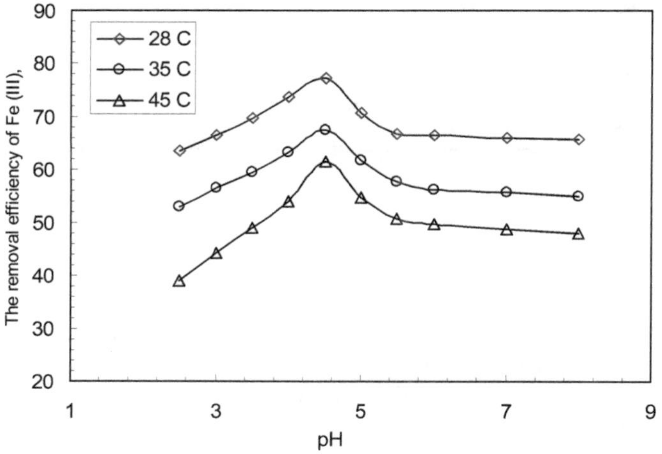

Fig. 4. Effect of initial pH on the removal efficiency, %, of Fe^{3+} ions at different temperatures. (Initial concentration of Fe^{3+} ions: 100 ppm, Agitation speed: 100 rpm, Mass of olive cake: 1 g, Dose: 5 g/l, Contact time: 24 hr).

2.9 Effect of initial concentration

At high-level concentrations, the available sites of adsorption become fewer. This behaviour is connected with the competitive diffusion process of the Fe^{3+} ions through the microchannel and pores in NB [20]. This competitive will lock the inlet of channel on the surface and prevents the metal ions to pass deeply inside the NB, *i.e.* the adsorption occurs on the surface only. These results indicate that energetically less favorable sites by increasing metal concentration in aqueous solution. This results are found matching with our recently studies using natural zeolite [19] and olive cake [21], in addition to other reported by Rao *et al.* [23] and Karthikeyan *et al.* [24]. The removal efficiency of Fe^{3+} ions on NQ and NB as well as zeolite at different initial concentrations (50, 100, 200, 300 and 400 ppm) is shown in the Figures 5-6. It is evident from the figure that the percentage removal of Fe^{3+} ions on NQ is slightly depended on the initial concentration. While the removal efficiency of Fe^{3+} ions using NB decreases as the initial concentration of Fe^{3+} ions increases. For example, the percentage removal is calculated 98 % using the initial concentration of 50 ppm, while it is found 28 % using high-level of 400 ppm [20].

On the other hand, it is clear from Figure 6 that the removal efficiency of Fe^{3+} ions using NQ is less affected by the initial concentration. For instance, the percentage removal using 50 ppm of the initial concentration is found 35 %, while is found 34.9 % using high-level concentrations (400 ppm). This means that the high concentration of Fe^{3+} ions will create and activate of some new activation sites on the adsorbent surface [20, 25].

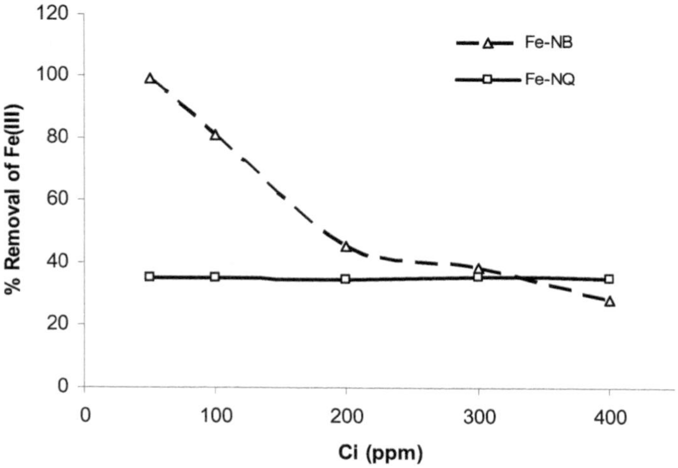

Fig. 5. The effect of initial concentration namely 50, 100, 200, 300 and 400 ppm of Fe^{3+} ions at constant contact time (2.5 hours), adsorbent dosage 4 g/L of natural NQ and NB, Temperature (30 °C) and agitation speed (300 rpm).

2.10 Dosage effect

The removal efficiency is generally increased as the concentration dose increases over these temperature values. This can be explained by the fact that more mass available, more the contact surface offered to the adsorption. The effect of the Jordanian Natural Zeolite (JNZ) dosage on the removal of Fe^{3+} ions is shown in Figure 6 [19]. The adsorbent dosage is varied from 10 to 40 g/l. The initial Fe^{3+} ions concentration, stirrer speed, initial pH and temperature are 1000 ppm, 300 rpm, 1% HNO_3, and 30 °C, respectively. This figure shows that the maximum removal of 69.15 % is observed with the dosage of 40 g/l. We observed that the removal efficiency of adsorbents generally improved with increasing amount of JNZ. This is expected because the higher dose of adsorbent in the solution, the greater availability of exchangeable sites for the ions, *i.e.* more active sites are available for binding of Fe^{3+} ions. Moreover, our recent studies using olive cake, natural quartz and natural bentonite and tripoli [19-22] are qualitatively in a good agreement with each other and with those found in the literatures [26].

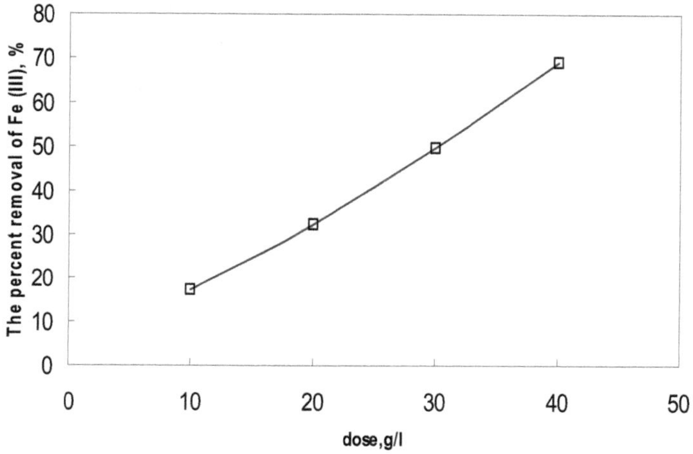

Fig. 6. The adsorbent dose of JNZ vs. Percent removal of Fe^{3+} ions: 1 g adsorbent/ 50 ml Fe^{3+} ions solution, 30 °C, initial pH of 1% HNO_3, 300 rpm, and constant initial concentration (1000 ppm).

3. Adsorption operation

Adsorption from solution is usually conducted using either the column or the batch operation. It should be possible to characterize the solution - adsorbent system by both technique operations and arrive at the same result. This is due to the physical and/or chemical forces applicable in each case must be identical. Furthermore, the results obtained from the batch experiment should be somewhat more reliable. Among the most serious objections of the column experiments are: (1)the inherent difficulties associated to maintain a constant flow rate; (2) the difficulty of ensuring a constant temperature throughout the column; (3) the appreciable probability of presence the channels within the packed column; and (4) the relatively large expenditure both in time and manpower required for a column experiment.

3.1 Batch operation

In a batch operation, fixed amount of adsorbent is mixed all at once with specific volume of adsorbate (with the range of initial concentration). Afterwards, the system kept in agitation for a convenient period of time. Separation of the resultant solution is accomplished by filtering, centrifuging, or decanting. The optimum pH, contact time, agitation speed and optimum temperature are fixed and used in this technique. For instance, the contact time study, the experiment are carried out at constant initial concentration, agitation speed, pH, and temperature. During the adsorption progress, the mixture container must be covered by

alumina foil to avoid the evaporation. The samples are withdrawn at different time intervals, for example, every 5 minuets or every 15 minutes.
The uptake of heavy metal ions was calculated from the mass balance, which was stated as the amount of solute adsorbed onto the solid. It equal the amount of solute removed from the solution. Mathematically can be expressed in equation 1 [27]:

$$q_e = \frac{(C_i - C_e)}{S} \tag{1}$$

q_e : Heavy metal ions concentration adsorbed on adsorbent at equilibrium (mg of metal ion/g of adsorbent).

C_i : Initial concentration of metal ions in the solution (mg/l).

C_e : Equilibrium concentration or final concentration of metal ions in the solution (mg/l). S : Dosage (slurry) concentration and it is expressed by equation 2:

$$S = \frac{m}{v} \tag{2}$$

Where v is the initial volume of metal ions solution used (L) and m is the mass of adsorbent. The percent adsorption (%) was also calculated using equation 3:

$$\% \text{ adsorption} = \frac{C_i - C_e}{C_i} \times 100\% \tag{3}$$

3.2 Column operation

In a column operation, the solution of adsorbate such as heavy metals (with the range of initial concentration) is allowed to percolate through a column containing adsorbent (ion exchange resin, silica, carbon, etc.) usually held in a vertical position. For instance, column studies were carried out in a column made of Pyrex glass of 1.5 cm internal diameter and 15 cm length. The column was filled with 1 g of dried PCA by tapping so that the maximum amount of adsorbent was packed without gaps. The influent solution was allowed to pass through the bed at constant flow rate of 2 mL/min, in down flow manner with the help of a fine metering valve. The effluent solution was collected at different time intervals.
The breakthrough adsorption capacity of adsorbate (heavy metal ions) was obtained in column at different cycles using the equation 4 [28].

$$q_e = [(C_i - C_e)/m] \times bv \tag{4}$$

Where *Ci* and *Ce* denote the initial and equilibrium (at breakthrough) of heavy metal ions concentration (mg/L) respectively. bv was the breakthrough volume of the heavy metal ions solution in liters, and m was the mass of the adsorbent used (g). After the column was exhausted, the column was drained off the remaining aqueous solution by pumping air. The adsorption percent is given by equation 5.

$$\% \text{ Desorption} = (C_e/C_i) \times 100 \tag{5}$$

4. Thermodynamic and adsorption isotherms

Adsorption isotherms or known as equilibrium data are the fundamental requirements for the design of adsorption systems. The equilibrium is achieved when the capacity of the adsorbent materials is reached, and the rate of adsorption equals the rate of desorption. The theoretical adsorption capacity of an adsorbent can be calculated with an adsorption isotherm. There are basically two well established types of adsorption isotherm the Langmuir and the Freundlich adsorption isotherms. The significance of adsorption isotherms is that they show how the adsorbate molecules (metal ion in aqueous solution) are distributed between the solution and the adsorbent solids at equilibrium concentration on the loading capacity at different temperatures. That mean, the amount of sorbed solute versus the amount of solute in solution at equilibrium.

4.1 Langmuir adsorption isotherm

Langmuir is the simplest type of theoretical isotherms. Langmuir adsorption isotherm describes quantitatively the formation of a monolayer of adsorbate on the outer surface of the adsorbent, and after that no further adsorption takes place. Thereby, the Langmuir represents the equilibrium distribution of metal ions between the solid and liquid phases [29]. The Langmuir adsorption is based on the view that every adsorption site is identical and energically equivalent (thermodynamically, each site can hold one adsorbate molecule). The Langmuir isotherm assume that the ability of molecule to bind and adsorbed is independent of whether or not neighboring sites are occupied. This mean, there will be no interactions between adjacent molecules on the surface and immobile adsorption. Also mean, trans-migration of the adsorbate in the plane of the surface is precluded. In this case, the Langmuir isotherms is valid for the dynamic equilibrium adsorption – desorption processes on completely homogeneous surfaces with negligible interaction between adsorbed molecules that exhibit the form:

$$q_e = (Q \times b \times C_e)/(1+b \times C_e) \quad (6)$$

C_e = The equilibrium concentration in solution
q_e = the amount adsorbed for unit mass of adsorbent
Q and b are related to standard monolayer adsorption capacity and the Langmuir constant, respectively.

$$q_{\max} = Q \times b \quad (7)$$

$q_{\max}$ = is the constant related to overall solute adsorptivity (l/g).
Equation 6 could be re-written as:

$$C_e/q_e = 1/(q_{\max} \times b) + (1/q_{\max}) \times C_e \quad (8)$$

In summary, the Langmuir model represent one of the the first theoretical treatments of non-linear sorption and suggests that uptake occurs on a homogenous surface by monolyer sorption without interaction between adsorbed molecules. The Langmuir isotherm assumes that adsorption sites on the adsorbent surfaces are occupied by the adsorbate in the solution. Therefore the Langmuir constant (b) represents the degree of adsorption affinity the adsorbate. The maximum adsorption capacity (Q) associated with complete monolayer cover is typically expressed in (mg/g). High value of b indicates for much stronger affinity of metal ion adsorption.

The shape of the isotherm (assuming the (x) axis represents the concentration of adsorbing material in the contacting liquid) is a gradual positive curve that flattens to a constant value.

A plot of C_e/q_e versus C_e gives a straight line of slope $1/q_{max}$ and intercept $1/(q_{max}\times b)$, for example, as shown in Figure 7.

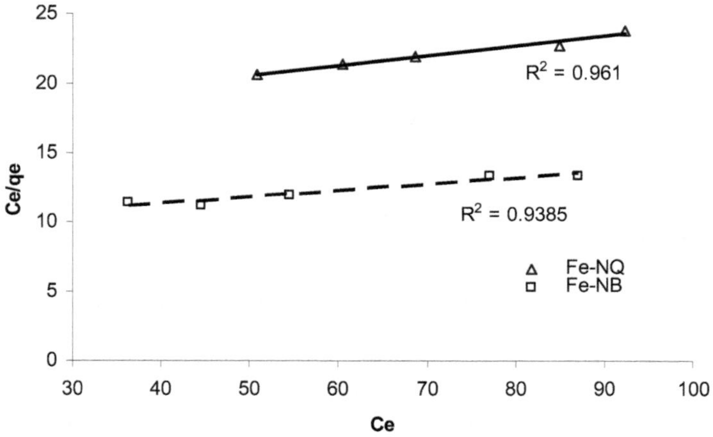

Fig. 7. The linearized Langmuir adsorption isotherms for Fe^{3+} ions adsorption by natural quartz (NQ) and bentonite (NB) at constant temperature 30 °C. (initial concentration: 400 ppm, 300 rpm and contact time: 2.5 hours).

The effect of isotherm shape is discussed from the direction of the predicting whether and adsorption system is "favorable" or "unfavorable". Hall et al (1966) proposed a dimensionless separation factor or equilibrium parameter, R_L, as an essential feature of the Langmuir Isotherm to predict if an adsorption system is "favourable" or "unfavourable", which is defined as [30]:

$$R_L = 1/(1+bC_i) \tag{9}$$

C_i = reference fluid-phase concentration of adsorbate (mg/l) (initial Fe^{3+} ions concentration)
b = Langmuir constant (ml mg^{-1})
Value of R_L indicates the shape of the isotherm accordingly as shown in Table 1 below. For a single adsorption system, C_i is usually the highest fluid-phase concentration encountered.

Value of R_L	Type of Isotherm
$0 < r < 1$	Favorable
$r > 1$	Unfavorable
$r = 1$	Linear
$R = 0$	Irreversible

Table 1. Type of isotherm according to value of R_L

4.2 Freundlich adsorption isotherms

Freundlich isotherm is commonly used to describe the adsorption characteristics for the heterogeneous surface [31]. It represents an initial surface adsorption followed by a condensation effect resulting from strong adsorbate-adsorbate interaction. Freundlich isotherm curves in the opposite way of Langmuir isotherm and is exponential in form. The heat of adsorption, in many instances, decreases in magnitude with increasing extent of adsorption. This decline in heat is logarithmic implying that the adsorption sites are distributed exponentially with respect to adsorption energy. This isotherm does not indicate an adsorption limit when coverage is sufficient to fill a monolayer ($\theta = 1$). The equation that describes such isotherm is the Freundlich isotherm, given as [31]:

$$q_e = K_f (C_e)^{1/n} \quad n > 1 \tag{10}$$

K_f = Freundlich constant related to maximum adsorption capacity (mg/g). It is a temperature-dependent constant.

n = Freundlich contestant related to surface heterogeneity (dimensionless). It gives an indication of how favorable the adsorption processes.

With $n = 1$, the equation reduces to the linear form: $q_e = k \times C_e$

The plotting q_e versus C_e yield a non-regression line, which permits the determination of $(1/n)$ and K_f values of $(1/n)$ ranges from 0 to 1, where the closer value to zero means the more heterogeneous the adsorption surface. On linearization, these values can be obtained by plotting $(ln\ q_e)$ versus $(ln\ C_e)$ as presented in equation 11. From the plot, the vales K_f and n can be obtained.

$$ln\ q_e = ln\ K_f + (1/n) ln\ C_e \tag{11}$$

where, the slop = $(1/n)$, and the intercept = $ln\ K_f$

4.3 Dubinin–Kaganer–Radushkevich (DKR)

The DKR isotherm is reported to be more general than the Langmuir and Freundlich isotherms. It helps to determine the apparent energy of adsorption. The characteristic porosity of adsorbent toward the adsorbate and does not assume a homogenous surface or constant sorption potential [32].

The Dubinin–Kaganer–Radushkevich (DKR) model has the linear form

$$\ln q_e = \ln X_m - \beta \varepsilon^2 \tag{12}$$

where X_m is the maximum sorption capacity, β is the activity coefficient related to mean sorption energy, and ε is the Polanyi potential, which is equal to

$$\varepsilon = RT \ln(1 + \frac{1}{C_e}) \tag{13}$$

where R is the gas constant (kJ/kmol- K) .

The slope of the plot of $\ln q_e$ versus ε^2 gives β (mol^2/J^2) and the intercept yields the sorption capacity, Xm (mg/g) as shown in Fig. 6. The values of β and Xm, as a function of temperature are listed in table 1 with their corresponding value of the correlation coefficient, R^2. It can be observed that the values of β increase as temperature increases while the values of Xm decrease with increasing temperature.

The values of the adsorption energy, E, was obtained from the relationship [33]

$$E = (-2\beta)^{-1/2}$$

4.4 Thermodynamics parameters for the adsorption

In order to fully understand the nature of adsorption the thermodynamic parameters such as free energy change (ΔG^o) and enthalpy change (ΔH^o) and entropy change (ΔS^o) could be calculated. It was possible to estimate these thermodynamic parameters for the adsorption reaction by considering the equilibrium constants under the several experimental conditions. They can calculated using the following equations [34]:

$$\Delta G = - R \ln K_d(T) \tag{14}$$

$$\ln K_d = \Delta S/R - \Delta H/RT \tag{15}$$

$$\Delta G = \Delta H - T\Delta S \tag{16}$$

The K_d value is the adsorption coefficient obtained from Langmuir equation. It is equal to the ratio of the amount adsorbed (x/m in mg/g) to the adsorptive concentration (y/a in mg/dm^3)

$$K_d = (x/m).\ (y/a) \tag{17}$$

These parameters are obtained from experiments at various temperatures using the previous equations. The values of ΔH^o and ΔS^o are determined from the slop and intercept of the linear plot of ($\ln K_d$) vs. (1/T).

In general these parameters indicate that the adsorption process is spontaneous or not and exothermic or endothermic. The standard enthalpy change (ΔH^o) for the adsorption process is: (i) positive value indicates that the process is endothermic in nature. (ii) negative value indicate that the process is exothermic in nature and a given amount of heat is evolved during the binding metal ion on the surface of adsorbent. This could be obtained from the plot of percent of adsorption (% C_{ads}) *vs.* Temperature (T). The percent of adsorption increase with increase temperature, this indicates for the endothermic processes and the opposite is correct [35]. The positive value of (ΔS^o) indicate an increase in the degree of freedom (or disorder) of the adsorbed species.

5. Motivation for the removal and sorption Fe^{3+} ions

In practice form recent studies, the natural zeolite [19], activated carbon[36], olive cake [21], quartz and bentonite [20] and jojoba seeds [37] are used as an adsorbent for the adsorption of mainly trivalent iron ions in aqueous solution. The Motivation for the removal and sorption Fe^{3+} ions is that iron ions causes serious problems in the aqueous streams especially at high levels concentration [38 - 39]. Usually, the iron ions dissolve from rocks and soils toward the ground water at low levels, but it can occur at high levels either through a certain geological formation or through the contamination by wastes effluent of the industrial processes such as pipeline corrosion, engine parts, metal finishing and galvanized pipe manufacturing [40 - 41].

The presence of iron at the high-levels in the aqueous streams makes the water in unusable for an several considerations: Firstly, Aesthetic consideration such as discoloration, the metallic taste even at low concentration (1.8 mg/l), odor, and turbidity, staining of laundry and plumbing fixtures. Secondly, the healthy consideration where the high level of iron ions precipitates as an insoluble Fe^{3+}-hydroxide under an aerobic conditions at neutral or alkaline pH [42]. This can generate toxic derivatives within the body by using drinking

water and can contribute to disease development in several ways. For instance, an excessive amounts of iron ions in specific tissues and cells (iron-loading) promote development of infection, neoplasia, cardiomyopathy, arthropathy, and various endocrine and possibly neurodegenerative disorders. Finally, the industrial consideration such as blocking the pipes and increasing of corrosion. In addition to that, iron oxides promote the growth of micro-organism in water which inhibit many industrial processes in our country [43].
In response to the human body health, its environmental problems and the limitation water sources especially in Jordan [44], the high-levels of Fe^{3+} ions must be removed from the aqueous stream to the recommended limit 5.0 and 0.3 ppm for both inland surface and drinking water, respectively. These values are in agreement with the Jordanian standard parameters of water quality[45]. For tracing Fe^{3+} ions into recommended limit, many chemical and physical processes were used such as supercritical fluid extraction, bioremediation, oxidation with oxidizing agent [46]. These techniques were found not effective due to either extremely expensive or too inefficient to reduce such high levels of ions from the large volumes of water [47 - 48]. Therefore, the effective process must be low cost-effective technique and simple to operate [49 - 52]. It found that the adsorption process using natural adsorbents realize these prerequisites. In addition to that, the natural adsorbents are environmental friend, existent in a large quantities and has good adsorption properties. The binding of Iron(III) ion with the surface of the natural adsorbent could change their forms of existence in the environment. In general they may react with particular species, change oxidation states and precipitate [53]. In spite of the abundant reported researches in the adsorption for the removal of the dissolved heavy metals from the aqueous streams, however the iron(III) ions still has limited reported studies. Therefore our studies are concentrated in this field. From our previous work, the natural zeolite [19], quartz and Bentonite [20], olive cake [21], in addition to the chitin [24], activated carbon [54 - 55] and alumina [56] have been all utilized for this aspect at low levels. The adsorption isotherm models (Langmuir and Freundlich) are used in order to correlate the experimental results.

5.1 Sorption Fe^{3+} ions using natural quartz (NQ) and bentonite (NB))

It is known from the chemistry view that surface of NB and NQ is ending with $(Si\text{-}O)^-$ negatively charged. These negative entities might bind metal ions *via* the coordination aspects especially at lower pH values as known in the literatures. Fe^{3+} ions are precipitated in the basic medium. Therefore, the 1 % HNO_3 stock solution is used to soluble Fe^{3+} ions and then achieving the maximum adsorption percentages [20].
The binding of metal ions might be influenced on the surface of NB more than NQ. This is due to the expected of following ideas: (i) NQ have pure silica entity with homogeneous negatively charged, therefore the binding will be homogeneous. (ii) The natural bentonite has silica surface including an inner-layer of alumina and iron oxide, which cause a heterogeneous negatively charged. Therefore, the binding Fe^{3+} ions on the surface on NB might be complicated.
The adsorption thermodynamics modelsof Fe^{3+} ions on NQ and NB at 30 °C are examined [20]. The calculated results of the Langmuir and Freundlich isotherm constants are given in Table 2. The high values of R^2 (>95%) indicates that the adsorption of Fe^{3+} ions onto both NQ and NB was well described by Freundlich isotherms.It can also be seen that the q_{max} and

the adsorption intensity values of NB are higher than that of NQ. The calculated *b* values indicate the interaction forces between NB surface and Fe^{3+} ions are stronger than in case of using NQ, this is in agreement with the higher ionic potential of Fe^{3+}. This means that the NB is more powerful adsorbent than NQ. Furthermore, based on this information, we found that the adsorption using NQ and NB is much higher as compared to carbon used by other authors [57].

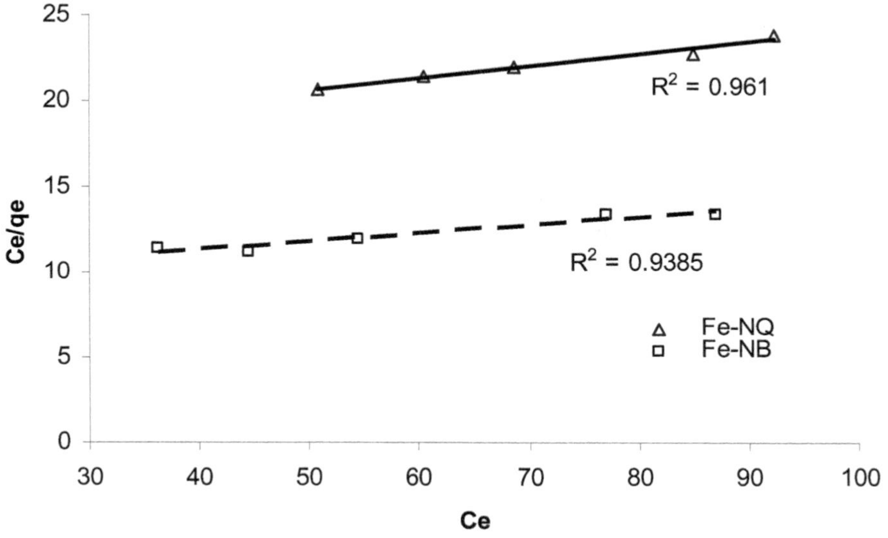

Fig. 8. The linearized Langmuir adsorption isotherms for Fe^{3+} ions adsorption by natural quartz (NQ) and bentonite (NB) at constant temperature 30 °C. (Initial concentration: 400 ppm, 300 rpm and contact time: 2.5 hours).

Langmuir Constants Adsorbent	q_{max}	**b (L/mol)**	**ΔG/1000 (kJ/mol)**
NQ	14.4	226.3	-13.4
NB	20.96	283.8	-13.9

NQ = Natural Quartz
NB = Natural Bentonite

Table 2. Langmuir constants for adsorption of Fe^{3+} ions on NQ and NB

The obtained experimental data also has well described by Freundlich isotherm model into both NQ and NB. The negative value of ΔG^o (- 13.4 and - 13.9 KJ/mol, respectively) confirms the feasibility of the process and the spontaneous nature of adsorption with a high preference for metal ions to adsorb onto NB more easily than NQ in pseudo second order rate reaction.

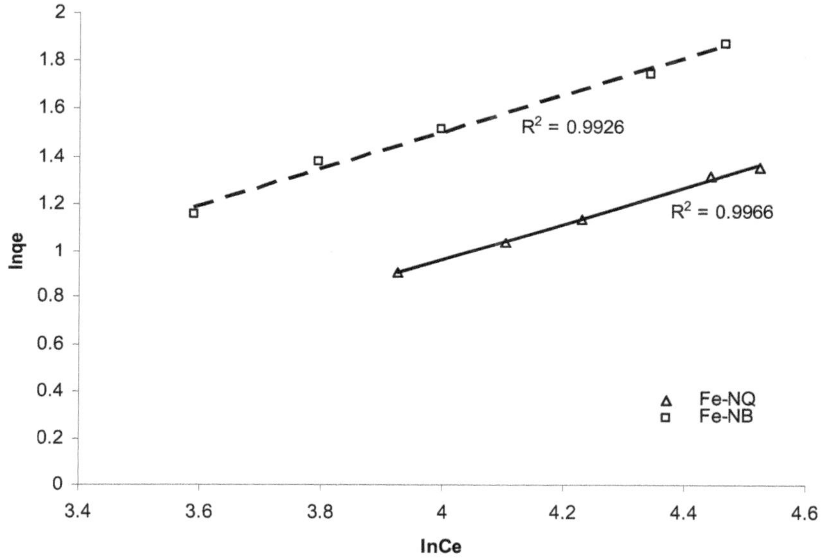

Fig. 9. The linearized Freundlich adsorption isotherms for Fe^{3+} ions adsorption by natural Quartz (NQ) and bentonite (NB) at constant temperature 30 °C. (initial concentration: 400 ppm, normal 1% HNO_3 aqueous solution, 300 rpm, contact time: 2.5 hours).

5.2 Sorption isotherms of Fe^{3+} ions processes using Jordanian natural zeolite (JNZ)

Isotherm studies are conducted at 30 °C by varying the initial concentration of Fe^{3+} ions [19]. Representative initial concentration (1000, 800, 600, 400 ppm) of Fe^{3+} ions are mixed with slurry concentrations (dose) of 20g/L for 150 min., which is the equilibrium time for the zeolite and Fe^{3+} ions reaction mixture. The equilibrium results are obtained at the 1% HNO_3 model solution of Fe^{3+} ions. The Langmuir isotherm model is applied to the experimental data as presented in Figure 10. Our experimental results give correlation regression coefficient, R^2, equals to 0.998, which are a measure of goodness-of-fit and the general empirical formula of the Langmuir model by

$$\frac{c_e}{q_e} = 0.136c_e + 8.72$$

Our results are in a good qualitatively agreements with those found from adsorption of Fe^{3+} on the palm fruit bunch and maize cob [58].

Figure 11 represents the fitting data into the Freundlich model. We observe that the empirical formula of this model is found as $\ln q_e = 0.1058 \ln c_e + 1.2098$ with R^2 value equals to 0.954. It can be seen that the Langmuir model has a better fitting model than Freundlich as the former have higher correlation regression coefficient than the latter.

The value of standard Gibbs free energy change calculated at 30 °C is found to be -16.98 kJ/mol. The negative sign for ΔG^0 indicates to the spontaneous nature of Fe^{3+} ions adsorption on the JNZ.

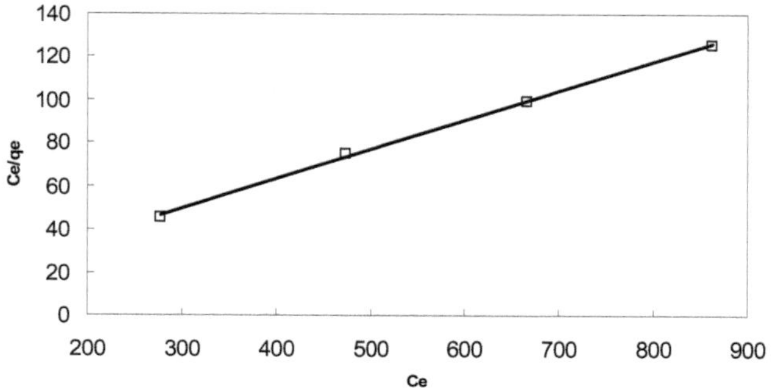

Fig. 10. Langmuir isotherm model plot of q_e/C_e of Fe^{3+} ions on JNZ *vs.* C_e (ppm) of Fe^{3+} ions.

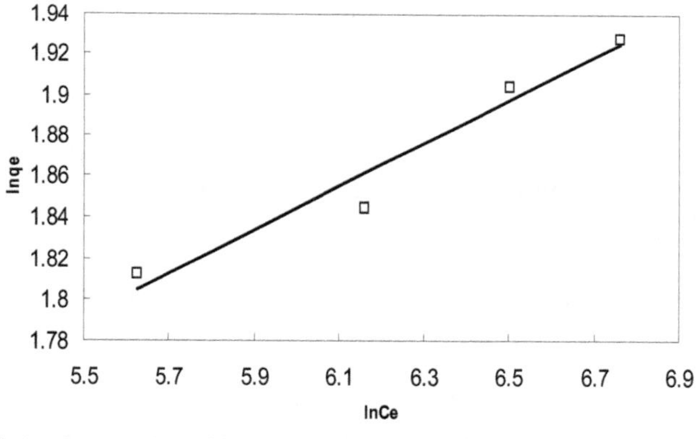

Fig. 11. Freundlich isotherm plot of $\ln q_e$ *vs.* $\ln C_e$, where C_e is the equilibrium concentration of Fe^{3+} ions concentration.

5.3 Sorption isotherms of Fe^{3+} ions processes using olive cake

To conduct the isotherm were studied at the initial pH of solution which was adjusted at the optimum value (pH = 4.5) and the mass of olive cake which was taken as 0.3, 0.5, 0.75 and 1.0 g at different temperatures of 28, 35 and 45 °C. Three adsorption isotherms models were used: Langmiur, Frendulich and Dubinin-Kaganer-Radushkevich (DKR) [21]. Figure 12 shows the experimental data that were fitted by the linear form of Langmiur model, (C_e / q_e) versus C_e, at temperatures of 28, 35 and 45 ^{0}C. The values of q_{max} and b were evaluated from the slope and intercept respectively for the three isothermal lines. These values of q_{max} and b are listed in Table 1 with their uncertainty and their regression coefficients, R^2. Table 3 shows that the values of q_{max} and b are decreased when the solution temperature increased from 28 to 45 ^{0}C. The decreasing in the values of q_{max} and b with increasing temperature indicates that the Fe^{3+} ions are favorably adsorbed by olive cake at lower temperatures, which shows that the adsorption process is exothermic. In order to justify the validity of olive cake as an adsorbent for Fe^{3+} ions adsorption, its adsorption potential must be compared with other adsorbents like eggshells [59] and chitin [24] used for this purpose. It may be observed that the maximum sorption of Fe^{3+} ions on olive cake is approximately greater 10 times than those on the chitin and eggshells.

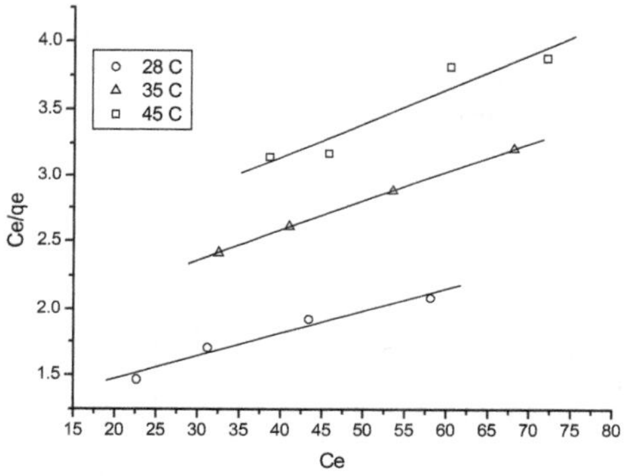

Fig. 12. Langmuir isotherm model plot of q_e/C_e of Fe^{3+} ions on olive cake *vs.* C_e (ppm) of Fe^{3+} ions.

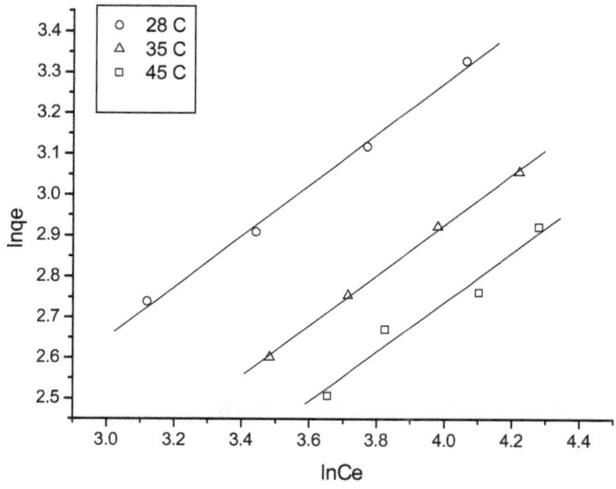

Fig. 13. The linearized Freundlich adsorption isotherms for Fe^{3+} ions adsorption by olive cake at different temperatures. (Initial concentration of Fe^{3+} ions: 100 ppm, Agitation speed: 100 rpm, pH: 4.5, Contact time: 24 hr).

The Freundlich constants K_f and n, which respectively indicating the adsorption capacity and the adsorption intensity, are calculated from the intercept and slope of plot $\ln q_e$ versus $\ln C_e$ respectively, as shown in Figure 13. These values of K_f and n are also listed in Table 3 with their regression coefficients. It can be observed that the values of K_f are decreased with increasing the temperature of solution from 28 to 45 ^{0}C. The decreasing in these values with temperature confirms also that the adsorption process is exothermic. It can be also seen that the values of *1/n* decreases as the temperature increases. Our experimental data of values K_f and *1/n* are considered qualitatively consistence with those that found in adsorption of Fe^{3+} ions on eggshells [59] and chitin [24].
The negative value of ΔG^0 (Table 4)confirms the feasibility of the process and the spontaneous nature of sorption. The values of ΔH^0 and ΔS^0 are found to be -10.83 kJ/mol and 19.9 J/mol-K, respectively (see Table 4). The negative values of ΔH^0 indicate and exothermic sorption reaction process. The positive value of ΔS^0 shows the increasing randomness at the solid/liquid interface during the sorption of Fe^{3+} ions onto olive cake.

T (°C)	Langmuir			Freundlich			DKR			
	b	q_{max}	R^2	$1/n$	k_f	R^2	β	X_m	E	R^2
28	0.0152 ±0.00011	58.479 ±3.44	0.91	0.626	2.164 ±0..98	0.99	-0.00006	28.321 ±1.04	91.287	0.91
35	0.0130 ±0.00013	45.249 ±1.17	0.99	0.618	1.578 ±0.15	0.99	-0.00009	23.539 ±2.12	74.535	0.97
45	0.012 ±0.00014	39.370 ±2.07	0.96	0.609	1.354 ±0.12	0.96	-0.0001	20.816 ±1.01	70.711	0.95

Table 3. Langmuir, Freundlich and DKR constants for adsorption of Fe^{3+} ions on olive cake

T(°C)	b(L/mol)	- ΔG^0 (kJ/mol)	- ΔH^0 (kJ/mol)	ΔS^0 (J/mol-K)
28	847.2174	16.8718		
35	727.8947	16.8755	10.83	19.9
45	667.7246	17.1953		

Table 4. Thermodynamics parameters for the adsorption of Fe^{3+} ions on olive cake.

5.4 Sorption isotherms of Fe^{3+} ions processes using chitosan and cross-linked chitosan beads

A batch adsorption system was applied to study the adsorption of Fe^{2+} and Fe^{3+} ions from aqueous solution by chitosan and cross-linked chitosan beads [60 - 61]. The adsorption capacities and rates of Fe^{2+} and Fe^{3+} ions onto chitosan and cross-linked chitosan beads were evaluatedas shown in Figure 14. Experiments were carried out as function of pH, agitation period, agitation rate and concentration of Fe^{2+} and Fe^{3+} ions. Langmuir and Freundlich

adsorption models were applied to describe the isotherms and isotherm constants. Equilibrium data agreed very well with the Langmuir model.
The calculated results of the Langmuir and Freundlich isotherm constants are given in Table 5. It is found that the adsorptions of Fe^{2+} and Fe^{3+} ions on the chitosan and cross-linked chitosan beads correlated well (R>0.99) with the Langmuir equation as compared to the Freundlich equation under the concentration range studied.

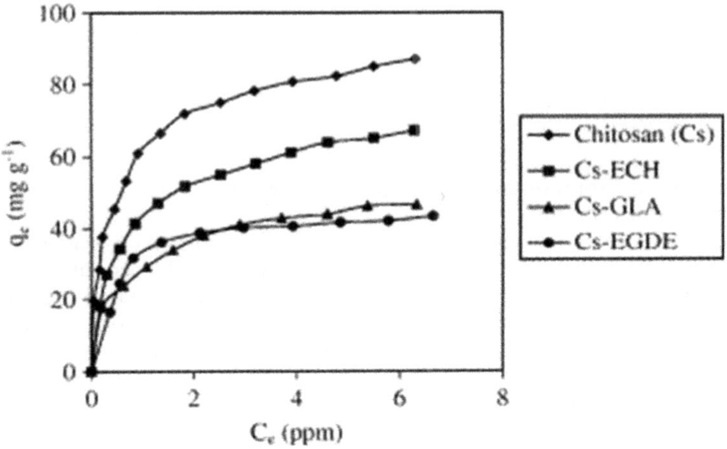

Fig. 14. Adsorption isotherms of Fe^{3+} ions on chitosan and cross-linked chitosan beads. chitosan, chitosan-GLA and chitosan-ECH bead

Iron	Beads	Langmuir			Freundlich		
		Q (mg g^{-1})	b (ml mg^{-1})	R	K_F (mg g^{-1})	n	R
Fe(II)	Chitosan	64.10	2197	0.9986	42.74	4.77	0.7730
	Chitosan-ECH	57.47	1891	0.9998	33.25	3.19	0.9597
	Chitosan-GLA	45.25	1023	0.9995	21.84	2.71	0.9744
	Chitosan-EGDE	38.61	762	0.9985	17.15	2.79	0.9985
Fe(III)	Chitosan	90.09	2413	0.9989	55.27	3.32	0.9824
	Chitosan-ECH	72.46	1550	0.9987	39.35	2.98	0.9788
	Chitosan-GLA	51.55	1405	0.9989	28.63	3.47	0.9881
	Chitosan-EGDE	46.30	2076	0.9991	28.36	3.61	0.8982

Table 5. Langmuir and Freundlich isotherm constants and correlation coefficients

Table 6 lists the calculated results. Based on the effect of separation factor on isotherm shape, the R_L values are in the range of $0<R_L<1$, which indicates that the adsorptions of Fe^{2+} and Fe^{3+} ions on chitosan and cross-linked chitosan beads are favourable. Thus, chitosan and cross-linked chitosan beads are favourable adsorbers. As mentioned earlier, chitosan and cross-linked chitosan beads are microporous biopolymers, therefore pores are large enough to let Fe^{2+} and Fe^{3+} ions through. The mechanism of ion adsorption on porous adsorbents may involve three steps [60 - 61]: (i) diffusion of the ions to the external surface of adsorbent; (ii) diffusion of ions into the pores of adsorbent; (iii) adsorption of the ions on the internal surface of adsorbent. The first step of adsorption may be affected by metal ion concentration and agitation period. The last step is relatively a rapid process.

Iron	Initial concentration (ppm)	R_L value			
		Chitosan	Chitosan-ECH	Chitosan-GLA	Chitosan-EGDE
Fe(II)	3	0.1317	0.1498	0.2457	0.3044
	6	0.0705	0.0810	0.1401	0.1795
	9	0.0481	0.0555	0.0980	0.1273
Fe(III)	3	0.1214	0.1769	0.1917	0.1383
	6	0.0646	0.0971	0.1060	0.0743
	9	0.0440	0.0669	0.0732	0.0508

Table 6. RL values based on the Langmuir equation

5.5 Adsorption of Fe^{3+} ions on activated carbons obtained from bagasse, pericarp of rubber fruit and coconut shell

The adsorptions of Fe^{3+} ions from aqueous solution at room temperature on activated carbons obtaining from bagasse, pericarp of rubber fruit and coconut shell have been studied [62]. The adsorption behavior of Fe^{3+} ions on these activated carbons could be interpreted by Langmuir adsorption isotherm as monolayer coverage. The maximum amounts of Fe^{3+} ions adsorbed per gram of these activated carbons were 0.66 mmol/g, 0.41 mmol/g and 0.18 mmol/g, respectively. The mechanism by which the adsorption of Fe^{3+} ions onto the activated carbon can be performed after being reduced to Fe^{3+} ions [63].
Figures 15 to 17 show the Langmuir plots that have the greatest values of iron adsorption on three types of activated carbons. The maximum adsorption at monolayer coverage on bagasse, pericarp of rubber fruit and coconut shellare in the range 0.25 - 0.66 mmol/g, 0.11 - 0.41 mmol/g and 0.12 - 0.19 mmol/g, respectively. The experimental result shows that the amount of iron ion adsorbed on activated carbons decreased with increasing adsorption temperature. This suggested that the adsorption mechanism was physical adsorption, in contrast to chemical adsorption in which the amount of adsorbate adsorbed on an adsorbent increases with increasing adsorption temperature [63].

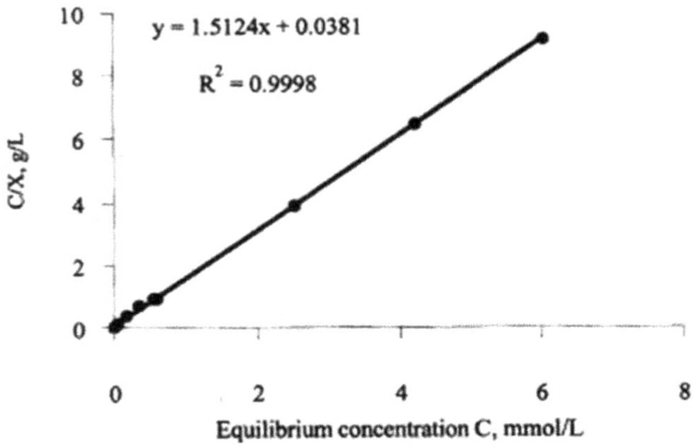

Fig. 15. Langmuir plot for the adsorption of Fe^{3+} ions on activated carbon obtained from bagasse.

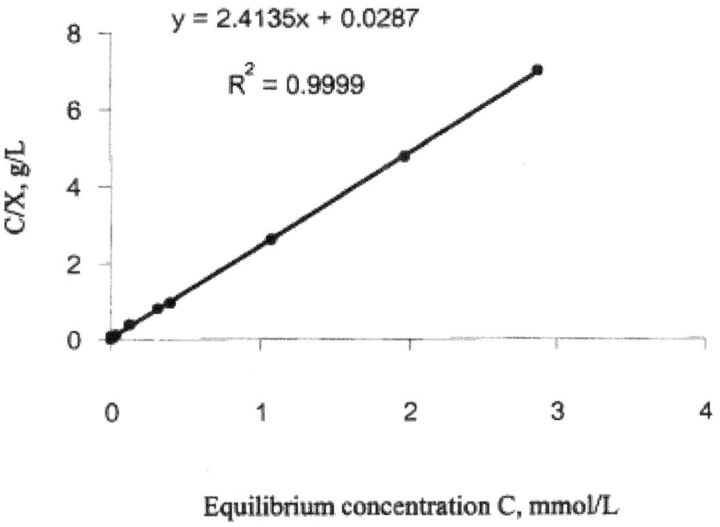

Fig. 16. Langmuir plot for the adsorption of Fe^{3+} ions on activated carbon obtained from pericarp of rubber fruit.

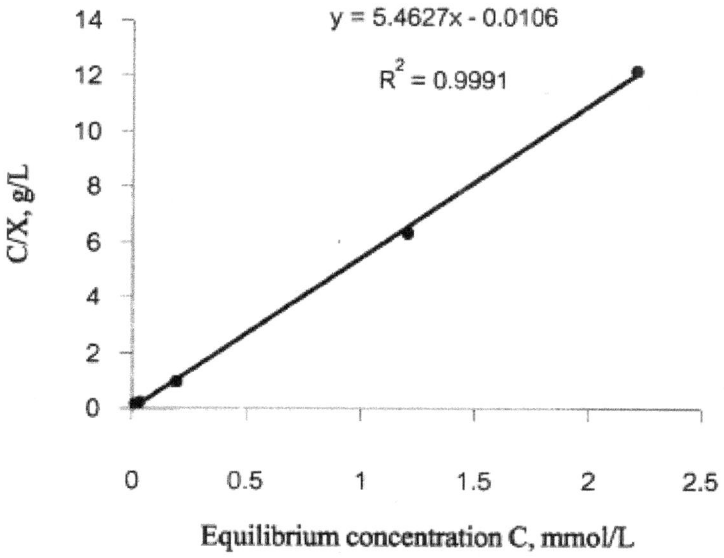

Fig. 17. Langmuir plot for the adsorption of Fe^{3+} ions on activated carbon obtained from coconut shell.

Study of the temperature dependence on these adsorptions has revealed them to be exothermic processes with the heats of adsorption of about -8.9 kJ/mol, -9.7 kJ/mol and -5.7 kJ/mol for bagasse, pericarp of rubber fruit and coconut shell, respectively. The value of Langmuir isotherm constant for the maximum adsorption at monolayer coverage (X_{max}) and the heats of adsorption (ΔH_{ads}) of Fe^{3+} ions on three types of activated carbons was summarized in Table 7.

Raw Materials	X_{max}	ΔH_{ads} (KJ/mol)
Bagasse	0.66	- 8.9
Pericarp of Rubber Fruit	0.41	- 9.7
Coconut Shell	0.19	- 5.7

Table 7. The maximum adsorption of iron ion at monolayer coverage (X_{max}) and the heats of adsorption (ΔH_{ads}) for iron ion on three types of activated carbons and activation temperature at 600 °C

5.6 Adsorption of Fe^{3+} ions on unmodified raphia palm (Raphia Hookeri) fruit endocarp

The adsorption of aqueous Fe^{3+} ions onto the surface of Raphia palm fruit endocarp (nut) was studied in a batch system [64 – 67]. The influence of initial Fe^{3+} ions concentration, temperature and particle size was investigated and the results showed that particle size and temperature affected the sorption rate and that the adsorption was fast with a maximum percentage adsorption of 98.7% in 20 min as initial metal ion concentration was increased. There is a general decrease in sorption efficiency as the particle size is increased. The increased sorption with smaller particle size means that there is higher external surface area available for adsorption with smaller particle at a constant total mass.

Four isotherms; Langmuir, Freundlich, Dubinin-Radushkevich (D-R) and Temkin were used to model the equilibrium sorption experimental data. The sorption process was found to follow chemisorption mechanism. From Dubinin-Radushkevich (D-R) isotherm, the apparent energy of adsorption was 353.55 Kj/mol. The apparent energy shows if the sorption process follows physisorption, chemisorption or ion exchange. It has been reported:

1. Physiasorption processes have adsorption energies <40 Kj/mol
2. Chemisorption processes have adsorption energies > 40 Kj/mol
3. Chemical ion exchange have adsorption energies between 8.0 and 16 Kj/mol
4. Adsorption is physical in nature have adsorption energies <8.0 Kj/mol

From the result obtained, the sorption of Fe^{3+} ions onto Raphia palm fruit endocarp (nut) was chemisorption process.

In order to describe the thermodynamic behaviuor of the sorption of Fe^{3+} ion onto Raphia palm fruit endocarp (nut) from aqueous solution, thermodynamic parameters including ΔG^{o}, ΔH^{o}, ΔS^{o}, were evaluated. The value of ΔH^{o} is negative indicating exothermic process. The standard Gibbs free energy indicates that the the sorption process is spontaneous in nature and also feasible. The decreasing in ΔG^{o} values with increasing temperature shows a decrease in feasibility of sorption at higher temperature.

Isotherm Constants	Fe(III)
Langmuir	- 16.3500
q_{max}	- 0.0800
b	0.9321
R^2	33.1000
$\Delta q(\%)$	
Freundlich	
n	0.6900
K_f	3.9800
R^2	0.9222
$\Delta q(\%)$	8.8100
Dubinin-Radushkevich	
β	$4.0*10^{-6}$
q_D	104.4800
E	353.5500
R^2	0.9461
Δq (%)	24.5200

Table 8. Isotherm constant for adsorption of Fe^{3+} ions onto Raphia palm fruit endocarp (nut) from aqueous solution

Constants (KJ/mol/K)	Fe(III)
ΔH	- 2560.4600
ΔS	19.1900
E_A	2094.8000
S^n	0.0900
R^2 (Vant Hoff)	0.4633
R^2 (Sticking Probability)	0.4747

Table 9. The activation energy of Fe(III)

The activation energy of any reaction process depicts the energy barrier which the reactants must overcome before any reaction could take place. High activation energy to react hence decrease in reaction rate.

6. Conclusion

High and low concentration level of Fe^{3+} ions can be adsorbed on different types of natural adsorbents. This process can be used to remove Fe^{3+} ions from aqueous solutions. The thermodynamic isotherms indicate the behavior picture of Fe^{3+} ions onto the surface of natural adsorbent as homogenous or heterogeneous monolayer coverage. It depends on the chemical nature of adsorbent surfaces. Mostly, the thermodynamic parameters show the spontaneous and exothermic adsorption processes of Fe^{3+} ions onto the surfaces of natural adsorbents, indicating of easier handling.

7. References

[1] Sawyer, C. N., and McCarty, P. L. (1978). Chemistry for environmental engineering, 3rd Ed., McGraw-Hill, Singapore, 85–90

[2] Lee, S. H. D., and Johnson, I. J. (1980). "Removal of gaseous alkali metal compounds from hot flue gas by particulate sorbents." J. Engrg. Power, 102, 397–402.

[3] Uberoi, M., and Shadman, F. (1990). "Sorbents for removal of lead compounds from hot flue gases." AICHE J., 36(2), 307–309.

[4] Ho, T. C., Chen, C., Hopper, J. R., and Oberacker, D. A. (1992). "Metal capture during fluidized bed incineration of wastes contaminated with lead chloride." Combustion Sci. and Technol., 85, 101–116.

[5] Ho, T. C., Chen, J. M., Shukla, S., and Hopper, J. R. (1990). "Metal capture during fluidized bed incineration of solid wastes." AICHE Symp. Ser., 276(86), 51–60.

[6] Ho, T. C., Chu, H. W., and Hopper, J. R. (1993). "Metal volatilization and separation during incineration." Waste Mgmt., 13, 455–466.

[7] Ho, T. C., Tan, T., Chen, C., and Hopper, J. R. (1991). "Characteristics of metal capture during fluidized bed incineration." Aiche Symp. Ser., 281(87), 118–126. a) M.C. Macias-Pbrez, C. Salinas-Martinez de Lecea, M.J. Muiloz- Guillena and A. Linares-Solano, Low Temeratijre SO_2 C A P m by Calcium Based Sorbents: Caaracterization of the Active Calcium. b) MuRoz-Guillena, M.J., Linares-Solano, A. and Salinas-Martinez, de Lecea, C. Appl. Surf. Sci.. 1995, 89, 197. Chau-Hwa Yang and James G. Goodwin, Jr.,Journal of Catalysis 78: 1, 1982, Pages 182-187A. Guerrero-Ruiz, Reaction Kinetics and Catalysis Letters, 49:1, 53-60, http://www.webref.org/chemistry/p/physisorption.htm, International Union of Pure and Applied Chemistry, http://wikichemistry.com/konfuciy.asp?tda=dt&t=13145&fs=physisorption+-+charactersitics

[8] P. Somasundaran, Somil C. Mehta, X. Yu, and S. Krishnakumar, handbook of Surface and Colloid Chemistry, Third Edition, Colloid Systems and Interfaces Stability of Dispersions through Polymer and Surfactant Adsorption, chapter 6.

[9] Deguo Kong 2009 , Master Thesis, Department of Land and Water Resources Engineering , Royal Institute of Technology (KTH), SE-100 44 Stockholm, Sweden.

[10] M. G. Lee, J.K. Cheon, S.K. Kam, J. Ind. Chem. Eng. 9(2) (2003) 174-180. a) M. Al-Anber and Z. Al-Anber, Utilization of natural zeolite as ion-exchange and sorbent material in the removal of iron. Desalination, 255, (2008) 70 – 81. b) M. Al-anber, Removal of Iron(III) from Model Solution UsingJordanian Natural Zeolite: Magnetic Study, Asian J. Chem. Vol. 19, No. 5 (2007), 3493-3501.

[11] A. Guerrero-Ruiz, Reaction Kinetics and Catalysis Letters, 49:1, 53-60.

[12] http://www.webref.org/chemistry/p/physisorption.htm, International Union of Pure and Applied Chemistry.

[13] http://wikichemistry.com/konfuciy.asp?tda=dt&t=13145&fs=physisorption+-+charactersitics

[14] P. Somasundaran, Somil C. Mehta, X. Yu, and S. Krishnakumar, handbook of Surface and Colloid Chemistry, Third Edition, Colloid Systems and Interfaces Stability of Dispersions through Polymer and Surfactant Adsorption, chapter 6.

[15] Deguo Kong 2009 , Master Thesis, Department of Land and Water Resources Engineering , Royal Institute of Technology (KTH), SE-100 44 Stockholm, Sweden.

a) R.C. Weast, CRC Handbook of Chemistry and Physics, CRC Press, Boca Raton, FL, 1979, p. F-214.]= b) Y. Onganer, C. Temur, Adsorption dynamics of Fe(III) from aqueous solution onto activated carbon, J. Colloid Interf. Sci. 205 (1998) 241–244.

[16] Adil Denizli, Ridvan Say and Yakup Arica Separation and Purification Technology, 21 (2000) 181-190.

[17] M. G. Lee, J.K. Cheon, S.K. Kam, J. Ind. Chem. Eng. 9(2) (2003) 174-180.

[18] a) M. Al-Anber and Z. Al-Anber, Utilization of natural zeolite as ion-exchange and sorbent material in the removal of iron. Desalination, 255, (2008) 70 – 81. b) M. Al-anber, Removal of Iron(III) from Model Solution Using Jordanian Natural Zeolite: Magnetic Study, Asian J. Chem. Vol. 19, No. 5 (2007), 3493-3501.

[19] Mohammad Al-Anber, Removal of High-level Fe^{3+} from Aqueous Solution using Jordanian Inorganic Materials: Bentonite and Quartz, Desalination 250 (2010) 885-891.

[20] Z. Al-Anber and M. Al-Anber, Thermodynamics and Kinetic Studies of Iron(III) Adsorption by Olive Cake in a Batch System, J. Mex. Chem. Soc. 2008, 52(2), 108-115.

[21] Tayel El-Hasan, Zaid A. Al-Anber, Mohammad Al-Anber*, Mufeed Batarseh, Farah Al-Nasr, Anf Ziadat, Yoshigi Kato and Anwar Jiries. Removal of Zn^{2+}, Cu^{2+} and Ni^{2+} Ions from Aqueous Solution via Tripoli: Simple Component with Single Phase Model, Current World Environment, 3(1) (2008) 01-14

[22] L.N. Rao, K.C.K. Krishnaiah, A. Ashutosh, Indian J. Chem. Technol. 1 (1994) 13.

[23] G. Karthikeyan, N.M. Andal and K Anbalagan, Adsorption studies of iron(III) on chitin, J. Chem. Sci. 117 (6) (2005) 663–672.

[24] a) N. Khalid, S.Ahmad, S.N. Kiani, J.Ahmed, Removal ofmercury fromaqueous solutions by adsorption to rice husks, Sep. Sci. Technol. 34 (16) (1999) 3139–3153. b) N. Khalid, S. Ahmed, S.N. Kiani, J. Ahmed, Removal of Lead from Aqueous Solutions Using Rice Husk, Sep. Sci. Technol. 33 (1998) 2349–2362.

[25] G. Karthikeyan, N.M. Andal and K Anbalagan, Adsorption studies of iron(III) on chitin, J. Chem. Sci. 117 (6) (2005) 663–672.

[26] Journal of Research of the National Bureau of Standards—A. Physics and Chemistry, 66A:6 (1962) 503-515

[27] N. Kannan and T. Veemaraj, Removal of Lead(II) Ions by Adsorption onto Bamboo Dust and Commercial Activated Carbons -A Comparative Study, E-Journal of Chemistry, 2009, 6(2), 247-256.

[28] I. Langmuir, adsorption of gases on plain surfaces of glass mica platinum, J. Am. Chem. Soc. 40 (1918) 136-403. b) J.M. Coulson, J.F. Richardson with J.R. Backhurst, J.H. Harker, (1991) Chemical Engineering, Vol 2, 4th Edt. "practical Technology and separation processes" Pergamon Press, Headington Hill Hall, Oxford. c) (Domenico, P.A. and Schwartz, F.W., "Physical and Chemical Hydrogeology", 1st Ed., John Wiley and Sons, New York, 1990). d) Reddi, L.N. and Inyang, H.I., "Geo-Environmental Engineering Principles and Applications", Marcel Decker Inc., New York, 2001. e) Nitzsche, O. and Vereecken, H., "Modelling Sorption and Exchange Processes in Column Experiments and Large Scale Field Studies". Mine Water and the Environment, 21, 15-23, 2002. f) F. Banat, S. Al-Asheh, D. Al-Rousan, A comparative study of copper and zinc ion adsorption on to activated date-pits. Adsorption Science Technology 2002, 20 (4) 319-335.

[29] Vermeulan et al., 1966T.H. Vermeulan, K.R. Hall, L.C. Eggleton and A. Acrivos Ind. Eng. Chem. Fundam. 5 (1966), pp. 212–223.
[30] a) Metcalf and Eddy. (2003). Wastewater Engineering, Treatment, Disposal and Reuse, 3ed Ed. McGraw-Hill: New York. b) Zeldowitsch, 1934 J. Zeldowitsch, Über den Mechanismus der katalytischen Oxydation von CO an MnO_2, Acta Physicochim. URSS 1 (1934), pp. 364–449. c) A. EDWIN VASU, E-Journal of Chemistry, 5:1, (2008) 1-9.
[31] Hutson N D, Yang R T, Adsorption. 1997, 3, 189. Krishna B S, Murty D S R, Jai Prakash B S, J. Colloid Interf Sci. 2000, 229, 230.
[32] a) Arivoli S, Venkatraman B R, Rajachandrasekar T and Hema M, Adsorption of ferrous ion from aqueous solution by low cost activated carbon obtained from natural plant material, Res J Chem Environ. 2007, 17, 70-78. b) Arivoli S, Kalpana K, Sudha R and Rajachandrasekar T, Comparative study on the adsorption kinetics and thermodynamics of metal ions onto acid activated low cost carbon, E J Chem, 2007, 4, 238-254. c) Renmin Gong, Yingzhi Sun, Jian Chen, Huijun Liu, Chao yang, Effect of chemical modification on dye adsorption capacity of peanut hull, Dyes and Pigments, 2005, 67, 179.
[33] Jj S Arivoli, P.Martin Deva Prasath and M Thenkuzhali, EJEAFChe, 6 (9), (2007) 2323-2340
[34] O. Sirichote, W. Innajitara, L. Chuenchom, D. Chunchit and K. Naweekan. Songklanakarin J. Sci. Technol., 2002, 24(2) : 235-242.
M. A. AL-Anber, Z. AL-Anber, I. AL-Momani, Thermodynamics and Kinetic Studies of Iron(III) Adsorption by Jojoba seeds in a Batch System, in preparation (2011).
[35] J. H. Duffus, ''Heavy metals``- A Meaningless Term? IUPAC Technical Report, Pure Appl. Chem. 74:5 (2002) 793-807.
[36] E. D. Weinberg, Iron Loading and Disease Surveillance. Emerging Infectious Diseases, 5 (3) 1999.
[37] B. Das, P. Hazarika, G. Saikia, H. Kalita, D.C. Goswami, H.B. Das, S.N. Dube, R.K. Dutta, Removal of iron from groundwater by ash: A systematic study of a traditional method, Journal of Hazardous Materials 141 (2007) 834–841.
[38] a) Gedge G (1992) Corrosion of cast Iron in potable water service. In: Proceedings of the Institute of Materials Conference. London, UK. b) Rice, O. Corrosion Control with Calgon. Journal AWWA, 39(6), (1947) 552. c) Hidmi, L.; Gladwell, D. & Edwards, M. Water Quality and Lead, Copper, and Iron Corrosion in Boulder Water. Report to the City of Boulder, CO (1994).
[39] J.D. Zuane, Handbook of Drinking Water Quality. Van Nostrand Reinhold, New York, 1990.
a) G. Chen, Electrochemical technologies in wastewater treatment. Sep. Purif. Technol. 38 (1), (2004) 11–41. b) M. Wessling-Resnick, Biochemistry of iron uptake. Crit. Rev. Biochem. Mol. Biol. 34, (1999) 285–314. c) G. J. Kontoghiorghes and E.D. Weinberg Iron: mammalian defense systems, mechanisms of disease, and chelation therapy approaches. Blood Rev. 9 (1995) 33-45. d) E.D. Weinberg and G.A. Weinberg. The role of iron in infection. Current Opinion in Infectious Diseases, 8 (1995) 164-9. e) E.D. Weinberg, The role of iron in cancer. Eur. J. Cancer. Prev. 5 (1996) 19-36. f) J.R. Connor and J.L. Beard, Dietary iron supplements in the elderly: to use or not to use? Nutrition Today, 32 (1997) 102-9. j) T-P. Tuomainen, K. Punnonen, K.

Nyyssonen, J.T. Salonen. Association between body iron stores and the risk of acute myocardial infarction in men. Circulation, 97 (1998) 1461-6. h) J.M. McCord, Effects of positive iron status at a cellular level. Nutr. Rev., 54 (1996) 85-8. i) D. Weinberg, Patho-ecological implications of microbial acquisition of host iron. Reviews in Medical Microbiology, 9 (1998) 171-8.

[40] P. Sarin, V.L. Snoeyink, J. Bebee, K.K. Jim, M. A. Beckett, W.M. Kriven, J. A. Clement, Iron release from corroded iron pipes in drinking water distribution systems, Water Research, 38,5, (2004)1259-1269.

[41] http://www.unu.edu/unupress/unupbooks/80858e/80858E02.htm

[42] F.R. Spellman, Handbook for Waterworks Operator Certification, Vol. 2, Technomic Publishing Company Inc., Lancaster, USA, 2001, pp 6–11, 81–83.

W.C. Andersen, T.J. Bruno, Application of gas–liquid entraining rotor to supercritical fluid extraction: removal of iron (III) from water, Anal. Chim. Acta 485 (2003) 1–8.

[43] a) D. Ellis, C. Bouchard, G. Lantagne, Removal of iron and manganese from groundwater by oxidation and microfiltration, Desalination 130 (2000) 255–264. b) Hikmet Katırcıog˘lu, Belma Aslım, Ali Rehber Tu¨rker, Tahir Atıc, Yavuz Beyatl, Removal of cadmium(II) ion from aqueous system by dry biomass, immobilized live and heat-inactivated Oscillatoria sp. H1 isolated from freshwater (Mogan Lake) Bioresource Technology 99 (2008) 4185–4191

[44] P. Berbenni, A. Pollice, R. Canziani, L. Stabile, F. Nobili, Removal of iron and manganese from hydrocarbon-contaminated groundwaters, Bioresour. Technol. 74 (2000) 109–114.

[45] a) M. Kalin, W.N. Wheeler, G. Meinrath, The removal of uranium from mining waste water using algal/microbial biomass. J. Environ. Radioact. 78 (2005) 151–177. b) F. Veglio, F. Beolchini, Removal of metals by biosorption: a review. Hydrometall. 44 (1997) 301–316.

[46] M. Uchida, S. Ito, N. Kawasaki, T. Nakamura, S. Tanada, Competitive adsorption of chloroform and iron ion onto activated carbon fiber, J. Colloid Interf. Sci. 220 (1999) 406–409.

[47] M. Pakula, S. Biniak, A. Swiatkowski, Chemical and electrochemical studies of interactions between iron (III) ions and activated carbon surface, Langmuir 14 (1998) 3082–3089.

[48] C. Huang, W.P. Cheng, Thermodynamic parameters of iron-cyanide adsorption onto - Al_2O_3, J. Colloid Interf. Sci. 188 (1997) 270–274.

[49] a) E. A. Sigworth and S. B. Smith, Adsorption of inorganic compound by activated carbon, J. Am. Water Works Assoc., 64, 386 (1972). b) E. Jackwerth, J. Lohmar and G. Wittler, Fresenius' Z. Anal. Chem., 266, 1(1973). c) A. Edwin Vasu, Adsorption of Ni(II), Cu(II) and Fe(III) from Aqueous Solutions Using Activated Carbon, E-Journal of Chemistry, 5 (1), 2008, 1-9. d) A. Ucer, A. Uyanik, S.F. Aygun, Adsorption of Cu^{2+}, Cd^{2+}, Zn^{2+}, Mn^{2+} and Fe^{3+} ions by tannic acid immobilised activated carbon, Separation and Purification Technology 47 (2006) 113–118.

[50] Hideko Koshima, Adsorption of Iron(III) on Activated Carbon from Hydrochloric Acid Solution, Analytical Sciences (1), 1985, 195.

[51] a) N. Cvjetićanin, D. Cvjetićanin, D. Golobočanin, and M. Pravica, Adsorption of colloidal trivalent iron on alumina, J. Radioanalytical and Nuclear Chem., 54(1-2), 1979, 149-158.

[52] A. Edwin Vasu, E-Journal of Chemistry, 5:1, (2008) 1-9.
[53] M.M. Nassar, K.T. Ewida, E.E. Ebrahiem, Y.H. Magdy and M.H. Mheaedi, Adsorption of iron and manganese ions using low-cost materials as adsorbents, Adsorp. Sci. Technol., 22(1) (2004) 25–37.
[54] Nacèra Yeddou, Aicha Bensmaili, Equilibrium and kinetic modeling of iron adsorption by eggshells in a batch system: effect of temperature, Desalination 206 (2007) 127–134.
[55] W. S. Wan Ngah, S. Ab Ghani and A. Kamari, Bioresource Technology Volume 96, Issue 4, March 2005, Pages 443-450.
[56] Peniche-Covas, C., Alvarez, L.W., Arguelles-Monal, W., 1992. The adsorption of mercuric ions by chitosan. J. Appl. Polym. Sci. 46, 1147–1150.
[57] O. Sirichote, W. Innajitara, L. Chuenchom, D. Chunchit and K. Naweekan. Song klanakarin J. Sci. Technol., 2002, 24(2) : 235-242.
[58] Uchida, M., Shinohara, O., Ito, S., Kawasaki, N., Nakamura, T. and Tanada, S. 2000. Reduction of iron(III) ion by activated carbon fiber. J. Colloid Interface Sci. 224: 347-350.
[59] C.Y. Abasi , A.A. Abia and J.C. Igwe, Environmental Research Journal, 5(3), 2011, Page No.: 104-113.
[60] Y. S. Ho, C. T. Huang, H. W. Huang, equlibrium sorption isotherm for metal ions on tree fern. Process biochem, 37 (2002) 1421-1430.
[61] B. A. Shah, A. V. Shah, R. R> Singh, Sorption isotherms and kinatics of chrmoium uptake from watewater using natural sorpbent materails. Intern. J. Environ. Sci. Technology. 6, (2009) 77-90.
[62] a) M. Horsfall, A.I. Spiff, A.A. Abia, studies the influence of mercaptoacetic acid (MAA) modification of cassava (Manihot esculenta cranz) waste biomass on the adsorption of Cu^{2+} and Cd^{2+} from aqueous solution. Bull Korean Chem. Soc. 25 (2004) 969-976. b) P. Loderio, J,L. Barriada, R. Herrero, M.E. Sastre-Vicente, The marine macroalge crstoserira baccata as biosorbent for Cd(II) and Pb(II) removal: kinetic and equilibrium studies. Environ. Pollution, 142 (2006) 264-273.

2

Thermodynamics and Reaction Rates

Miloslav Pekař
Brno University of Technology
Czech Republic

1. Introduction

Thermodynamics has established in chemistry principally as a science determining possibility and direction of chemical transformations and giving conditions for their final, equilibrium state. Thermodynamics is usually thought to tell nothing about rates of these processes, their velocity of approaching equilibrium. Rates of chemical reactions belong to the domain of chemical kinetics. However, as thermodynamics gives some restriction on the course of chemical reactions, similar restrictions on their rates are continuously looked for. Similarly, because thermodynamic potentials are often formulated as driving forces for various processes, a thermodynamic driving force for reactions rates is searched for.

Two such approaches will be discussed in this article. The first one are restrictions put by thermodynamics on values of rate constants in mass action rate equations. The second one is the use of the chemical potential as a general driving force for chemical reactions and also "directly" in rate equations. These two problems are in fact connected and are related to expressing reaction rate as a function of pertinent independent variables.

Relationships between chemical thermodynamics and kinetics traditionally emerge from the ways that both disciplines use to describe equilibrium state of chemical reactions (chemically reacting systems or mixtures in general). Equilibrium is the main domain of classical, equilibrium, thermodynamics that has elaborated elegant criteria (or, perhaps, definitions) of equilibria and has shown how they naturally lead to the well known equilibrium constant. On the other hand, kinetics describes the way to equilibrium, i.e. the nonequilibrium state of chemical reactions, but also gives a clear idea on reaction equilibrium. Combining these two views various results on compatibility between thermodynamics and kinetics, on thermodynamic restrictions to kinetics etc. were published. The main idea can be illustrated on the trivial example of decomposition reaction AB = A + B with rate (kinetic) equation $r = \vec{k}c_{AB} - \overleftarrow{k}c_A c_B$ where r is the reaction rate, $\vec{k}, \overleftarrow{k}$ are the forward and reverse rate constants, and c_α are the concentrations. In equilibrium, the reaction rate is zero, consequently $\vec{k} / \overleftarrow{k} = (c_A c_B / c_{AB})_{eq}$. Because the right hand side corresponds to the thermodynamic equilibrium constant (K) it is concluded that $K = \vec{k} / \overleftarrow{k}$. However, this is simplified approach not taking into account conceptual differences between the true thermodynamic equilibrium constant and the ratio of rate constants that is called here the kinetic equilibrium constant. This discrepancy is sometimes to be removed by restricting this approach to ideal systems of elementary reactions but even then some questions remain.

Chemical potential (μ) is introduced into chemical kinetics by similar straightforward way (Qian & Beard, 2005). If it is expressed by $\mu_\alpha = \mu_\alpha^\circ + RT\ln c_\alpha$, multiplied by stoichiometric coefficients, summed and compared with rate equation it is obtained for the given example that:

$$\Delta\mu \equiv -\mu_{AB} + \mu_A + \mu_B = RT\ln\frac{c_A c_B}{Kc_{AB}} = RT\ln\left(\overleftarrow{r} / \overrightarrow{r}\right) \tag{1}$$

(note that the equivalence of thermodynamic and kinetic equilibrium constants is supposed again; $\overrightarrow{r}, \overleftarrow{r}$ are the forward and reverse rates). Equation (1) used to be interpreted as determining the (stoichiometric) sum of chemical potentials (Δμ) to be some (thermodynamic) "driving force" for reaction rates. In fact, there is "no kinetics", no kinetic variables in the final expression $\Delta\mu = RT\ln\left(\overleftarrow{r} / \overrightarrow{r}\right)$ and reaction rates are directly determined by chemical potentials what is questionable and calls for experimental verification.

2. Restrictions put by thermodynamics on values of rate constants

2.1 Basic thermodynamic restrictions on rate constants coming from equilibrium

Perhaps the only one work which clearly distinguishes kinetic and thermodynamic equilibrium constant is the kinetic textbook by Eckert and coworkers (Eckert et al., 1986); the former is in it called the empirical equilibrium constant. This book stresses different approaches of thermodynamics and kinetics to equilibrium. In thermodynamics, equilibrium is defined as a state of minimum free energy (Gibbs energy) and its description is based on stoichiometric equation and thermodynamic equilibrium constant containing activities. Different stoichiometric equations of the same chemical equation can give different values of thermodynamic equilibrium constant, however, equilibrium composition is independent on selected stoichiometric equation. Kinetic description of equilibrium is based on zero overall reaction rate, on supposed reaction mechanism or network (reaction scheme) and corresponding kinetic (rate) equation. Kinetic equilibrium constant usually contains concentrations. According to that book, thermodynamic equilibrium data should be introduced into kinetic equations indirectly as shown in the Scheme 1.

Simple example reveals basic problems. Decomposition of carbon monoxide occurs (at the pressure p) according to the following stoichiometric equation:

$$2\,CO = CO_2 + C \tag{R1}$$

Standard state of gaseous components is selected as the ideal gas at 101 kPa and for solid component as the pure component at the actual pressure (due to negligible effects of pressure on behavior of solid components, the dependence of the standard state on pressure can be neglected here). Ideal behavior is supposed. Then $a_\alpha = p_\alpha/p^\circ = p_{rel}\, n_\alpha/n_\Sigma$ for α = CO, CO_2, where $p_{rel} = p/p^\circ$, and $a_C = 1$; a_α is the activity, p_α is the partial pressure, p° the standard pressure, n_α is the number of moles, and n_Σ the total number of moles. Thermodynamic equilibrium constant is then given by

$$K = \left(\frac{n_\Sigma n_{CO_2}}{p_{rel} n_{CO}^2}\right)_{eq} = \left[\frac{(c_{CO} + c_{CO_2})c_{CO_2}}{p_{rel} c_{CO}^2}\right]_{eq} \tag{2}$$

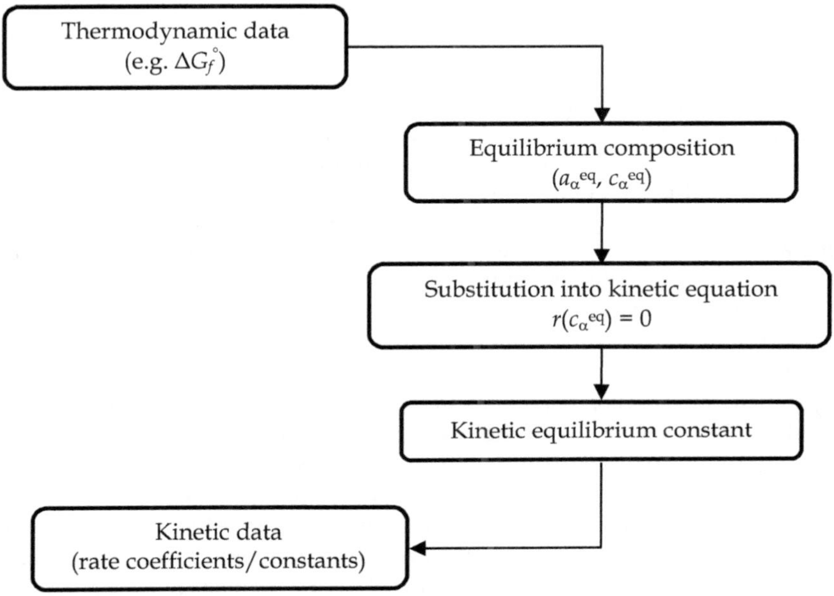

Scheme 1. Connecting thermodynamics and kinetics correctly (Eckert et al., 1986)

On contrary, the ratio of rate constants is given by

$$\left(\frac{\vec{k}}{\overleftarrow{k}}\right)_{eq} = \left(\frac{c_{CO_2} c_C}{c_{CO}^2}\right)_{eq} \tag{3}$$

It is clear that thermodynamic and kinetic equilibrium constants need not be equivalent even in ideal systems. For example, the former does not contain concentration of carbon and though this could be remedied by stating that carbon amount does not affect reaction rate and its concentration is included in the reverse rate constant, even then the kinetic equilibrium constant could depend on carbon amount in contrast to the thermodynamic equilibrium constant. Some discrepancies could not be remedied by restricting on elementary reactions only – in this example the presence of p_{rel} and of the total molar amount, generally, the presence of quantities transforming composition variables into standard state-related (activity-related) variables, and, of course, discrepancy in dimensionalities of the two equilibrium constants.

Let us use the same example to illustrate the procedure suggested by Eckert et al. (1986). At 1300 K and 202 kPa the molar standard Gibbs energies are (Novák et al., 1999): $G_m^\circ(CO) = -395.3$ kJ/mol, $G_m^\circ(CO_2) = -712.7$ kJ/mol, $G_m^\circ(C) = -20.97$ kJ/mol and from them the value of thermodynamic equilibrium constant is calculated: K = 0.00515. Equilibrium molar balance gives $(n_{CO_2})_{eq} = (n_C)_{eq} = x$, $(n_{CO})_{eq} = 1-2x$, $n_\Sigma = 1-x$. Then from (2) follows x = 0.0107 (Novák et al., 1999). Equilibrium composition is substituted into (3):

$$\left(\frac{\vec{k}}{\overleftarrow{k}}\right)_{\text{eq}} = \frac{0.0107 \times 0.0107}{0.09786^2} = 0.012 \tag{4}$$

and this is real and true result of thermodynamic restriction on values of rate constants valid at given temperature. More precisely, this is a restriction put on the ratio of rate constants, values of which are supposed to be independent on equilibrium, in other words, dependent on temperature (and perhaps on pressure) only and therefore this restriction is valid also out of equilibrium at given temperature. The numerical value of this restriction is dependent on temperature and should be recalculated at every temperature using the value of equilibrium constant at that temperature.

Thus, simple and safe way how to relate thermodynamics and kinetics, thermodynamic and kinetic equilibrium constants, and rate constants is that shown in Scheme 1. However, it gives no general equations and should be applied specifically for each specific reaction (reacting system) and reaction conditions (temperature, at least). There are also works that try to resolve relationship between the two types of equilibrium constant more generally and, in the same time, correctly and consistently. They were reviewed previously and only main results are presented here, in the next section. But before doing so, let us note that kinetic equilibrium constant can be used as a useful indicator of the distance of actual state of reacting mixture from equilibrium and to follow its approach to equilibrium. In the previous example, actual value of the fraction $c_{\mathrm{CO_2}} c_{\mathrm{C}} / c_{\mathrm{CO}}^2$ can be compared with the value of the ratio $\vec{k} / \overleftarrow{k}$ and relative distance from equilibrium calculated, for more details and other examples see our previous work (Pekař & Koubek, 1997, 1999, 2000).

2.2 General thermodynamic restrictions on rate constants

As noted in the preceding section there are several works that do not rely on simple identification of thermodynamic and kinetic equilibrium constants. Hollingsworth (1952a, 1952b) generalized restriction on the ratio of forward and reverse reaction rates (*f*) defined by

$$f(c_\alpha, T) = \vec{f}(c_\alpha, T) / \overleftarrow{f}(c_\alpha, T) \equiv \vec{r} / \overleftarrow{r} \tag{5}$$

Hollingsworth showed that sufficient condition for consistent kinetic and thermodynamic description of equilibrium is

$$F(Q_r, T) = \Phi(Q_r/K) \text{ and } \Phi(1) = 1 \tag{6}$$

where F is the function f with transformed variables, $F(Q_r, T) = f(c_\alpha, T)$, and Q_r is the well known reaction quotient. The first equality in (6) says that function F should be expressible as a function Φ of Q_r/K. This is too general condition saying explicitly nothing about rate constants. Identifying kinetic equilibrium constant with thermodynamic one, condition (6) is specialized to

$$\Phi(Q_r/K) = (Q_r/K)^{-z} \tag{7}$$

where z is a positive constant. Equation (7) is a generalization of simple identity $K = \vec{k} / \overleftarrow{k}$ from introduction. Hollingsworth also derived the necessary consistency condition:

$$f - 1 = (Q_r/K - 1)\,\Psi(c_\alpha, T, u_j) \tag{8}$$

in the neighbourhood of $Q_r/K = 1$ (i.e., of equilibrium); u_j stands for a set of non-thermodynamic variables. Example of practical application of Hollingsworth's approach in an ideal system is given by Boyd (Boyd, 1977).
Blum (Blum & Luus, 1964) considered a general mass action rate law formulated as follows:

$$r = \vec{k}\phi\prod_{\alpha=1}^{m} a_\alpha^{\omega_\alpha} - \overleftarrow{k}\phi\prod_{\alpha=1}^{m} a_\alpha^{\omega'_\alpha} \tag{9}$$

where ϕ is some function of activities, a_α, of reacting species, ω_α and ω'_α are coefficients which may differ from the stoichiometric coefficients (ν_α), in fact, reaction orders. Supposing that both the equilibrium constant and the ratio of the rate constants are dependent only on temperature, they proved that

$$\vec{k} / \overleftarrow{k} = K^z \tag{10}$$

where

$$z = (\omega'_\alpha - \omega_\alpha) / \nu_\alpha;\ \alpha = 1,\ldots,n \tag{11}$$

General law (9) is rarely used in chemical kinetics, in reactions of ions it probably does not work (Laidler, 1965; Boudart, 1968). It can be transformed, particularly simply in ideal systems, to concentrations. Samohýl (personal communication) pointed out that criteria (11) may be problematic, especially for practically irreversible reactions. For example, reaction orders for reaction 4 NH_3 + 6 NO = 5 N_2 + 6 H_2O were determined as follows: $\omega_{NH_3} = 1$, $\omega_{NO} = 0.5$, $\omega_{N_2} = \omega_{H_2O} = 0$. Orders for reversed direction are unknown, probably because of practically irreversible nature of the reaction. Natural selection could be, e.g., $\omega'_{NO} = 0$ (reaction is not inhibited by reactant), then $z = 1/12$ and from this follows $\omega'_{NH_3} = 2/3$ which seems to be improbable (rather strong inhibition by reactant).

2.3 Independence of reactions, Wegscheider conditions

Wegscheider conditions belong also among "thermodynamic restrictions" on rate constants and have been introduced more than one hundred years ago (Wegscheider, 1902). In fact, they are also based on equivalence between thermodynamic and kinetic equilibrium constants disputed in previous sections. Recently, matrix algebra approaches to find these conditions were described (Vlad & Ross, 2009). Essential part of them is to find (in)dependent chemical reactions. Problem of independent and dependent reactions is an interesting issue sometimes found also in studies on kinetics and thermodynamics of reacting mixtures. As a rule, a reaction scheme, i.e. a set of stoichiometric equations (whether elementary or nonelementary), is proposed, stoichiometric coefficients are arranged into stoichiometric matrix and linear (matrix) algebra is applied to find its rank which determines the number of linearly (stoichiometrically) independent reactions; all other reactions can be obtained as linear combinations of independent ones. This procedure can be viewed as an a posteriori analysis of the proposed reaction mechanism or network. Bowen has shown (Bowen, 1968) that using not only matrix but also vector algebra interesting results can be obtained on the basis of knowing only components of reacting mixture, i.e. with no reaction scheme. This is a priori type of analysis and is used in continuum nonequilibrium (rational) thermodynamics. Because Bowen's results are important for this article they are briefly reviewed now for reader's convenience.

Let a reacting mixture be composed from n components (compounds) which are formed by z different atoms. Atomic composition of each component is described by numbers $T_{\sigma\alpha}$ that indicate the number of atoms σ (= 1, 2,..., z) in component α (= 1, 2,..., n). Atomic masses M_{a}^{σ} in combination with these numbers determine the molar masses M_{α}:

$$M_{\alpha} = \sum_{\sigma=1}^{z} M_{\mathrm{a}}^{\sigma} T_{\sigma\alpha} \tag{12}$$

Although compounds are destroyed or created in chemical reactions the atoms are preserved. If J^{α} denotes the number of moles of the component α formed or reacted per unit time in unit volume, i.e. the reaction rate for the component α (component rate in short), then the persistence of atoms can be formulated in the form

$$\sum_{\alpha=1}^{n} T_{\sigma\alpha} J^{\alpha} = 0; \qquad \sigma = 1,2,\ldots,z \tag{13}$$

This result expresses, in other words, the mass conservation.

Atomic numbers can be arranged in matrix $\| T_{\sigma\alpha} \|$ of dimension $z \times n$. Chemical reactions are possible if its rank (h) is smaller than the number of components (n), otherwise the system (13) has only trivial solution, i.e. is valid only for zero component rates. If $h < z$ then a new $h \times n$ matrix $\| S_{\sigma\alpha} \|$ with rank h can be constructed from the original matrix $\| T_{\sigma\alpha} \|$ and used instead of it:

$$\sum_{\alpha=1}^{n} S_{\sigma\alpha} J^{\alpha} = 0; \qquad \sigma = 1,2,\ldots,h \tag{14}$$

In this way only linearly independent relations from (13) are retained and from the chemical point of view it means that instead of (some) atoms with masses M_{a}^{σ} only some their linear combinations with masses M_{e}^{σ} should be considered as elementary building units of components:

$$M_{\alpha} = \sum_{\sigma=1}^{h} M_{\mathrm{e}}^{\sigma} S_{\sigma\alpha} \tag{15}$$

Example. Mixture of NO_2 and N_2O_4 has the matrix $\| T_{\sigma\alpha} \|$ of dimension 2×2 and rank 1; the matrix $\| S_{\sigma\alpha} \|$ is of dimension 1×2 and can be selected as $(1 \quad 2)$ which means that the elementary building unit is NO_2 and $M_{\mathrm{e}}^{1} = M_{\mathrm{a}}^{1} + 2M_{\mathrm{a}}^{2} \equiv M_{\mathrm{a}}^{\mathrm{N}} + 2M_{\mathrm{a}}^{\mathrm{O}}$.

Multiplying each of the z relations (13) by corresponding M_{a}^{σ} and summing the results for all σ it follows that $\sum_{\alpha=1}^{n} M_{\alpha} J^{\alpha} = 0$. This fact can be much more effectively formulated in vector form because further important implications than follow. The last equality indicates that component molar masses and rates should form two perpendicular vectors, i.e. vectors with vanishing scalar product. Let us introduce n-dimensional vector space, called the component space and denoted by U, with base vectors $\mathbf{e}_\alpha$ and reciprocal base vectors $\mathbf{e}^\alpha$ (α = 1, 2,..., n). Then the vector of molar masses **M** and the vector of reaction rates **J** are defined in this space as follows:

$$\mathbf{M} = \sum_{\alpha=1}^{n} M_\alpha \mathbf{e}^\alpha, \quad \mathbf{J} = \sum_{\alpha=1}^{n} J^\alpha \mathbf{e}_\alpha \tag{16}$$

To proceed further we use relations (14) and (15) because in contrast to relations (12) and (13) the matrix $\|S_{\sigma\alpha}\|$ is of "full rank" (does not contain linearly dependent rows). The product of the two vectors can be then expressed in the following form:

$$\mathbf{M}.\mathbf{J} = \left(\sum_{\alpha=1}^{n}\sum_{\sigma=1}^{h} M_\mathrm{e}^\sigma S_{\sigma\alpha}\mathbf{e}^a\right).\left(\sum_{\alpha=1}^{n} J^\alpha \mathbf{e}_\alpha\right) = \left(\sum_{\sigma=1}^{h} M_\mathrm{e}^\sigma \sum_{\alpha=1}^{n} S_{\sigma\alpha}\mathbf{e}^a\right).\left(\sum_{\alpha=1}^{n} J^\alpha \mathbf{e}_\alpha\right) = 0 \tag{17}$$

where the latter equality follows using (14). Because the matrix $\|S_{\sigma\alpha}\|$ has rank h, the vectors

$$\mathbf{f}_\sigma = \sum_{\alpha=1}^{n} S_{\sigma\alpha}\mathbf{e}^a; \quad \sigma = 1,2,\ldots,h \tag{18}$$

that appear in (17) are linearly independent and thus form a basis of a h-dimensional subspace W of the space U (remember that $h < n$). This subspace unambiguously determines complementary orthogonal subspace V (of dimension n–h), i.e. U = V $\oplus$ W, V $\perp$ W. From (17) follows:

$$\mathbf{M} = \sum_{\sigma=1}^{h} M_\mathrm{e}^\sigma \mathbf{f}_\sigma \tag{19}$$

which shows that **M** can be expressed in the basis of the subspace W or **M** $\in$ W. From (14) and $(16)_2$ follows:

$$\mathbf{J}.\mathbf{f}_\sigma = 0; \quad \sigma = 1,2,\ldots,h \tag{20}$$

which means that **J** is perpendicular to all basis vectors of the subspace W, consequently, **J** lies in the complementary orthogonal subspace V, **J** $\in$ V.

Let us now select basis vectors in the subspace V and denote them $\mathbf{d}^p$, p = 1, 2,..., n–h. Of course, these vectors lie also in the (original) space U and can be expressed using its basis vectors analogically to (16):

$$\mathbf{d}^p = \sum_{\alpha=1}^{n} P^{p\alpha}\mathbf{e}_\alpha \tag{21}$$

Because of orthogonality of subspaces V and W, their bases conform to equation

$$\mathbf{f}_\sigma.\mathbf{d}^p = \sum_{\alpha=1}^{n} S_{\sigma\alpha}P^{p\alpha} = 0 \tag{22}$$

which can be alternatively written in matrix form as

$$\| P^{p\alpha} \| \times \| S_{\sigma\alpha} \|^{\mathrm{T}} = \| 0 \| \tag{23}$$

Meaning of the matrix $\| P^{p\alpha} \|$ can be deduced from two consequences. First, because the reaction vector **J** lies in the subspace V, it can be expressed also using its basis vectors, $\mathbf{J} = \sum_{p=1}^{n-h} J_p \mathbf{d}^p$. Substituting for **J** from $(16)_2$ and for $\mathbf{d}^p$ from (21), it follows:

$$J^\alpha = \sum_{p=1}^{n-h} J_p P^{p\alpha}; \ \alpha = 1, 2, \ldots, n \tag{24}$$

Second, because the vector of molar masses **M** is in the subspace W, it is perpendicular to all vectors $\mathbf{d}^p$ and thus

$$0 = \mathbf{d}^p.\mathbf{M} = \sum_{\alpha=1}^{n} P^{p\alpha} M_\alpha ; p = 1, 2, \ldots, n-h \tag{25}$$

as follows after substitution from (19), (21), (220. Eq. (25) shows that matrix $\| P^{p\alpha} \|$ enables to express component rates in n–h quantities J_p which are, in fact, rates of n–h independent reactions shown by (25) if instead of molar masses M_α the corresponding chemical symbols are used. In other words $\| P^{p\alpha} \|$ is the matrix of stoichiometric coefficients of component α in (independent) reaction p.

Vector algebra thus shows that chemical transformations fulfilling persistence of atoms (mass conservation) can be equivalently described either by component reaction rates or by rates of independent reactions. The number of the former is equal to the number of components (n) whereas the number of the latter is lower (n–h) which could decrease the dimensionality of the problem of description of reaction rates. In kinetic practice, however, changes in component concentrations (amounts) are measured, i.e. data on component rates and not on rates of individual reactions are collected. Reactions, in the form of reaction schemes, are suggested a posteriori on the basis of detected components, their concentrations changing in time and chemical insight. Then dependencies between reactions can be searched. Vector analysis offers rather different procedure outlined in Scheme 2. Dependencies are revealed at the beginning and then only independent reactions are included in the (kinetic) analysis. Vector analysis also shows how to transform (measured) component rates into (suggested, selected) rates of independent reactions. This transformation is made by standard procedure for interchange between vector bases or between vector coordinates in different bases. First, the contravariant metric tensor with components $d^{rp} = \mathbf{d}^r.\mathbf{d}^p$ is constructed and then its inversion (covariant metric tensor) with components d_{rp} is found. From $\mathbf{J} = \sum_{p=1}^{n-h} J_p \mathbf{d}^p$ it follows that $\mathbf{J}.\mathbf{d}_r = \sum_{p=1}^{n-h} J_p \mathbf{d}^p.\mathbf{d}_r \equiv J_r$. Using in the latter equation the well known relationship between metric tensors and corresponding base vectors and the definition of base vectors (21) it finally follows:

$$J_p = \sum_{\alpha=1}^{n} \left(J^\alpha \sum_{r=1}^{n-h} P^{r\alpha} d_{rp} \right); p = 1, 2, \ldots, n-h \tag{26}$$

Of course, so far we have seen only relationships between reaction rates and no explicit equations for them like, e.g., the kinetic mass action law. Analysis based only on permanence of atoms cannot give such equations – they belong to the domain of chemical kinetics although they can also be devised by thermodynamics, see Section 4.
Simple example on Wegscheider conditions was presented by Vlad and Ross (Vlad & Ross, 2009) – isomerization taking place in two ways:

$$A = B, \quad 2A = A + B \tag{R2}$$

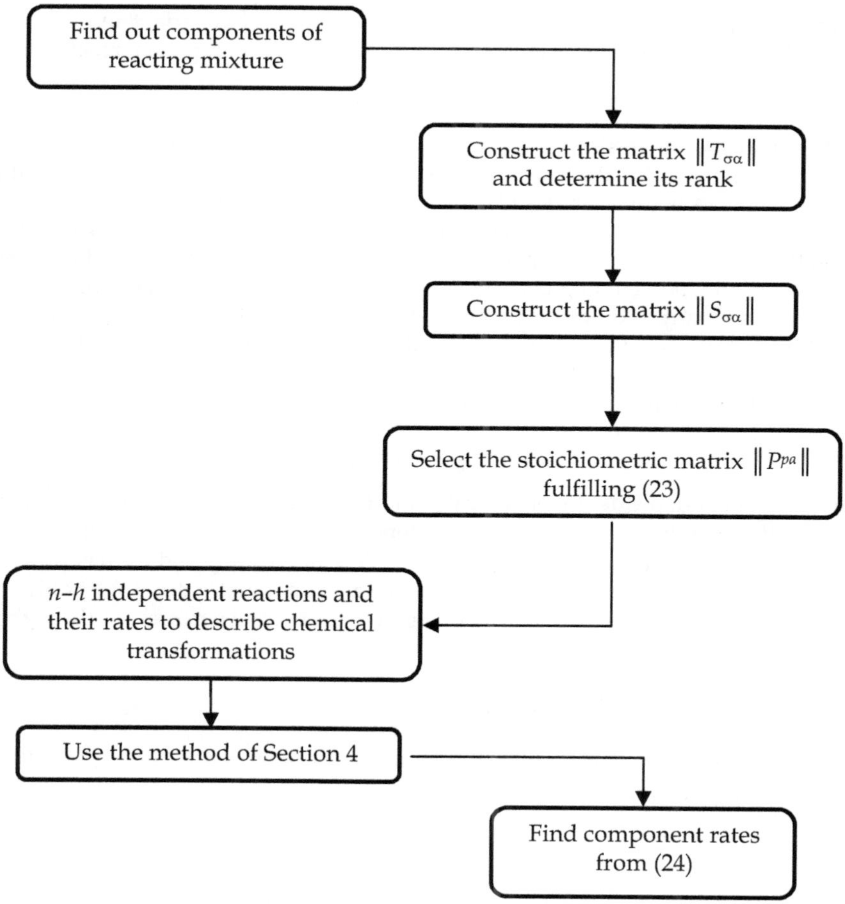

Scheme 2. Alternative procedure to find reaction rates

Vlad and Ross note that if the (thermodynamic) equilibrium constant is $K = (c_B / c_A)_{eq}$ and if kinetic equations are expressed e.g. $r_1 = \vec{k}_1 c_A - \overleftarrow{k}_1 c_B$ then the consistency between thermodynamic and kinetic description of equilibrium is achieved only if the following (Wegscheider) condition holds:

$$\vec{k}_1 / \overleftarrow{k}_1 = \vec{k}_2 / \overleftarrow{k}_2 = K \tag{27}$$

It can be easily checked that in this mixture of one kind of atom and two components the rank of the matrix $\| T_{\sigma\alpha} \|$ (dimension 1×2) is 1 and there is only one independent reaction. The matrix $\| S_{\sigma\alpha} \|$ can be selected as equal to the matrix $\| T_{\sigma\alpha} \|$ and then the stoichiometric matrix can be selected as $\begin{pmatrix} -1 & 1 \end{pmatrix}$ which corresponds to the first reaction (A = B) selected as the independent reaction. There is one base vector $\mathbf{d}^1 = -\mathbf{e}_1 + \mathbf{e}_2$ giving one component contravariant tensor $d^{11} = 2$ and corresponding component of covariant tensor $d_{11} = 1/2$. Consequently, the rate of the independent reaction is related to component reaction rates by:

$$J^{\mathrm{A}} = J_1 P^{11} = -J_1, \quad J^{\mathrm{B}} = J_1 P^{12} = J_1 \tag{28}$$

and $J^{\mathrm{A}} = -J^{\mathrm{B}}$ which follows also from (14). Kinetics of transformations in a mixture of two isomers can be thus fully described by one reaction rate only - either from the two component rates can be measured and used for this purpose, the other component rate is then determined by it, can be calculated from it. At this stage of analysis there is no indication that two reactions should be considered and this should be viewed as some kind of "external" information coming perhaps from experiments. At the same time this analysis does not provide any explicit expression for reaction rate and its dependence on concentration - this is another type of external information coming usually from kinetics. Let us therefore suppose the two isomerization processes given above and their rates formulated in the form of kinetic mass action law:

$$r_1 = \vec{k}_1 c_{\mathrm{A}} - \overleftarrow{k}_1 c_{\mathrm{B}}, \quad r_2 = \vec{k}_2 c_{\mathrm{A}}^2 - \overleftarrow{k}_2 c_{\mathrm{A}} c_{\mathrm{B}} \tag{29}$$

Then the only one independent reaction rate is in the form $J_1 = r_1 + r_2$. Note, that although the first reaction has been selected as the independent reaction, the rate of independent reaction is not equal to (its mass action rate) r_1. This interesting finding has probably no specific practical implication. However, individual traditional rates (r_i) should not be independent. Let us suppose that r_2 is dependent on r_1, i.e. can be expressed through it: $r_2 = br_1$; then

$$\left(b\vec{k}_1 - \vec{k}_2 c_{\mathrm{A}}\right) c_{\mathrm{A}} - \left(b\overleftarrow{k}_1 - \overleftarrow{k}_2 c_{\mathrm{A}}\right) c_{\mathrm{B}} = 0 \tag{30}$$

should be valid for any concentrations. Sufficient conditions for this are $b = \vec{k}_2 c_{\mathrm{A}} / \vec{k}_1 = \overleftarrow{k}_2 c_{\mathrm{A}} / \overleftarrow{k}_1$ and from them follows:

$$\vec{k}_1 \overleftarrow{k}_2 = \overleftarrow{k}_1 \vec{k}_2 \tag{31}$$

i.e. "kinetic part" of Wegscheider condition (27). Substituting derived expressions for b into br_1 it can be easily checked that r_2 really results. Although the derivation is rather straightforward and is not based on linear dependency with constant coefficients it points to assumption that Wegscheider conditions are not conditions for consistency of kinetics with thermodynamics but results of dependencies among reaction rates. Moreover, this derivation need not suppose equality of thermodynamic and kinetic equilibrium constant.

There is a thermodynamic method giving kinetic description in terms of independent reactions as noted in Scheme 2, see Section 4.
More complex reaction mixture and scheme was discussed by Ederer and Gilles (Ederer & Gilles, 2007). Their mixture was composed from six formal components (A, B, C, AB, BC, ABC) formed by three atoms (A, B, C). Three independent reactions are possible in this mixture while four reactions were considered by Ederer and Gilles (Ederer & Gilles, 2007) $r_4 = b_1r_1 + b_2r_2 + b_3r_3$ with following mass action rate equations:

$$r_1 = \vec{k}_1 c_A c_B - \overleftarrow{k}_1 c_{AB},\quad r_2 = \vec{k}_2 c_{AB} c_C - \overleftarrow{k}_2 c_{ABC},\quad r_3 = \vec{k}_3 c_B c_C - \overleftarrow{k}_3 c_{BC},\quad r_4 = \vec{k}_4 c_A c_{BC} - \overleftarrow{k}_4 c_{ABC} \quad (32)$$

Let us suppose that the fourth reaction rate can be expressed through the other three rates: $b_1r_1 + b_2r_2 + b_3r_3$. By similar procedure as in the preceding example we arrive at conditions $b_2 = \overleftarrow{k}_4 / \overleftarrow{k}_2$, $b_3 = -\vec{k}_4 c_A / \overleftarrow{k}_3$, and $b_1 = b_2 \vec{k}_2 c_C / \overleftarrow{k}_1 = -b_3 \vec{k}_3 c_C / \vec{k}_1 c_A$ from which it follows that

$$\frac{\overleftarrow{k}_1 \overleftarrow{k}_2 \vec{k}_3 \vec{k}_4}{\vec{k}_1 \vec{k}_2 \overleftarrow{k}_3 \overleftarrow{k}_4} = 1 \quad (33)$$

i.e., Wegscheider condition derived in (Ederer & Gilles, 2007) from equilibrium considerations. Thus also here Wegscheider condition seems to be a result of mutual dependence of reaction rates and not a necessary consistency condition between thermodynamics and kinetics.
If reactions A + B = AB, AB + C = ABC, and B + C = BC are selected as independent ones then (24) gives

$$J^A = -J_1,\ J^B = -J_1 - J_3,\ J^C = -J_2 - J_3,\ J^{AB} = J_1 - J_2,\ J^{BC} = J_3,\ J^{ABC} = J_2 \quad (34)$$

Remember that, e.g., $J_1 \neq r_1$ but that the relationships between rates of independent reactions and mass action rates (32) follow from (34):

$$J_1 = r_1 + r_4,\ J_2 = r_2 + r_4,\ J_3 = r_3 - r_4 \quad (35)$$

Eq. (26) gives more complex expressions for independent rates, e.g. $J_1 = -J^A/2 - J^B/4 + J^{AB}/4 - J^{BC}/4 + J^{ABC}/4$, whereas from (24), i.e. (34), simply follow: $J_1 = -J^A$, $J_2 = J^{ABC}$, $J_3 = J^{BC}$. This is because the rates are considered as vector components - components J^α of six dimensional space are transformed to components J_i in three dimensional subspace. Consequently, in practical applications (24) should be preferred in favor of (26) also to express J_i in terms of J^α.
Message from the analysis of independence of reactions in this example is that it is sufficient to measure three component rates only (J^A, J^{ABC}, J^{BC}); the remaining three component rates are determined by them. Although concentrations, i.e. component rates, are measured in kinetic experiments, results are finally expressed in reaction rates, rates of reactions occurring in suggested reaction scheme. Component rates are simply not sufficient in kinetic analysis and they are (perhaps always) translated into rates of reaction steps. However, from the three independent rates there cannot be unambiguously determined rates of four reactions in suggested reaction schemes as (35) demonstrates (three equations for four unknown r_i). One equation more is needed and this is the above equation relating r_4 to the remaining three rates. Equations containing r_i are too general and in practice are replaced by mass action expressions shown in (32) - eight parameters (rate constants) are thus

introduced in this example. They can be in principle determined from three equations (35) with the three measured independent reactions, four equations relating equilibrium composition (or thermodynamic equilibrium constant) and kinetic equilibrium constant and one Wegscheider condition (33), i.e. eight equations in total. Alternative thermodynamic method is described in Section 4.

Algebraically more rigorous is this analysis in the case of first order reactions as was illustrated on a mixture of three isomers and their triangular reaction scheme which is traditional example used to discuss consistency between thermodynamics and kinetics. Here, Wegscheider relations are consequences of linear dependence of traditional mass action reaction rates (Pekař, 2007).

2.4 Note on standard states

Preceding sections demonstrated that one of the main problems to be solved when relating thermodynamics and kinetics is the transformation between activities and concentration variables. This is closely related to the selection of standard state (important and often overlooked aspect of relating thermodynamic and kinetic equilibrium constants) and to chemical potential. Standard states are therefore briefly reviewed in this section and chemical potential is subject of the following section.

Rates of chemical reactions are mostly expressed in terms of concentrations. Among standard states introduced and commonly used in thermodynamics there is only one based on concentration – the standard state of nonelectrolyte solute on concentration basis. Only this standard state can be directly used in kinetic equations. Standard state in gaseous phase or mixture is defined through (partial) pressure or fugacity. As shown above even in mixture of ideal gases it is impossible to simply use this standard state in concentration based kinetic equations. Although kinetic equations could be reformulated into partial pressures there still remains problem with the fact that standard pressure is fixed (at 1 atm or, nowadays, at 10^5 Pa) and its recalculation to actual pressure in reacting mixture may cause incompatibility of thermodynamic and kinetic equilibrium constants (see the factor p_{rel} in the example above in Section 2.1). This opens another problem – the very selection of standard state, particularly in relation to activity discussed in subsequent section. In principle, it can be selected arbitrarily, as dependent only on temperature or on temperature and pressure. Standard states strictly based on the (fixed) standard pressure are of the former type and only such will be considered in this article. All other states, including states dependent also on pressure, will be called the reference state; the same approach is used, e.g. by de Voe (de Voe, 2001).

The value of thermodynamic equilibrium constant and its dependence or independence on pressure is thus dependent on the selected standard (or reference) state. This is quite uncommon in chemical kinetics where the dependence of rate constants is not a matter of selection of standard states but result of experimental evidence or some theory of reaction rates. As a rule, rate constant is always function of temperature. Sometimes also the dependence on pressure is considered but this is usually the case of nonelementary reactions. Consequently, attempts to relate thermodynamic and kinetic equilibrium constants should select standard state consistently with functional dependence of rate constants. On the other hand, the method of Scheme 1 is self-consistent in this aspect because equilibrium composition is independent of the selection of standard state.

3. Chemical potential and activity revise

Chemical potential is used in discussions on thermodynamic implications on reaction rates, particularly in the form of (stoichiometric) difference between chemical potentials of reaction products and reactants and through its explicit relationship to concentrations (activities, in general). Before going into this type of analysis basic information is recapitulated.
Chemical potential is in classical, equilibrium thermodynamics defined as a partial derivative of Gibbs energy (G):

$$\mu_\alpha = (\partial G / \partial n_\alpha)_{T,p,n_{j\neq\alpha}} \tag{36}$$

Although another definitions through another thermodynamic quantities are possible (and equivalent with this one), the definition using the Gibbs energy is the most useful for chemical thermodynamics. Chemical potential expresses the effect of composition and this effect is also essential in chemical kinetics. To make the mathematical definition of the chemical potential applicable in practice its relationship to composition (concentration) should be stated explicitly. Practical chemical thermodynamics suggests that this is an easy task but we must be very careful and bear all (tacit) presumptions in mind to arrive at proper conclusions. Generally the explicit relationship between chemical composition and chemical potential is stated defining the activity of a component α:

$$a_\alpha = \exp\left(\frac{\mu_\alpha - \mu_\alpha^\circ}{RT}\right) \tag{37}$$

which can be transformed to

$$\mu_\alpha = \mu_\alpha^\circ + RT\ln a_\alpha \tag{38}$$

but this still lacks direct interconnection/linkage to measurable concentrations. Just this is the main problem of applying chemical potential (and activities) in rate equations which systematically use molar concentrations. Even when reaction rates would be expressed using activities in place of concentrations the activities should be properly calculated from the measured concentrations, in other words, the concentrations should be correctly transformed to the activities. Activity is very easily related to measurable composition variable in the case of mixture of ideal gases. Providing that Gibbs energy is a function of temperature, pressure and molar amounts, following relation is well known from thermodynamics for the partial molar volume: $\overline{V}_\alpha = (\partial \mu_\alpha / \partial p)_{T,n_{j\neq\alpha}}$. In a mixture of ideal gases partial molar volumes are equal to the molar volume of the mixture, V_m (Silbey et al., 2005). Because $V_m = RT/p$ we can write:

$$RT / p = (\partial \mu_{\alpha,g} / \partial p_\alpha)(\partial p_\alpha / \partial p) = x_\alpha (\partial \mu_{\alpha,g} / \partial p_\alpha) \tag{39}$$

and

$$RT / p_\alpha = (\partial \mu_{\alpha,g} / \partial p_\alpha) \tag{40}$$

Integration from the standard state to some actual state then yields

$$\mu_{\alpha,g} = \mu_{\alpha,g}^{\circ} + RT\ln\left(p_\alpha / p^\circ\right) \tag{41}$$

Comparing with the definition of activity it follows

$$a_\alpha = p_\alpha / p^\circ \text{ (mixture of ideal gases)} \tag{42}$$

Application of this relationship was illustrated in the example given above. Note that (42) was not derived from the definition of activity but comparing the properties of chemical potential in the ideal gas mixture (41) with the definition of activity. Note also that the partial derivative in the original definition of chemical potential is in general a function of molar amounts (contents) of all components but eq. (42) states that the chemical potential of a component α is a function only of the content of that component.
In a real gas mixture, non-idealities should be taken into account, usually by substituting fugacity (f_α) for the partial pressure:

$$\mu_{\alpha,g} = \mu_{\alpha,g}^{\circ} + RT\ln\left(f_\alpha / p^\circ\right) \tag{43}$$

The fugacity can be eliminated in favor of directly measurable quantities using the fugacity coefficient ϕ_α

$$f_\alpha = \phi_\alpha p_\alpha \tag{44}$$

and its relationship to the partial molar volume and the total pressure (de Voe, 2001):

$$\mu_{\alpha,g} = \mu_{\alpha,g}^{\circ} + RT\ln\left(p_\alpha / p^\circ\right) + \int_0^p \left(\overline{V}_\alpha - RT / p\right)\mathrm{d}p \tag{45}$$

It should be stressed that in derivation of the expression for the fugacity coefficient it was assumed that the Gibbs energy is a function of (only) temperature, pressure, and molar amounts of all components. Comparing with the definition of activity we have

$$a_\alpha = f_\alpha / p^\circ \text{ (mixture of gases)} \tag{46}$$

If kinetic equations for mixture of real gases are written in partial pressures then thermodynamic and kinetic equilibrium constants are incompatible due to the presence of fugacity coefficient or the integral in eq. (45). Kinetic equations for mixture of real gases could be formulated in terms of fugacities instead of concentrations (or partial pressures) to achieve compatibility between thermodynamic and kinetic equilibrium constants but even than the same problem remains with the presence of the standard pressure in thermodynamic relations. Kinetic equations formulated in fugacities are really rare – some success in this way was demonstrated by Eckert and Boudart (Eckert & Boudart, 1963) while Mason (Mason, 1965) showed, using the same data, that fugacities need not remedy the whole situation.
Similar derivation for liquid state (solutions) has different basis. It stems from the equilibrium between liquid and gaseous phase in which the following identity holds: $\mu_{\alpha,g} = \mu_{\alpha,l}$. Introducing expression (41) or (43) and using either Raoult's or Henry's law for the

relationship between compositions of equilibrated liquid and gaseous phases final form of $\mu_{\alpha,l}$ dependence on the composition of liquid is obtained. For example, with Raoult's law $p_\alpha = x_\alpha p_\alpha^*$ and ideal gas phase we have this equation

$$\mu_{\alpha,l} = \mu_{\alpha,g}^{\circ} + RT\ln\left(x_\alpha p_\alpha^* / p^{\circ}\right) \equiv \mu_\alpha^{\text{ref}} + RT\ln x_\alpha \tag{47}$$

which has, in fact, inspired the definition of an ideal (liquid, solid, or gas) mixture as a mixture with the chemical potential defined, at a given T and p, as $\mu_\alpha = \mu_\alpha^{\text{ref}} + RT\ln x_\alpha$ where μ_α^{ref} is a function of both T and p. This definition, as well as the identity in (47), can be simply related to the definition of activity only if the standard state is selected consistently with the reference state, i.e. if the former is a function of both T and p. If the standard state is selected as dependent on temperature, as it should be, than the pressure factor (Γ_α) should be introduced (see, e.g., de Voe, 2001)

$$\Gamma_\alpha = \exp\left(\frac{\mu_\alpha^{\text{ref}} - \mu_\alpha^{\circ}}{RT}\right) \tag{48}$$

Then the activity of a (non-electrolyte) component in real solution is written as $a_\alpha = \Gamma_\alpha \gamma_\alpha x_\alpha$ where γ_α is the activity coefficient introduced by the equation $\mu_\alpha = \mu_\alpha^{\text{ref}} + RT\ln(\gamma_\alpha x_\alpha)$. Introducing activities in place of concentrations means in this case to know the pressure factor and to transform molar fractions into molar concentrations to be consistent with thermodynamics.

The main problems with using activities defined for liquid systems can be summarized as follows. Activity is based on molar fractions whereas kinetic uses concentrations. Although there are formulas for the conversion of these variables they do not allow direct substitution, they introduce other variables (e.g., solution density) and lead to rather complex expression of thermodynamic equilibrium constant in concentrations. Whereas concentrations of all species are independent (variables) this is not true for molar fractions - value of one from them is unambiguously determined by values of remaining ones. Chemical potential in liquid and activity based on it are introduced on the basis of (liquid-gas) equilibrium while kinetics essentially works with reactions out of equilibrium. Applicability of equilibrium-based formulated in fugacities are really rare in nonequilibrium states deserves further study. The problem with molar fractions can be resolved by the use of molar concentration based Henry's law giving for ideal-dilute solution $\mu_{\alpha,l} = \mu_{\alpha,c}^{\text{ref}} + RT\ln c_\alpha / c^{\circ}$, however, rate equations should be formulated with the standard concentration. Sometimes following relationship is used: $\mu_{\alpha,l} = \mu_{\alpha,c}^{\text{ref}} + RT\ln c_\alpha / c_\Sigma$ (Ederer & Gilles, 2007) where c_Σ is the sum of all concentrations. In this case, the invertibility for c_α is problematic because it is included in c_Σ; reaction rates should be then formulated in c_α / c_Σ instead of concentrations that is quite unusual. Of course, the value of activity is dependent on the selected standard state, anyway. All attempts to relate thermodynamic and kinetic equilibrium constants should pay great attention to the selection of standard state and its consequences to be really rigorous and correct.

It is clear from this basic overview that chemical potential, activity and their interrelation are in principle equilibrium quantities which, in kinetic applications, are to be used for non-equilibrium situations. Let us now trace one relatively simple non-equilibrium approach to description of chemically reacting systems and its results regarding the chemical potential. Samohýl has developed rational thermodynamic approach for chemically reacting fluids with linear transport properties (henceforth called briefly linear fluids) and these fluids seem to include many (non-electrolyte) systems encountered in chemistry (Samohýl & Malijevský, 1976; Samohýl, 1982, 1987). This is a continuum mechanics based approach working with densities of quantities and specific quantities (considered locally, in other words, as fields but this is not crucial for the present text) therefore it primarily uses densities of components (more precisely, the density of component mass) instead of their molar concentrations or fractions that are common in chemistry. This density, in fact, is known in chemistry as a mass concentration with dimension of mass per (unit) volume and can be thus easily recalculated to concentration quantities more common in chemistry. Chemical potential of a reacting component α is defined in this theory as follows:

$$g_\alpha = \partial\left(\rho \overline{f}\right) / \partial \rho_\alpha \tag{49}$$

Here ρ is the density of mixture, i.e. the sum of all component densities ρ_α, and $\overline{f}$ is the specific free energy of (reacting) mixture as a function of relevant independent variables (the value of this function is denoted by f). Inspiration for this definition came from the entropic inequality (the "second law" of thermodynamics) as formulated in rational thermodynamics generally for mixtures and from the fact that this definition enabled to derive classical (equilibrium) thermodynamic relations in the special case that is covered by classical theory. The chemical potential g_α thus has the dimensions of energy per mass. The product $\rho\overline{f}$ essentially transforms the specific quantity to its density and the definition (49) can be viewed as a generalization of the classical definition (36) – partial derivative of mixture free energy (as a function) with respect to an independent variable expressing the amount of a component.

The specific free energy $\overline{f}$ is function of various (mostly kinematic and thermal) variables but here it is sufficient to note that component densities are among them, of course.

In the case of linear fluids it can be proved that free energy is function of densities and temperature only, $f = \overline{f}(\rho_1, \rho_2, \ldots, \rho_n, T)$. The same result is proved also for chemical potentials g_α and also for reaction rates expressed as component mass created or destroyed by chemical reactions at a given place and time in unit volume, $r_\alpha = \overline{r}_\alpha(\rho_1, \rho_2, \ldots, \rho_n, T)$. These rates can be easily transformed to molar basis much more common in chemistry using the molar mass M_α: $J^\alpha = r_\alpha / M_\alpha$. Component densities are directly related to molar concentration by a similar equation: $c_\alpha = \rho_\alpha / M_\alpha$. In this way, the well known kinetic empirical law – the law of mass action – is derived theoretically in the form: $J^\alpha = \overline{J}^\alpha(c_1, c_2, \ldots, c_n, T)$. Apparently, activities could be introduced into this function as independent variables controlling reaction rates by means of relations as $a_\alpha = \Gamma_\alpha \gamma_\alpha c_\alpha / c^\circ$ but this is not rigorous because these relations are consequences of chemical potential and its explicit dependence on mixture composition and not definitions per se. Therefore, chemical potentials should be introduced as independent variables at first. This could be done providing that component densities can be expressed as functions of chemical potential, i.e. providing that functions $g_\alpha = \overline{g}_\alpha(\rho_1, \rho_2, \ldots, \rho_n, T)$ are invertible (with respect to densities). This invertibility is not self-evident and the best way would be to prove it. Samohýl has proved (Samohýl, 1982, 1987) that if mixture of linear fluids fulfils Gibbs' stability conditions then the matrix with elements $\partial\overline{g}_\alpha / \partial\rho_\gamma$ (α, γ = 1,..., n) is regular which ensures the invertibility. This stability is a standard requirement for reasonable behavior of many reacting systems of chemist's interest, consequently the invertibility can be considered to be guaranteed and we can transform the rate functions as follows:

$$J^\alpha = \overline{J}^\alpha(\rho_1, \rho_2, \ldots, \rho_n, T) = \hat{J}^\alpha(g_1, g_2, \ldots, g_n, T) = \breve{J}^\alpha(\mu_1, \mu_2, \ldots, \mu_n, T) \tag{50}$$

where the last transformation was made using the following transformation of (specific) chemical potential into the traditional chemical potential (which will be called the molar chemical potential henceforth): $\mu_\alpha = g_\alpha M_\alpha$. Using the definition of activity (37) another transformation, to activities, can be made providing that the standard state is a function of temperature only:

$$\breve{J}^\alpha(\mu_1, \mu_2, \ldots, \mu_n, T) = \tilde{J}^\alpha(a_1, a_2, \ldots, a_n, T) \tag{51}$$

It should be stressed that chemical potential of component α as defined by (49) is a function of densities of all components, i.e. of ρ_γ, γ = 1,..., n, therefore also the molar chemical potential is following function of composition: $\mu_\alpha = \overline{\mu}_\alpha(c_1, c_2, \ldots, c_n, T)$. Note that generally any rate of formation or destruction (J^α) is a function of densities, or chemical potentials, or activities, etc. of all components.
Although the functions (dependencies) given above were derived for specific case of linear fluids they are still too general. Yet simpler fluid model is the simple mixture of fluids which is defined as mixture of linear fluids constitutive (state) equations of which are independent on density gradients. Then it can be shown (Samohýl, 1982, 1987) that

$$\partial\overline{f}_\alpha / \partial\rho_\gamma = 0 \quad \text{for } \alpha \neq \gamma;\ \alpha, \gamma = 1, \ldots, n \tag{52}$$

and, consequently, also that $g_\alpha = \overline{g}_\alpha(\rho_\alpha, T)$, i.e. the chemical potential of any component is a function of density of this component only (and of temperature). Mixture of ideal gases is defined as a simple mixture with additional requirement that partial internal energy and enthalpy are dependent on temperature only. Then it can be proved (Samohýl, 1982, 1987) that chemical potential is given by

$$g_\alpha = g_\alpha^\circ(T) + R_\alpha T \ln\left(p_\alpha / p^\circ\right) \tag{53}$$

that is slightly more general than the common model of ideal gas for which $R_\alpha = R/M_\alpha$. Thus the expression (41) is proved also at nonequilibrium conditions and this is probably only one mixture model for which explicit expression for the dependence of chemical potential on composition out of equilibrium is derived. There is no indication for other cases while the function $g_\alpha = \overline{g}_\alpha(\rho_\alpha, T)$ should be just of the logarithmic form like (47). Let us check conformity of the traditional ideal mixture model with the definition of simple mixture. For solute in an ideal-dilute solution following concentration-based expression is used:

$$\mu_\alpha = \mu_\alpha^{\text{ref}} + RT\ln\left(c_\alpha / c^\circ\right) \tag{54}$$

where μ_α^{ref} includes (among other) the gas standard state and concentration-based Henry's constant. Changing to specific quantities and densities we obtain:

$$g_\alpha = \mu_\alpha^{\text{ref}} / M_\alpha + (RT / M_\alpha)\ln\left(\rho_\alpha / M_\alpha c^\circ\right) \tag{55}$$

which looks like a function of ρ_α and T only, i.e. the simple mixture function $g_\alpha = \overline{g}_\alpha(\rho_\alpha, T)$. However, the referential state is a function of pressure so this is not such function rigorously. Except ideal gases there is probably no proof of applicability of classical expressions for dependence of chemical potential on composition out of equilibrium and no proof of its logarithmic point. There are probably also no experimental data that could help in resolving this problem.

4. Solution offered by rational thermodynamics

Rational thermodynamics offers certain solution to problems presented so far. It should be stressed that this is by no means totally general theory resolving all possible cases. But it clearly states assumptions and models, i. e. scope of its potential application.

The first assumption, besides standard balances and entropic inequality (see, e.g., Samohýl, 1982, 1987), or model is the mixture of linear fluids in which the functional form of reaction rates was proved: $J^\alpha = \overline{J}^\alpha(c_1, c_2, \ldots, c_n, T)$ (Samohýl & Malijevský, 1976; Samohýl, 1982, 1987). Only independent reaction rates are sufficient that can be easily obtained from component rates, cf. (26) from which further follows that they are function of the same variables. This function, $J_i = \overline{J}_i(c_1, c_2, \ldots, c_n, T)$, is approximated by a polynomial of suitable degree (Samohýl & Malijevský, 1976; Samohýl, 1982, 1987). Equilibrium constant is defined for each independent reaction as follows:

$$-RT\ln K_p = \sum_{\alpha=1}^{n} \mu_\alpha^\circ P^{p\alpha}; \quad p = 1, 2, \ldots, n-h \tag{56}$$

Activity (37) is supposed to be equal to molar concentrations (divided by unit standard concentration), which is possible for ideal gases, at least (Samohýl, 1982, 1987). Combining this definition of activity with the proved fact that in equilibrium $\sum_{\alpha=1}^{n} (\mu_\alpha)_{\text{eq}} P^{p\alpha} = 0$ (Samohýl, 1982, 1987) it follows

$$K_p = \prod_{\alpha=1}^{n} \left[(c_\alpha)_{\text{eq}}\right]^{P^{p\alpha}} \tag{57}$$

Some equilibrium concentrations can be thus expressed using the others and (57) and substituted in the approximating polynomial that equals zero in equilibrium. Equilibrium polynomial should vanish for any concentrations what leads to vanishing of some of its coefficients. Because the coefficients are independent of equilibrium these results are valid also out of it and the final simplified approximating polynomial, called thermodynamic polynomial, follows and represents rate equation of mass action type. More details on this method can be found elsewhere (Samohýl & Malijevský, 1976; Pekař, 2009, 2010). Here it is illustrated on two examples relevant for this article.

First example is the mixture of two isomers discussed in Section 2. 3. Rate of the only one independent reaction, selected as A = B, is approximated by a polynomial of the second degree:

$$J_1 = k_{00} + k_{10}c_A + k_{01}c_B + k_{20}c_A^2 + k_{02}c_B^2 + k_{11}c_Ac_B \tag{58}$$

The concentration of B is expressed from the equilibrium constant, $(c_B)_{eq} = K(c_A)_{eq}$ and substituted into (58) with $J_1 = 0$. Following form of the polynomial in equilibrium is obtained:

$$0 = k_{00} + \left(k_{10} + Kk_{01}\right)(c_A)_{eq} + \left(k_{20} + K^2k_{02} + Kk_{11}\right)(c_A)_{eq}^2 \tag{59}$$

Eq. (59) should be valid for any values of equilibrium concentrations, consequently

$$k_{00} = 0; \quad k_{10} = -Kk_{01}; \quad k_{20} = -K^2k_{02} - Kk_{11} \tag{60}$$

Substituting (60) into (58) the final thermodynamic polynomial (of the second degree) results:

$$J_1 = k_{10}\left(-Kc_A + c_B\right) + k_{02}\left(-K^2c_A^2 + c_B^2\right) + k_{11}\left(-Kc_A^2 + c_Ac_B\right) \tag{61}$$

Note, that coefficients k_{ij} are functions of temperature only and can be interpreted as mass action rate constants (there is no condition on their sign, if some k_{ij} is negative then traditional rate constant is k_{ij} with opposite sign). Although only the reaction A = B has been selected as the independent reaction, its rate as given by (61) contains more than just traditional mass action term for this reaction. Remember that component rates are given by (28). Selecting $k_{02} = 0$ two terms remain in (61) and they correspond to the traditional mass action terms just for the two reactions supposed in (R2). Although only one reaction has been selected to describe kinetics, eq. (61) shows that thermodynamic polynomial does not exclude other (dependent) reactions from kinetic effects and relationship very close to $J_1 = r_1 + r_2$, see also (29), naturally follows. No Wegscheider conditions are necessary because there are no reverse rate constants. On contrary, thermodynamic equilibrium constant is directly involved in rate equation; it should be stressed that because no reverse constant are considered this is not achieved by simple substitution of K for $\overleftarrow{k}_j$ from (27). Eq. (61) also extends the scheme (R2) and includes also bimolecular isomerization path: 2A = 2B.

This example illustrated how thermodynamics can be consistently connected to kinetics considering only independent reactions and results of nonequilibrium thermodynamics with no need of additional consistency conditions.

Example of simple combination reaction A + B = AB will illustrate the use of molar chemical potential in rate equations. In this mixture of three components composed from two atoms only one independent reaction is possible. Just the given reaction can be selected with equilibrium constant defined by (56): $\ln K = \left(-\mu_A^\circ - \mu_B^\circ + \mu_{AB}^\circ\right)/(-RT)$ and equal to

$K = (c_{AB} / c_A c_B)_{eq}$, cf. (57). The second degree thermodynamic polynomial results in this case in following rate equation:

$$J_1 = k_{110}(c_A c_B - K^{-1} c_{AB}) \tag{62}$$

that represents the function $J_1 = \bar{J}_1(T, c_A, c_B, c_{AB})$. Its transformation to the function $J_1 = \breve{J}_1(T, \mu_A, \mu_B, \mu_{AB})$ gives:

$$J_1 = k_{110} \exp\left(\frac{-\mu_A^\circ - \mu_B^\circ}{RT}\right)\left[\exp\left(\frac{\mu_A + \mu_B}{RT}\right) - \exp\frac{\mu_{AB}}{RT}\right] \tag{63}$$

This is thermodynamically correct expression (for the supposed thermodynamic model) of the function $\breve{J}$ discussed in Section 3 and in contrast to (1). It is clear that proper "thermodynamic driving force" for reaction rate is not simple (stoichiometric) difference in molar chemical potentials of products and reactants. The expression in square brackets can be considered as this driving force. Equation (63) also lucidly shows that high molar chemical potential of reactants in combination with low molar chemical potential of products can naturally lead to high reaction rate as could be expected. On the other hand, this is achieved in other approaches, based on $\sum \nu_i \mu_i$, due to arbitrary selection of signs of stoichiometric coefficients. In contrast to this straightforward approach illustrated in introduction, also kinetic variable (k_{110}) is still present in eq. (63), explaining why some "thermodynamically highly forced" reactions may not practically occur due to very low reaction rate. Equation (63) includes also explicit dependence of reaction rate on standard state selection (cf. the presence of standard chemical potentials). This is inevitable consequence of using thermodynamic variables in kinetic equations. Because also the molar chemical potential is dependent on standard state selection, it can be perhaps assumed that these dependences are cancelled in the final value of reaction rate.

Rational thermodynamics thus provides efficient connection to reaction kinetics. However, even this is not totally universal theory; on the other hand, presumptions are clearly stated. First, the procedure applies to linear fluids only. Second, as presented here it is restricted to mixtures of ideal gases. This restriction can be easily removed, if activities are used instead of concentrations, i.e. if functions $\tilde{J}$ are used in place of functions $\bar{J}$ - all equations remain unchanged except the symbol a_α replacing the symbol c_α. But then still remains the problem how to find explicit relationship between activities and concentrations valid at non equilibrium conditions. Nevertheless, this method seems to be the most carefully elaborated thermodynamic approach to chemical kinetics.

5. Conclusion

Two approaches relating thermodynamics and chemical kinetics were discussed in this article. The first one were restrictions put by thermodynamics on the values of rate constants in mass action rate equations. This can be also formulated as a problem of relation, or even equivalence, between the true thermodynamic equilibrium constant and the ratio of forward and reversed rate constants. The second discussed approach was the use of chemical potential as a general driving force for chemical reaction and "directly" in rate equations.

Both approaches are closely connected through the question of using activities, that are common in thermodynamics, in place of concentrations in kinetic equations and the problem of expressing activities as function of concentrations.
Thermodynamic equilibrium constant and the ratio of forward and reversed rate constants are conceptually different and cannot be identified. Restrictions following from the former on values of rate constants should be found indirectly as shown in Scheme 1.
Direct introduction of chemical potential into traditional mass action rate equations is incorrect due to incompatibility of concentrations and activities and is problematic even in ideal systems.
Rational thermodynamic treatment of chemically reacting mixtures of fluids with linear transport properties offers some solution to these problems whenever its clearly stated assumptions are met in real reacting systems of interest. No compatibility conditions, no Wegscheider relations (that have been shown to be results of dependence among reactions) are then necessary, thermodynamic equilibrium constants appear in rate equations, thermodynamics and kinetics are connected quite naturally. The role of ("thermodynamically") independent reactions in formulating rate equations and in kinetics in general is clarified.
Future research should focus attention on the applicability of dependences of chemical potential on concentrations known from equilibrium thermodynamics in nonequilibrium states, or on the related problem of consistent use of activities and corresponding standard states in rate equations.
Though practical chemical kinetics has been successfully surviving without special incorporation of thermodynamic requirements, except perhaps equilibrium results, tighter connection of kinetics with thermodynamics is desirable not only from the theoretical point of view but may be of practical importance considering increasing interest in analyzing of complex biochemical network or increasing computational capabilities for correct modeling of complex reaction systems. The latter when combined with proper thermodynamic requirements might contribute to more effective practical, industrial exploitation of chemical processes.

6. Acknowledgment

The author is with the Centre of Materials Research at the Faculty of Chemistry, Brno University of Technology; the Centre is supported by project No. CZ.1.05/2.1.00/01.0012 from ERDF. The author is indebted to Ivan Samohýl for many valuable discussions on rational thermodynamics.

7. References

Blum, L.H. & Luus, R. (1964). Thermodynamic Consistency of Reaction Rate Expressions. *Chemical Engineering Science*, Vol.19, No.4, pp. 322-323, ISSN 0009-2509

Boudart, M. (1968). *Kinetics of Chemical Processes*, Prentice-Hall, Englewood Cliffs, USA

Bowen, R.M. (1968). On the Stoichiometry of Chemically Reacting Systems. *Archive for Rational Mechanics and Analysis*, Vol.29, No.2, pp. 114-124, ISSN 0003-9527

Boyd, R.K. (1977). Macroscopic and Microscopic Restrictions on Chemical Kinetics. *Chemical Reviews*, Vol.77, No.1, pp. 93-119, ISSN 0009-2665

De Voe, H. (2001). *Thermodynamics and Chemistry*, Prentice Hall, ISBN 0-02-328741-1, Upper Saddle River, USA

Eckert, C.A. & Boudart, M. (1963). Use of Fugacities in Gas Kinetics. *Chemical Engineering Science*, Vol.18, No.2, 144-147, ISSN 0009-2509

Eckert, E.; Horák, J.; Jiráček, F. & Marek, M. (1986). *Applied Chemical Kinetics*, SNTL, Prague, Czechoslovakia (in Czech)

Ederer, M. & Gilles, E.D. (2007). Thermodynamically Feasible Kinetic Models of Reaction Networks. *Biophysical Journal*, Vol.92, No.6, pp. 1846-1857, ISSN 0006-3495

Hollingsworth, C.A. (1952a). Equilibrium and the Rate Laws for Forward and Reverse Reactions. *Journal of Chemical Physics*, Vol.20, No.5, pp. 921-922, ISSN 0021-9606

Hollingsworth, C.A. (1952b). Equilibrium and the Rate Laws. *Journal of Chemical Physics*, Vol.20, No.10, pp. 1649-1650, ISSN 0021-9606

Laidler, K.J. (1965). *Chemical Kinetics*, McGraw-Hill, New York, USA

Mason, D.M. (1965). Effect of Composition and Pressure on Gas Phase Reaction Rate Coefficient. *Chemical Engineering Science*, Vol.20, No.12, pp. 1143-1145, ISSN 0009-2509

Novák, J.; Malijevský, A.; Voňka, P. & Matouš, J. (1999). *Physical Chemistry*, VŠCHT, ISBN 80-7080-360-6, Prague, Czech Republic (in Czech)

Pekař, M. & Koubek, J. (1997). Rate-limiting Step. Does It Exist in the Non-Steady State? *Chemical Engineering Science*, Vol.52, No.14 , pp. 2291-2297, ISSN 0009-2509

Pekař, M. & Koubek, J. (1999). Concentration Forcing in the Kinetic Research in Heterogeneous Catalysis. *Applied Catalysis A*, Vol.177, No.1, pp. 69-77, ISSN 0926-860X

Pekař, M. & Koubek, J. (2000). On the General Principles of Transient Behaviour of Heterogeneous Catalytic Reactions. *Applied Catalysis A*, Vol.199, No.2, pp. 221-226, ISSN 0926-860X

Pekař, M. (2007). Detailed Balance in Reaction Kinetics – Consequence of Mass Conservation? *Reaction Kinetics and Catalysis Letters*, Vol. 90, No. 2, p. 323-329, ISSN 0133-1736

Pekař, M. (2009). Thermodynamic Framework for Design of Reaction Rate Equations and Schemes. *Collection of the Czechoslovak Chemical Communications*, Vol.74, No.9, pp. 1375–1401, ISSN 0010-0765

Pekař, M. (2010). Macroscopic Derivation of the Kinetic Mass-Action Law. *Reaction Kinetics, Mechanisms and Catalysis*, Vol.99, No. 1, pp. 29-35, ISSN 1878-5190

Qian, H. & Beard, D.A. (2005). Thermodynamics of Stoichiometric Biochemical Networks in Living Systems Far From Equilibrium. *Biophysical Chemistry*, Vol.114, No.3, pp. 213-220, ISSN 0301-4622

Samohýl, I. (1982). *Rational Thermodynamics of Chemically Reacting Mixtures*, Academia, Prague, Czechoslovakia (in Czech)

Samohýl, I. (1987). *Thermodynamics of Irreversible Processes in Fluid Mixtures*, Teubner, Leipzig, Germany

Samohýl, I. & Malijevský, A. (1976). Phenomenological Derivation of the Mass Action LAw of homogeneous chemical kinetics. *Collection of the Czechoslovak Chemical Communications*, Vol.41, No.8, pp. 2131-2142, ISSN 0010-0765

Silbey, R.J.; Alberty, R.A. & Bawendi M.G. (2005). *Physical Chemistry*, 4th edition, J.Wiley, ISBN 0-471-21504-X, Hoboken, USA

Vlad, M.O. & Ross, J. (2009). Thermodynamically Based Constraints for Rate Coefficients of Large Biochemical Networks. *WIREs Systems Biology and Medicine*, Vol.1, No.3, pp. 348-358, ISSN 1939-5094

Wegscheider, R. (1902). Über simultane Gleichgewichte und die Beziehungen zwischen Thermodynamik und Reaktionskinetik. *Zeitschrift für physikalische Chemie*, Vol. XXXIX, pp. 257-303

3

The Thermodynamics *in* Planck's Law

Constantinos Ragazas[1]
The Lawrenceville School
USA

1. Introduction

Quantum Physics has its historical beginnings with Planck's derivation of his formula for blackbody radiation, more than one hundred years ago. In his derivation, Planck used what latter became known as *energy quanta*. In spite of the best efforts at the time and for decades later, a more *continuous approach* to derive this formula had not been found. Along with Einstein's *Photon Hypothesis,* the *Quantization of Energy Hypothesis* thus became the foundations for much of the Physics that followed. This *physical view* has shaped our understanding of the Universe and has resulted in mathematical certainties that are counter-intuitive and contrary to our experience.

Physics provides *mathematical models* that seek to describe *what is* the Universe. We believe mathematical models of *what is* -- as with past metaphysical attempts -- are a never ending search getting us deeper and deeper into the 'rabbit's hole' [Frank 2010]. We show in this Chapter that a *quantum-view* of the Universe is not necessary. We argue that *a world without quanta* is not only possible, but desirable. We do not argue, however, with the mathematical formalism of Physics -- just the *physical view* attached to this.

We will present in this Chapter a mathematical derivation of *Planck's Law* that uses simple continuous processes, without needing *energy quanta* and *discrete statistics*. This *Law* is not true by Nature, but by Math. In our view, *Planck's Law* becomes a *Rosetta Stone* that enables us to translate known physics into simple and sensible formulations. To this end the quantity *eta* we introduce is fundamental. This is the time integral of energy that is used in our mathematical derivation of *Planck's Law.* In terms of this *prime physis* quantity *eta* (acronym for *e*nergy-*t*ime-*a*ction), we are able to define such physical quantities as energy, force, momentum, temperature and entropy. Planck's constant h (in units of energy-time) is such a quantity *eta*. Whereas currently h is thought as *action,* in our derivation of *Planck's Law* it is more naturally viewed as *accumulation of energy*. And while h is a constant, the quantity *eta* that appears in our formulation is a variable. Starting with *eta,* Basic Law can be mathematically derived and not be physically posited.

Is the Universe *continuous* or *discrete*? In my humble opinion this is a false dichotomy. It presents us with an impossible choice between two absolute views. And as it is always the case, making one side *absolute* leads to endless fabrications denying the opposite side. The Universe is neither *continuous* nor *discrete* because the Universe is both *continuous* and *discrete.* Our *view* of the Universe is *not* the Universe. The Universe simply *is*. In The Interaction of

---

[1] cragaza@lawrenceville.org

Measurement [Ragazas, 2010h] we argue with mathematical certainty that we cannot know through direct measurements what a physical quantity *E(t)* is as a function of time.

Since we are limited by our measurements of *'what is'*, we should consider these as the beginning and end of our knowledge of *'what is'*. Everything else is just *'theory'*. There is nothing real about theory! As the ancient Greeks knew and as the very word 'theory' implies. In Planck's Law is an Exact Mathematical Identity [Ragazas 2010f] we show *Planck's Law* is a mathematical truism that describes the *interaction of measurement*. We show that *Planck's Formula* can be *continuously* derived. But also we are able to explain *discrete* 'energy quanta'. In our view, *energy propagates continuously but interacts discretely*. Before there is *discrete manifestation* we argue there is *continuous accumulation* of energy. And this is based on the *interaction of measurement*.

Mathematics is a tool. It is a language of objective reasoning. But mathematical 'truths' are always 'conditional'. They depend on our presuppositions and our premises. They also depend, in my opinion, on the mental images we use to think. We phrase our explanations the same as we frame our experiments. In the single electron emission double-slit experiment, for example, it is assumed that the electron emitted at the source is the same electron detected at the screen. Our explanation of this experiment considers that these two electrons may be separate events. Not directly connected by some trajectory from *source* to *sensor*. [Ragazas 2010j]

We can have beautiful mathematics based on *any* view of the Universe we have. Consider the Ptolemy with their epicycles! But if the view leads to physical explanations which are counter-intuitive and defy common sense, or become too abstract and too removed from life and so not support life, than we must not confuse mathematical deductions with *physical realism*. Rather, we should change our view! And just as we can write bad literature using good English, we can also write bad physics using good math. In either case we do not fault the language for the story. We can't fault Math for the failings of Physics.

The failure of Modern Physics, in my humble opinion, is in not providing us with *physical explanations* that make sense; a *physical view* that is consistent with our experiences. A *view* that will not put us at odds with ourselves, with our understanding of our world and our lives. Math may not be adequate. Sense may be a better guide.

2. Mathematical results

We list below the main mathematical derivations that are the basis for the results in physics in this Chapter. The proofs can be found in the Appendix at the end. These mathematical results, of course, do not depend on Physics and are not limited to Physics. In *Stocks and Planck's Law* [Ragazas 2010l] we show how the same 'Planck-like' formula we derive here also describes a simple comparison model for stocks.

Notation. $E(t)$ is a real-valued function of the real-variable t

$\tau = \Delta t = t - s$ is an interval of t

$\Delta E = E(t) - E(s)$ is the change of E

$\eta = P = \int_s^t E(u)du$ is the accumulation of E

$\overline{E} = E_{av} = \frac{1}{t-s}\int_s^t E(u)du$ is the average of E

$\mathcal{T} = \mathcal{T}_\eta = \left(\frac{1}{\kappa}\right)\frac{\eta}{\tau}$ where κ is a scalar constant

D_x indicates differentiation with respect to x

r , ν are constants, often a rate of growth or frequency

Characterization 1: $E(t) = E_0 e^{rt}$ *if and only if* $\Delta E = Pr$

Characterization 2: $E(t) = E_0 e^{rt}$ *if and only if* $\frac{Pr}{e^{r(t-s)} - 1} = E(s)$

Characterization 2a: $E(t) = E_0 e^{rt}$ *if and only if* $\frac{Pr}{e^{Pr/E_{av}} - 1} = E(s)$

Characterization 3: $E(t) = E_0 e^{rt}$ *if and only if* $\frac{\Delta E}{e^{\Delta E/E_{av}} - 1} = E(s)$

Characterization 4: $E(t) = E_0 e^{rt}$ *if and only if* $\frac{\Delta E}{E_{av}} = r\Delta t$

Theorem 1a: $E(t) = E_0 e^{rt}$ *if and only if* $\frac{Pr}{e^{Pr/E_{av}} - 1}$ *is invariant with t*

Theorem 2: *For any integrable function* $E(t)$, $\lim_{t \to s} \frac{Pr}{e^{r\Delta t} - 1} = E(s)$

2.1 'Planck-like' characterizations [Ragazas 2010a]

Note that $E_{av} = \kappa\mathcal{T}_\eta$. We can re-write *Characterization 2a* above as,

$$E(t) = E_0 e^{\nu t} \quad \textit{if and only if} \quad E_0 = \frac{\eta\nu}{e^{\eta\nu/\kappa\mathcal{T}} - 1} \tag{1}$$

Planck's Law for blackbody radiation states that,

$$E_0 = \frac{h\nu}{e^{h\nu/kT} - 1} \tag{2}$$

where E_0 is the intensity of radiation, ν is the frequency of radiation and T is the (Kelvin) temperature of the blackbody, while h is Planck's constant and k is Boltzmann's constant. [Planck 1901, *Eqn 11*]. Clearly (1) and (2) have the exact same mathematical form, including the type of quantities that appear in each of these equations. We state the main results of this section as,

Result I: A 'Planck-like' characterization of simple exponential functions

$$E(t) = E_0 e^{\nu t} \quad \textit{if and only if} \quad E_0 = \frac{\eta\nu}{e^{\eta\nu/\kappa\mathcal{T}} - 1}$$

Using *Theorem 2* above we can drop the condition that $E(t) = E_0 e^{\nu t}$ and get,

Result II: A 'Planck-like' limit of any integrable function

For any *integrable* function $E(t)$, $E_0 = \lim_{t \to 0} \frac{\eta\nu}{e^{\eta\nu/\kappa\mathcal{T}} - 1}$

We list below for reference some helpful variations of these mathematical results that will be used in this Chapter.

$$E_0 = \frac{\Delta E}{e^{\Delta E/E_{av}} - 1} = \frac{\eta \nu}{e^{\eta \nu / \kappa \mathcal{T}_\eta} - 1} \qquad (\text{if } E(t) = E_0 e^{\nu t}) \tag{3}$$

$$E_0 \approx \frac{\Delta E}{e^{\Delta E/E_{av}} - 1} \approx \frac{\eta \nu}{e^{\eta \nu / \kappa \mathcal{T}_\eta} - 1} \qquad (\text{if } E(t) \text{ is integrable}) \tag{4}$$

$$E_0 = \frac{\eta \nu}{e^{\eta \nu / \kappa \mathcal{T}_\eta} - 1} \text{ is } \textit{exact} \text{ if and only if } \frac{\eta \nu}{e^{\eta \nu / \kappa \mathcal{T}_\eta} - 1} \text{ is } \textit{independent} \text{ of } \eta \tag{5}$$

Note that in order to avoid using limit approximations in (4) above, by (3) we will assume an *exponential of energy* throughout this Chapter. This will allow us to explore the underlying ideas more freely and simply. Furthermore in **Section 10.0** of this Chapter, we will be able to justify such an exponential time-dependent local representation of energy [Ragazas 2010i]. Otherwise, all our results (with the exception of **Section 8.0**) can be thought as pertaining to a blackbody with perfect emission, absorption and transmission of energy.

3. Derivation of Planck's law without *energy quanta* [Ragazas 2010f]

Planck's Formula as originally derived describes what physically happens at the *source*. We consider instead what happens at the *sensor* making the measurement. Or, equivalently, what happens at the *site of interaction* where energy exchanges take place. We assume we have a blackbody medium, with perfect emission, absorption and transmission of energy. We consider that measurement involves an *interaction* between the *source* and the *sensor* that results in energy exchange. This interaction can be mathematically described as a functional relationship between $E(s)$, the energy locally at the *sensor* at time s; ΔE, the energy absorbed by the *sensor* making the measurement; and $\overline{E}$, the average energy at the *sensor* during measurement. Note that *Planck's Formula* (2) has the exact same mathematical form as the mathematical equivalence (3) and as the limit (4) above. By letting $E(s)$ be an *exponential*, however, from (3) we get an *exact* formula, rather than the limit (4) if we assume that $E(s)$ is only an integrable function. The argument below is one of several that can be made. The *Assumptions* we will use in this very simple and elegant derivation of *Planck's Formula* will themselves be justified in later **Sections 5.0, 6.0 and 10.0** of this Chapter.

Mathematical Identity. For any integrable function $E(t)$, $\eta = \int_s^{s+\eta/E_{av}} E(u)du$ (6)

Proof: (see Fig. 1)

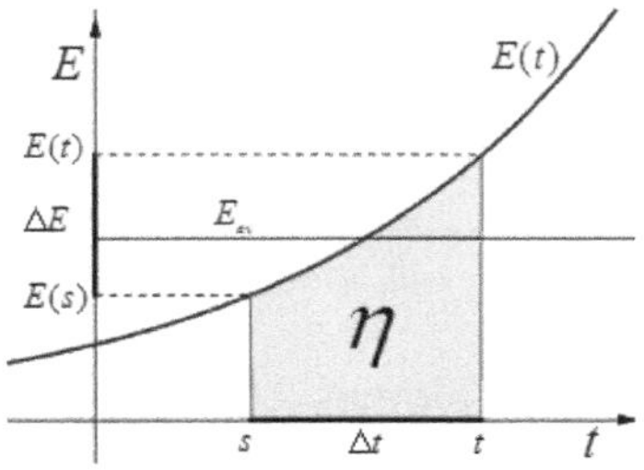

Fig. 1.

Assumptions: 1) Energy locally at the sensor at $t = s$ *can be represented by* $E(s) = E_0 e^{\nu s}$ *, where* E_0 *is the intensity of radiation and* ν *is the frequency of radiation. 2) When measurement is made, the source and the sensor are in equilibrium. The average energy of the source is equal to the average energy at the sensor. Thus,* $\overline{E} = kT$ *. 3) Planck's constant* h *is the minimal 'accumulation of energy' at the sensor that can be manifested or measured. Thus we have* $\eta = h$ *.*
Using the above *Mathematical Identity* (6) and *Assumptions* we have *Planck's Formula,*

$$h = \int_0^{\frac{h}{kT}} E_0 e^{\nu u} du = \frac{E_0}{\nu}\left[e^{h\nu/kT} - 1 \right] \text{ and so, } E_0 = \frac{h\nu}{e^{h\nu/kT} - 1}$$

Planck's Formula is a mathematical truism that describes the interaction of energy. That is to say, it gives a mathematical relationship between the energy locally at the *sensor,* the energy absorbed by the *sensor,* and the average energy at the *sensor* during measurement. Note further that when an amount of energy ΔE is absorbed by the *sensor,* $E(t)$ resets to E_0 .

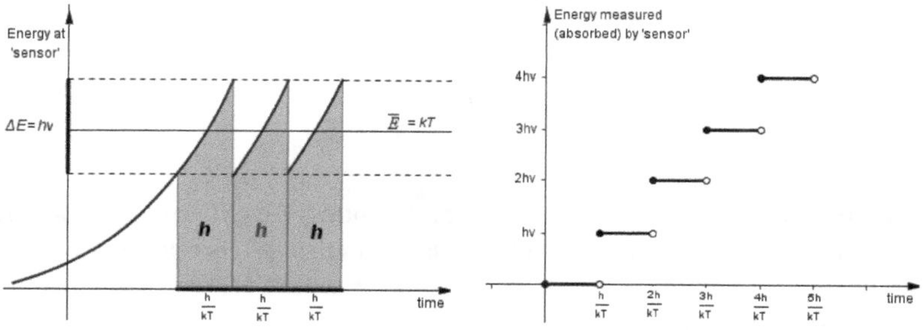

(a) (b)

Fig. 2.

Note: Our derivation, showing that *Planck's Law* is a mathematical truism, can now clearly explain why the experimental blackbody spectrum is so indistinguishable from the theoretical curve. **(http://en.wikipedia.org/wiki/File:Firas_spectrum.jpg)**
Conclusions:

1. Planck's Formula is an *exact mathematical truism* that describes the interaction of energy.
2. Energy propagates continuously but interacts discretely. The absorption or measurement of energy is made in discrete 'equal size sips'(energy quanta).
3. Before *manifestation of energy* (when an amount ΔE is absorbed or emitted) there is an *accumulation of energy* that occurs over a duration of time Δt .
4. The absorption of energy is proportional to frequency, $\Delta E = h\nu$ *(The Quantization of Energy Hypothesis).*
5. There exists a *time-dependent local representation of energy,* $E(t) = E_0 e^{\nu t}$, where E_0 is the intensity of radiation and ν is the frequency of radiation. [Ragazas 2011a]
6. The energy measured ΔE vs. Δt is linear with slope νkT for constant temperature T .
7. The time Δt required for an *accumulation of energy* h to occur at temperature T is given by $\Delta t = \dfrac{h}{kT}$.

4. Prime physis eta and the derivation of Basic Law [Ragazas 2010d]

In our derivation of *Planck's Formula* the quantity η played a prominent role. In this derivation η is the *time-integral of energy*. We consider this quantity η as *prime physis,* and define in terms of it other physical quantities. And thus mathematically derive Basic Law. Planck's constant h is such a quantity η, measured in units of *energy-time*. But whereas h is a constant, η is a variable in our formulation.

Definitions: For fixed $(\vec{\mathbf{x}}_0, t_0)$ *and along the x-axis for simplicity,*

Prime physis: η = *eta (energy-time-action)*

$$\text{Energy:} \quad E = \frac{\partial \eta}{\partial t} \tag{7}$$

$$\text{Momentum:} \quad p_x = \frac{\partial \eta}{\partial x} \tag{8}$$

$$\text{Force:} \quad F_x = \frac{\partial^2 \eta}{\partial x \partial t} \tag{9}$$

Note that the quantity *eta* is undefined. But it can be thought as 'energy-time-action' in units of *energy-time*. *Eta* is both *action* as well as *accumulation of energy*. We make only the following assumption about η.

Identity of Eta Principle: For the same physical process, the quantity η *is one and the same.*

Note: This *Principle* is somewhat analogous to a physical system being described by the *wave function*. Hayrani Öz has also used originally and consequentially similar ideas in [Öz 2002, 2005, 2008, 2010].

4.1 Mathematical derivation of Basic Law

Using the above definitions, and known mathematical theorems, we are able to derive the following Basic Law of Physics:

- Planck's Law, $E_0 = \dfrac{h\nu}{e^{h\nu/kT} - 1}$, is a mathematical truism **(Section 3.0)**
- The Quantization of Energy Hypothesis, $\Delta E = nh\nu$ **(Section 3.0)**
- *Conservation of Energy and Momentum.* The gradient of $\eta(\vec{\mathbf{x}},t)$ is $\vec{\nabla}\eta = \left\langle \dfrac{\partial \eta}{\partial x}, \dfrac{\partial \eta}{\partial t} \right\rangle = \langle p_x, E \rangle$. Since all gradient vector fields are *conservative,* we have the *Conservation of Energy and Momentum.*
- *Newton's Second law of Motion.* The second Law of motion states that $F = ma$. From definition (9) above we have,

$$F = \frac{\partial^2 \eta}{\partial x \partial t} = \frac{\partial^2 \eta}{\partial t \partial x} = \frac{\partial p_x}{\partial t} = \frac{\partial}{\partial t}(mv) = ma \text{, since } p_x = mv \text{ .}$$

- *Energy-momentum Equivalence.* From the definition of energy $E = \frac{\partial \eta}{\partial t}$ and of momentum $p_x = \frac{\partial \eta}{\partial x}$ we have that, $\eta = \int_{t_0}^{t} E(u)du$ and $\eta = \int_{x_0}^{x} p_x(u)du$.
Using the *Identity of Eta Principle,* the quantity η in these is one and the same. Therefore, $\int_{t_0}^{t} E(u)du = \int_{x_0}^{x} p_x(u)du$. Differentiating with respect to *t*, we obtain,
$E(t) = p_x(x) \cdot \frac{dx}{dt}$ or more simply, $E = p_x v$ *(energy-momentum equivalence)*
- *Schroedinger Equation:* Once the extraneous constants are striped from Schroedinger's equation, this in essence can be written as $\frac{\partial \psi}{\partial t} = H\psi$, where ψ is the *wave function* , H is the *energy operator,* and $H\psi$ is the energy at *any* $(\vec{\mathbf{x}},t)$. The definition (7) of energy $\frac{\partial \eta}{\partial t} = E$ given above is for a *fixed* $(\vec{\mathbf{x}}_0,t_0)$. Comparing these we see that whereas our definition of energy is for *fixed* $(\vec{\mathbf{x}}_0,t_0)$, Schroedinger equation is for *any* $(\vec{\mathbf{x}},t)$. But otherwise the two equations have the same form and so express the same underlying idea. Now (7) *defines* energy in terms of the more primary quantity η (which can be viewed as *accumulation of energy* or *action*) and so we can view Schroedinger Equation as in essence *defining* the energy of the system at *any* $(\vec{\mathbf{x}},t)$ while the *wave function* ψ can be understood to express the *accumulation of energy* at *any* $(\vec{\mathbf{x}},t)$. This suggests that the *wave function* ψ is the same as the quantity η. We have the following interesting interpretation of the wave function.
- The wave function gives the distribution of the accumulation of energy of the system.
- *Uncertainty Principle:* Since $\Delta E = \eta \nu$, for $\Delta t > \frac{1}{\nu}$ (a 'wavelength') we have $\Delta E \cdot \Delta t > \eta \nu \cdot \frac{1}{\nu} = \eta > h$. Or equivalently, for $\frac{\Delta E}{E_{av}} > 1$, we again have $\Delta E \cdot \Delta t > \eta > h$, since h is the *minimal eta that can be manifested.* Note that since $\frac{\Delta E}{E_{av}} = \nu \Delta t$ *(Characteristic 5),* we have $\frac{\Delta E}{E_{av}} > 1$ *if and only if* $\Delta t > \frac{1}{\nu}$. Since $E_{av} = kT$ and entropy is defined as $\Delta S = \frac{\Delta E}{T}$, we have that
- $\Delta E \cdot \Delta t > h$ if and only if $\Delta S > k$.
- *Planck's Law and Boltzmann's Entropy Equation Equivalence:*

Starting with our *Planck's Law* formulation, $E_0 = \frac{\Delta E}{e^{\Delta E/E_{av}} - 1}$ in (3) above and re-writing this equivalently we have, $e^{\Delta E/E_{av}} = 1 + \frac{\Delta E}{E_0} = \frac{E}{E_0}$ and so, $\frac{\Delta E}{E_{av}} = \ln\left(\frac{E}{E_0}\right)$. Using the definition of thermodynamic entropy we get $\Delta S_{\Theta} = \frac{\Delta E}{T} = k \cdot \frac{\Delta E}{E_{av}} = k \cdot \ln\left(\frac{E}{E_0}\right)$. If $\Omega(t)$ represents the number of microstates of the system at time t, then $E(t) = A\Omega(t)$, for some constant A. Thus, we get *Boltzmann's Entropy Equation,* $S_{\Theta} = k \ln \Omega$.

Conversely, starting with *Boltzmann's Entropy Equation,* $\Delta S_{\Theta} = k \ln\left(\frac{\Omega}{\Omega_0}\right) = k \ln\left(\frac{E}{E_0}\right)$.

Since $\Delta S_{\Theta} = \frac{\Delta E}{T}$ we can rewrite this equivalently as $\frac{\Delta E}{E_{av}} = \ln\left(\frac{E}{E_0}\right)$ and so

$e^{\Delta E/E_{av}} = 1 + \frac{\Delta E}{E_0} = \frac{E}{E_0}$. From this we have, *Planck's Law,* $E_0 = \frac{\Delta E}{e^{\Delta E/E_{av}} - 1}$ in (3) above.

- *Entropy-Time Relationship:* $\Delta S = k\nu\Delta t$ where ν is the *rate of evolution* of the system and Δt is the *time duration* of evolution, since $E_{av} = \frac{\eta}{\Delta t}$ and $\Delta E = \eta\nu$.
- *The Fundamental Thermodynamic Relation:* It is a well known fact that the internal energy U, entropy S, temperature T, pressure P and volume V of a system are related by the equation $dU = TdS - PdV$. By using increments rather than differentials, and using the fact that work performed by the system is given by $W = \int PdV$ this can be re-written as $\Delta S = \frac{\Delta U}{T} + \frac{\Delta W}{T}$. All the terms in this equation are various entropy quantities. The fundamental thermodynamic relation can be interpreted thus as saying, *"the total change of entropy of a system equals the sum of the change in the internal (unmanifested) plus the change in the external (manifested) entropy of the system"*. Considering the *entropy-time relationship* above, this can be rephrased more intuitively as saying *"the total lapsed time for a physical process equals the time for the 'accumulation of energy' plus the time for the 'manifestation of energy' for the process"*. This relationship along with *The Second Law of Thermodynamics* establish a *duration of time* over which there is *accumulation of energy before manifestation of energy* – one of our main results in this Chapter and a premise to our explanation of the double-slit experiment. [Ragazas 2010j]

5. The temperature of radiation [Ragazas 2010g]

Consider the energy $E(t)$ at a fixed point at time t. We define the *temperature of radiation* to be given by $\mathcal{T} = \mathcal{T}_{\eta} = \left(\frac{1}{\kappa}\right)\frac{\eta}{\tau}$ where κ is a scalar constant. Though in defining *temperature* this way the *accumulation of energy* η can be any value, when considering a *temperature scale* η is fixed and used as a *standard for measurement*. To distinguish *temperature* and *temperature scale* we will use $\mathcal{T}$ and $\mathcal{T}_{\eta}$ respectively. We assume that *temperature* is characterized by the following property:

Characterization of temperature: For a fixed η, *the temperature is inversely proportional to the duration of time for an accumulation of energy* η *to occur.*

Thus if *temperature* is twice as high, the accumulation of energy will be twice as fast, and visa-versa. This *characterization of temperature* agrees well with our physical sense of temperature. It is also in agreement with *temperature* as being the average kinetic energy of the motion of molecules.

For fixed η, we can define $\mathcal{T}_\eta = \left(\frac{1}{\kappa}\right)\frac{\eta}{\tau}$, which will be unique up to an arbitrary scalar constant κ. *Conversely,* for a given $\mathcal{T}$ as characterized above, we will have $\mathcal{T} = \rho \cdot \frac{1}{\tau}$, where ρ is a proportionality constant. By setting $\rho = \frac{\eta}{\kappa}$ we get $\mathcal{T} = \mathcal{T}_\eta = \left(\frac{1}{\kappa}\right)\frac{\eta}{\tau}$. We have the following *temperature-eta* correspondence:

Temperature-eta Correspondence: Given η, *we have* $\mathcal{T}_\eta = \left(\frac{1}{\kappa}\right)\frac{\eta}{\tau}$, *where* κ *is some arbitrary scalar constant. Conversely, given* $\mathcal{T}$ *we have* $\mathcal{T} = \mathcal{T}_\eta = \left(\frac{1}{\kappa}\right)\frac{\eta}{\tau}$, *for some fixed* η *and arbitrary scalar constant* κ. *Any temperature scale. therefore, will have some fixed* η *and arbitrary scalar constant* κ *associated with it.*

6. The meaning and existence of Planck's constant h [Ragazas 2010c]

Planck's constant h is a fundamental universal constant of Physics. And although we can experimentally determine its value to great precision, the reason for its existence and what it really means is still a mystery. Quantum Mechanics has adapted it in its mathematical formalism. But QM does not explain the meaning of h or prove why it must exist. Why does the Universe need h and *energy quanta?* Why does the mathematical formalism of QM so accurately reflect physical phenomena and predict these with great precision? Ask any physicists and uniformly the answer is "that's how the Universe works". The units of h are in *energy-time* and the conventional interpretation of h is as a *quantum of action*. We interpret h as *the minimal accumulation of energy* that can be manifested. Certainly the units of h agree with such interpretation. Based on our results above we provide an explanation for the existence of Planck's constant -- what it means and how it comes about. We show that the existence of *Planck's constant* is not necessary for the Universe to exist but rather h exists by Mathematical necessity and inner consistency of our system of measurements.

Using *eta* we defined in **Section 5.0** above the *temperature of radiation* as being proportional to the ratio of *eta/time*. To obtain a *temperature scale,* however, we need to fix *eta* as a standard for measurement. We show below that the fixed *eta* that determines the Kelvin *temperature scale* is Planck's constant h.

In The Interaction of Measurement [Ragazas 2010h] we argue that direct measurement of a physical quantity $E(t)$ involves a physical interaction between the *source* and the *sensor*. For measurement to occur an interval of time Δt must have lapsed and an incremental amount ΔE of the quantity will be absorbed by the *sensor*. This happens when there is an *equilibrium* between the *source* and the *sensor*. At *equilibrium,* the 'average quantity E_{av} from the source' will equal to the 'average quantity E_{av} at the sensor'. *Nothing in our observable World can exist without time, when the entity 'is' in equilibrium with its environment and its 'presence' can be observed and measured.* Furthermore as we showed above in **Section 3.0** the *interaction of measurement* is described by *Planck's Formula.*

From the mathematical equivalence (5) above we see that η can be *any* value and $\frac{\eta\nu}{e^{\eta\nu/\kappa\mathcal{T}_\eta}-1}$ will be invariant and will continue to equal to E_0. We can in essence (Fig. 3) 'reduce' the formula $E_0=\frac{\eta\nu}{e^{\eta\nu/\kappa\mathcal{T}_\eta}-1}$ by reducing the value of η and so the value of $E_{av}=\kappa\mathcal{T}_\eta$ will correspondingly adjust, and visa versa. Thus we see that η and $\mathcal{T}_\eta$ go *hand-in-hand* to maintain $E_0=\frac{\eta\nu}{e^{\eta\nu/\kappa\mathcal{T}_\eta}-1}$ invariant. And though the mathematical equivalence (5) above allows these values to be anything, the calibrations of these quantities in Physics require their value to be specific. Thus, for $\eta=h$ (Planck's constant) and $\kappa=k$ (Boltzmann's constant), we get $\mathcal{T}_\eta=T$ (Kelvin temperature) (see Fig. 3). Or, conversely, if we start with $\mathcal{T}_\eta=T$ and set the arbitrary constant $\kappa=k$, then this will force $\eta=h$. Thus we see that *Planck's constant* h*, Boltzmann's constant* k*, and Kelvin temperature* T are so defined and calibrated to fit Planck's Formula. Simply stated, when $\eta=h$, $\mathcal{T}_h=T$.

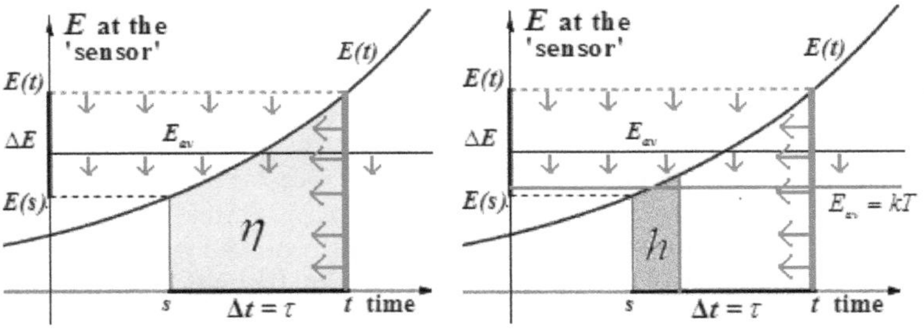

Fig. 3. $E_0=\frac{\Delta E}{e^{\Delta E/E_{av}}-1}=\frac{\eta\nu}{e^{\eta\nu/\kappa\mathcal{T}_\eta}-1}$, $\Delta E=\eta\nu$, $E_{av}=\kappa\mathcal{T}_\eta$, $\mathcal{T}_\eta=\left(\frac{1}{\kappa}\right)\frac{\eta}{\tau}$, $E(t)=E_0e^{\nu t}$

Conclusion: Physical theory provides a conceptual lens through which we 'see' the world. And based on this theoretical framework we get a measurement methodology. Planck's constant h is just that 'theoretical focal point' beyond which we cannot 'see' the world through our theoretical lens. Planck's constant h is the minimal eta that can be 'seen' in our measurements. Kelvin temperature scale requires the measurement standard eta to be h.

Planck's Formula is a mathematical identity that describes the interaction of measurement. It is invariant with time, accumulation of energy or amount of energy absorbed. Planck's constant exists because of the time-invariance of this mathematical identity. The calibration of Boltzmann's constant k *and Kelvin temperature* T*, with* kT *being the average energy, determine the specific value of Planck's constant* h.

7. Entropy and the second law of thermodynamics [Ragazas 2010b]

The quantity $\frac{\Delta E}{E_{av}}$ that appears in our *Planck's Law* formulation (3) is *'additive over time'*. This is so because under the assumption that *Planck's Formula* is *exact* we have that $\frac{\Delta E}{E_{av}}=\nu\Delta t$, by

Characterization 4. Interestingly, this quantity is essentially *thermodynamic entropy,* since $E_{av} = kT$, and so $\Delta S = \frac{\Delta E}{T} = k\nu\Delta t$. Thus entropy is *additive over time.* Since ν can be thought as the *evolution rate* of the system (both positive or negative), entropy is a measure of the *amount of evolution* of the system over a duration of time Δt . Such connection between *entropy* as *amount of evolution* and *time* makes eminent intuitive sense, since *time* is generally thought in terms of *change.* But, of course, this is *physical time* and not some mathematical abstract parameter as in *spacetime continuum.*

Note that in the above, *entropy* can be both positive or negative depending on the *evolution rate* ν . That the *duration of time* Δt is positive, we argue, is postulated by *The Second Law of Thermodynamics.* It is amazing that the most fundamental of all physical quantities *time* has no fundamental Basic Law pertaining to its nature. We argue *the Basic Law pertaining to time* **is** *The Second Law of Thermodynamics.* Thus, a more revealing rewording of this Law should state that *all physical processes take some positive duration of time to occur.* Nothing happens *instantaneously.* Physical time is really *duration* Δt (or dt) and not *instantiation* $t = s$.

8. The photoelectric effect without photons [Ragazas 2010k]

Photoelectric emission has typically been characterized by the following experimental facts *(some of which can be disputed, as noted)*:

1. For a given metal surface and frequency of incident radiation, the rate at which photoelectrons are emitted (the photoelectric current) is directly proportional to the intensity of the incident light.
2. The energy of the emitted photoelectron is independent of the intensity of the incident light but depends on the frequency of the incident light.
3. For a given metal, there exists a certain minimum frequency of incident radiation below which no photoelectrons are emitted. This frequency is called the threshold frequency. *(see below)*
4. The time lag between the incidence of radiation and the emission of photoelectrons is very small, less than 10^{-9} second.

Explanation of the Photoelectric Effect without the Photon Hypothesis: Let ν be the rate of radiation of an incident light on a metal surface and let α be the rate of absorption of this radiation by the metal surface. The combined rate locally at the surface will then be $\nu - \alpha$. The radiation energy at a point on the surface can be represented by $E(t) = E_0 e^{(\nu-\alpha)t}$, where E_0 is the intensity of radiation of the incident light. If we let η be the *accumulation of energy* locally at the surface over a time pulse τ , then by *Characterization 1* we'll have that $\Delta E = \eta(\nu - \alpha)$. If we let Planck's constant h be the *accumulation of energy* for an electron, the number of electrons n_e over the pulse of time τ will then be $n_e = \frac{\eta}{h}$ and the energy of an electron ΔE_e will be given by

$$\Delta E_e = \frac{\Delta E}{n_e} = h(\nu - \alpha) \tag{10}$$

Since $\eta = \int_0^\tau E_0 e^{(\nu-\alpha)u} du = E_0 \left[\frac{e^{(\nu-\alpha)\tau} - 1}{(\nu - \alpha)} \right]$, we get the *photoelectric current* I ,

$$I = \frac{n_e}{\tau} = \frac{\eta}{h\tau} = E_0 \left[\frac{e^{(\nu-\alpha)\tau} - 1}{h(\nu-\alpha)\tau} \right] \tag{11}$$

The absorption rate α is a characteristic of the metal surface, while the pulse of time τ is assumed to be constant for fixed experimental conditions. The quantity $\left[\frac{e^{(\nu-\alpha)\tau} - 1}{h(\nu-\alpha)\tau} \right]$ in equation (10) would then be *constant*.

Combining the above and using (10) and (11) we have *The Photoelectric Effect:*

1. For incident light of fixed frequency ν and fixed metal surface, the photoelectric current I is proportional to the intensity E_0 of the incident light. (by (11) above)
2. The energy ΔE_e of a photoelectron depends only on the frequency ν and not on the intensity E_0 of the incident light. It is given by the equation $\Delta E_e = h(\nu - \alpha)$ where h is Planck's constant and the absorption rate α is a property of the metal surface. (by (10) above)
3. If ΔE_e is taken to be the kinetic energy of a photoelectron, then for incident light with frequency ν less than the 'threshold frequency' α the kinetic energy of a photoelectron would be negative and so there will be no photoelectric current. (by (10) above) *(see Note below)*
4. The photoelectric current is almost instantaneous ($< 10^{-9}$ sec.), since for a single photoelectron we have that $\Delta t = \frac{h}{kT} < 10^{-9}$ sec. by *Conclusion* 7 **Section 3**.

Note: Many experiments since the classic 1916 experiments of Millikan have shown that there is photoelectric current even for frequencies below the threshold, contrary to the explanation by Einstein. In fact, the original experimental data of Millikan show an asymptotic behavior of the (photocurrent) vs (voltage) curves along the energy axis with no clear 'threshold frequency'. The photoelectric equations (10) and (11) we derived above agree with these experimental anomalies, however.

In an article Richard Keesing of York University, UK , states,

> *I noticed that a reverse photo-current existed ... and try as I might I could not get rid of it.*
> *My first disquieting observation with the new tube was that the I/V curves had high energy tails on them and always approached the voltage axis asymptotically. I had been brought up to believe that the current would show a well defined cut off, however my curves just refused to do so.*
> *Several years later I was demonstrating in our first year lab here and found that the apparatus we had for measuring Planck's constant had similar problems.*
> *After considerable soul searching it suddenly occurred on me that there was something wrong with the theory of the photoelectric effect ... [Keesing 2001]*

In the same article, taking the original experimental data from the 1916 experiments by Millikan, Prof. Keesing plots the graphs in Fig. 4.

In what follows, we analyze the asymptotic behavior of equation (11) by using a function of the same form as (11).

$$f(x) = \frac{A\left(e^{b(x-c)} - 1\right)}{x - c} + d \tag{12}$$

Note: We use d since some graphs typically are shifted up a little for clarity.

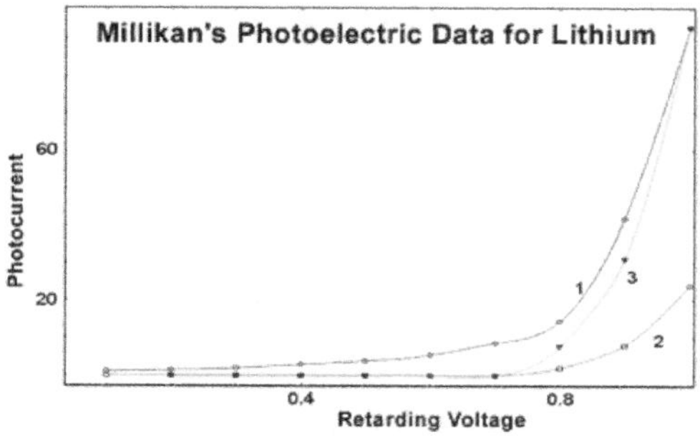

Fig. 4.

The graphs in Fig. 5 match the above experimental data to various graphs (in red) of equation (12)

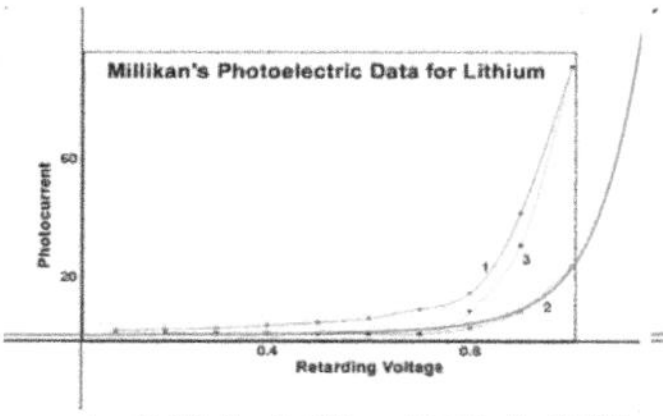

A=0.13 b=1.98 c=5.95 d=0,07
(a)

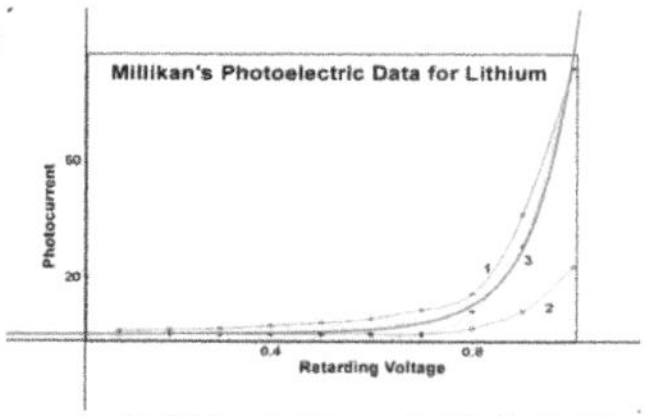

A=0.09 b=2.07 c=4.88 d=0.09
(b)

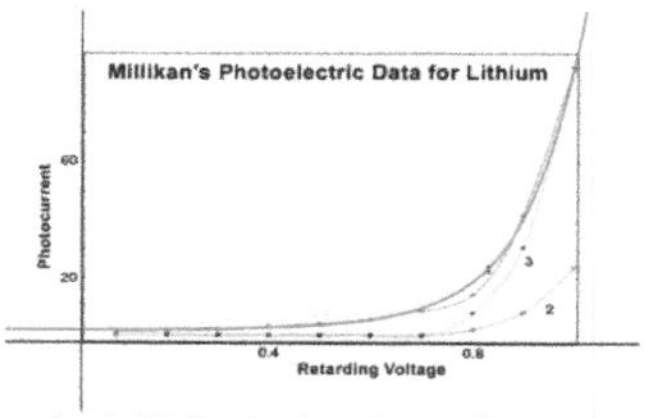

A=0.05 b=1.41 c=3.04 d=0.18
(c)

Fig. 5.

The above graphs (Fig. 5) seem to suggest that Eq. (11) agrees well with the experimental data showing the asymptotic behavior of the (photocurrent) v (energy) curves. But more systematic experimental work is needed.

9. Meaning and derivation of the De Broglie equations [Ragazas 2011a]

Consider $\eta_0(x_0,t_0)\rightarrow\eta(x,t)$. We can write $\frac{\Delta\eta}{\eta_0}$ = %-change of η = 'cycle of change'. For corresponding Δx and Δt we can write, $\lambda=\frac{\Delta x}{\Delta\eta/\eta_0}$ = "distance per cycle of change" and $\nu=\frac{\Delta\eta/\eta_0}{\Delta t}$ = "cycle of change per time". We can rewrite these as $\lambda=\frac{\eta_0}{\Delta\eta/\Delta x}$ and $\nu=\frac{\Delta\eta/\Delta t}{\eta_0}$.

Taking limits and letting $\eta_0=h$ (Planck's constant being the minimal η that can be measured) we get the *de Broglie equations:*

$$\lambda=\frac{\eta_0}{\Delta\eta/\Delta x}\rightarrow\frac{h}{\partial\eta/\partial x}=\frac{h}{p} \quad \text{and} \quad \nu=\frac{\Delta\eta/\Delta t}{\eta_0}\rightarrow\frac{\partial\eta/\partial t}{h}=\frac{E}{h}$$

Note: Since %-change in η can be both positive or negative, λ and ν can be both positive or negative.

10. The 'exponential of energy' $E(t)=E_0e^{\nu t}$ [Ragazas 2010i, 2011a]

From **Section 9.0** above we have that ν equals "%-change of η per unit of time". If we consider *continuous change*, we can express this as $\eta=\eta_0e^{\nu t}$. Differentiating with respect to t we have, $E(t)=\frac{\partial\eta}{\partial t}=\eta_0\nu e^{\nu t}$ and $E_0=\eta_0\nu$. Thus, $E(t)=E_0e^{\nu t}$

11. Proposition: *"If the speed of light is constant, then light is a wave"* [Ragazas 2011b]

Proof: We have that $\lambda=\frac{h}{p}$, $\nu=\frac{E}{h}$ and $\lambda\nu=c$. Since $p=\frac{\partial\eta}{\partial x}$ and $E=\frac{\partial\eta}{\partial t}$, we have that $\lambda\nu=\frac{\partial\eta/\partial t}{\partial\eta/\partial x}$. Differentiating, we get

$$D_t(\lambda\nu)=\frac{\frac{\partial^2\eta}{\partial t^2}\cdot\frac{\partial\eta}{\partial x}-\frac{\partial\eta}{\partial t}\cdot\frac{\partial^2\eta}{\partial t\partial x}}{\left(\frac{\partial\eta}{\partial x}\right)^2}\quad\text{and}\quad D_x(\lambda\nu)=\frac{\frac{\partial^2\eta}{\partial x\partial t}\cdot\frac{\partial\eta}{\partial x}-\frac{\partial\eta}{\partial t}\cdot\frac{\partial^2\eta}{\partial x^2}}{\left(\frac{\partial\eta}{\partial x}\right)^2}$$

Since $\lambda\nu=c$, we have that $D_t(\lambda\nu)=0$ and $D_x(\lambda\nu)=0$. Therefore,

$$\frac{\partial^2\eta}{\partial t^2}\cdot\frac{\partial\eta}{\partial x}-\frac{\partial\eta}{\partial t}\cdot\frac{\partial^2\eta}{\partial t\partial x}=0\quad\text{and}\quad\frac{\partial^2\eta}{\partial x\partial t}\cdot\frac{\partial\eta}{\partial x}-\frac{\partial\eta}{\partial t}\cdot\frac{\partial^2\eta}{\partial x^2}=0$$

Using $\lambda\nu=\frac{\partial\eta/\partial t}{\partial\eta/\partial x}$ and $\lambda\nu=c$, these can be written as,

$$\frac{\partial^2\eta}{\partial t^2}=c\cdot\frac{\partial^2\eta}{\partial t\partial x}\quad\text{and}\quad\frac{\partial^2\eta}{\partial x\partial t}=c\frac{\partial^2\eta}{\partial x^2}$$

Since 'mixed partials are equal', these equations combine to give us,

$$\frac{\partial^2\eta}{\partial t^2}=c^2\cdot\frac{\partial^2\eta}{\partial x^2},\quad\textit{the wave equation in one dimension}$$

Thus, for the speed of light to be constant the 'propagation of light' η must be a solution to the wave equation. *q.e.d*

12. The double-slit experiment [Ragazas 2011a]

The 'double-slit experiment' (where a beam of light passes through two narrow parallel slits and projects onto a screen an interference pattern) was originally used by Thomas Young in 1803, and latter by others, to demonstrate the wave nature of light. This experiment later came in direct conflict, however, with Einstein's *Photon Hypothesis* explanation of the Photoelectric Effect which establishes the particle nature of light. Reconciling these logically antithetical views has been a major challenge for physicists. The double-slit experiment embodies this quintessential mystery of Quantum Mechanics.

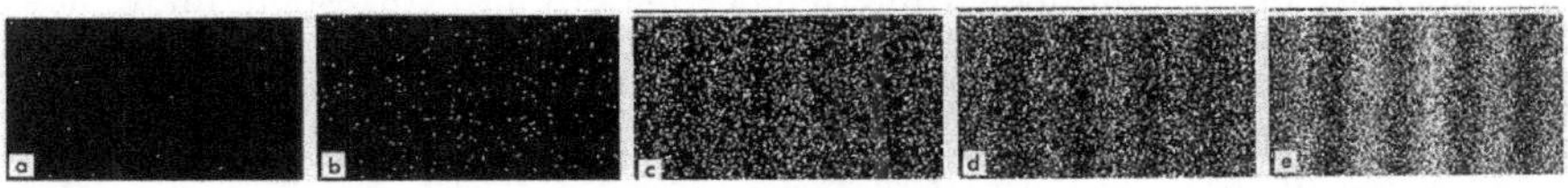

Fig. 6.

There are many variations and strained explanations of this simple experiment and new methods to prove or disprove its implications to Physics. But the 1989 Tonomura 'single electron emissions' experiment provides the clearest expression of this wave-particle enigma. In this experiment single emissions of electrons go through a simulated double-slit barrier and are recorded at a detection screen as 'points of light' that over time randomly fill in an interference pattern. The picture frames in Fig. 6 illustrate these experimental results. We will use these results in explaining the *double-slit experiment.*

12.1 Plausible explanation of the double-slit experiment

The basic logical components of this *double-slit experiment* are the 'emission of an electron at the source' and the subsequent 'detection of an electron at the screen'. It is commonly assumed that these two events are directly connected. The electron emitted at the source is assumed to be the same electron as the electron detected at the screen. We take the view that this may not be so. Though the two events (emission and detection) are related, they may not be directly connected. That is to say, there may not be a 'trajectory' that directly connects the electron emitted with the electron detected. And though many explanations in Quantum Mechanics do not seek to trace out a trajectory, nonetheless in these interpretations the detected electron is tacitly assumed to be the same as the emitted electron. This we believe is the source of the dilemma. We further adapt the view that while energy propagates continuously as a wave, the measurement and manifestation of energy is made in discrete units (*equal size sips*). This view is supported by all our results presented in this Chapter. And just as we would never characterize the nature of a vast ocean as consisting of discrete 'bucketfuls of water' because that's how we draw the water from the ocean, similarly we should not conclude that energy consists of discrete energy quanta simply because that's how energy is absorbed in our measurements of it.

The 'light burst' at the detection screen in the Tonomura *double-slit experiment* may not signify the arrival of "the" electron emitted from the source and going through one or the other of the two slits as a particle strikes the screen as a 'point of light'. The 'firing of an electron' at the source and the 'detection of an electron' at the screen are two separate events. What we have at the detection screen is a separate event of a light burst at some point on the screen, having absorbed enough energy to cause it to 'pop' (like popcorn at seemingly random manner once a seed has absorbed enough heat energy). The parts of the detection screen that over time are illuminated more by energy will of course show more 'popping'. The emission of an electron at the source is a separate event from the detection of a light burst at the screen. Though these events are connected they are not directly connected. There is no trajectory that connects these two electrons as being one and the same. The electron 'emitted' is not the same electron 'detected'.

What is emitted as an electron is a burst of energy which propagates continuously as a wave and going through both slits illuminates the detection screen in the typical interference pattern. This interference pattern is clearly visible when a large beam of energy illuminates the detection screen all at once. If we systematically lower the intensity of such electron beam the intensity of the illuminated interference pattern also correspondingly fades. For small bursts of energy, the interference pattern illuminated on the screen may be undetectable as a whole. However, when at a point on the screen *local equilibrium* occurs, we get a 'light burst' that in effect discharges the screen of an amount of energy equal to the energy burst that illuminated the screen. These points of discharge will be more likely to occur at those areas on the screen where the illumination is greatest. Over time we would get these dots of light filling the screen in the interference pattern.

We have a 'reciprocal relation' between 'energy' and 'time'. Thus, 'lowering energy intensity' while 'increasing time duration' is equivalent to 'increasing energy intensity' and 'lowering time duration'. But the resulting phenomenon is the same: the interference pattern we observe. This explanation of the *double-slit* experiment is logically consistent with the 'probability distribution' interpretation of Quantum Mechanics. The view we have of energy propagating continuously as a wave while manifesting locally in discrete units (*equal size sips)* when *local equilibrium* occurs, helps resolve the *wave-particle dilemma*.

12.2 Explanation *summary*

The argument presented above rests on the following ideas. These are consistent with all our results presented in this Chapter.

1. The 'electron emitted' is not be the same as the 'electron detected'.
2. Energy 'propagates continuously' but 'interacts discretely' when equilibrium occurs
3. We have 'accumulation of energy' before 'manifestation of energy'.

Our thinking and reasoning are also guided by the following attitude of *physical realism*:

a. Changing our detection devices while keeping the experimental setup the same can reveal something 'more' of the examined phenomenon but not something 'contradictory'.

b. If changing our detection devices reveals something 'contradictory', this is due to the detection device design and not to a change in the physics of the phenomenon examined.

Thus, using *physical realism* we argue that if we keep the experimental apparatus constant but only replace our 'detection devices' and as a consequence we detect something contradictory, the physics of the double slit experiment does not change. The experimental behavior has not changed, just the display of this behavior by our detection device has changed. The 'source' of the beam has not changed. The effect of the double slit barrier on that beam has not changed. So if our detector is now telling us that we are detecting 'particles' whereas before using other detector devices we were detecting 'waves', *physical realism* should tell us that this is entirely due to the change in our methods of detection. For the same input, our instruments may be so designed to produce different outputs.

13. Conclusion

In this Chapter we have sought to present a thumbnail sketch of *a world without quanta*. We started at the very foundations of Modern Physics with a simple and continuous mathematical derivation of *Planck's Law*. We demonstrated that *Planck's Law is an exact mathematical identity that describes the interaction of energy*. This fact alone explains why *Planck's Law* fits so exceptionally well the experimental data.

Using our derivation of *Planck's Law* as a *Rosetta Stone* (linking Mechanics, Quantum Mechanics and Thermodynamics) we considered the *quantity eta* that naturally appears in our derivation as *prime physis.* Planck's constant h is such a quantity. Energy can be defined as the time-rate of *eta* while momentum as the space-rate of *eta*. Other physical quantities can likewise be defined in terms of *eta*. Laws of Physics can and must be mathematically derived and not physically posited as Universal Laws chiseled into cosmic dust by the hand of God.

We postulated the *Identity of Eta Principle,* derived the Conservation of Energy and Momentum, derived Newton's Second Law of Motion, established the intimate connection between entropy and time, interpreted Schoedinger's equation and suggested that the *wave-function* ψ is in fact *prime physis* η. We showed that The Second Law of Thermodynamics pertains to *time* (and not entropy, which can be both positive and negative) and should be reworded to state that '*all physical processes take some positive duration of time to occur*'. We also showed the unexpected mathematical equivalence between *Planck's Law and Boltzmann's Entropy Equation* and proved that "*if the speed of light is a constant, then light is a wave*".

14. Appendix: Mathematical derivations

The proofs to many of the derivations below are too simple and are omitted for brevity. But the propositions are listed for purposes of reference and completeness of exposition.

Notation. We will consistently use the following notation throughout this APPENDIX:

$E(t)$ is a real-valued function of the real-variable t

$\Delta t = t - s$ is an 'interval of t'

$\Delta E = E(t) - E(s)$ is the 'change of E'

$P = \int_s^t E(u)du$ is the 'accumulation of E'

$\overline{E} = E_{av} = \frac{1}{t-s}\int_s^t E(u)du$ is the 'average of E'

D_x indicates 'differentiation with respect to x'

r is a constant, often an 'exponential rate of growth'

14.1 Part I: Exponential functions

We will use the following characterization of exponential functions without proof:

Basic Characterization: $E(t) = E_0e^{rt}$ *if and only if* $D_tE = rE$

Characterization 1: $E(t) = E_0e^{rt}$ *if and only if* $\Delta E = Pr$

Proof: Assume that $E(t) = E_0e^{rt}$. We have that $\Delta E = E(t) - E(s) = E_0e^{rt} - E_0e^{rs}$,

while $P = \int_s^t E_0e^{ru}du = \frac{1}{r}\left[E_0e^{rt} - E_0e^{rs}\right] = \frac{\Delta E}{r}$. Therefore $\Delta E = Pr$.

Assume next that $\Delta E = Pr$. Differentiating with respect to t, $D_tE = rD_tP = rE$.

Therefore by the *Basic Characterization,* $E(t) = E_0e^{rt}$. *q.e.d*

Theorem 1: $E(t) = E_0 e^{rt}$ *if and only if* $\frac{Pr}{e^{r\Delta t}-1}$ *is invariant with respect to t*

Proof: Assume that $E(t) = E_0 e^{rt}$. Then we have, for *fixed s*,

$$P = \int_s^t E_0 e^{ru} du = \frac{E_0}{r}\left[e^{rt} - e^{rs}\right] = \frac{E_0 e^{rs}}{r}\left[e^{r(t-s)} - 1\right] = \frac{E(s)}{r}\left(e^{r(t-s)} - 1\right)$$

and from this we get that $\frac{Pr}{e^{r\Delta t}-1} = E(s) =$ constant. Assume next that $\frac{Pr}{e^{r\Delta t}-1} = C$ is constant with respect to *t*, for *fixed s*.

Therefore, $D_t\left[\frac{Pr}{e^{r\Delta t}-1}\right] = \frac{rE(t)\cdot\left[e^{r\Delta t}-1\right] - rP\cdot\left[re^{r\Delta t}\right]}{\left(e^{r\Delta t}-1\right)^2} = 0$ and so, $E(t) = \left(\frac{Pr}{e^{r\Delta t}-1}\right)e^{r\Delta t} = C\cdot e^{r\Delta t}$

where C is constant. Letting $t = s$ we get $E(s) = C$. We can rewrite this as $E(t) = E(s)e^{r(t-s)} = E_0 e^{rt}$. *q.e.d*

From the above, we have

Characterization 2: $E(t) = E_0 e^{rt}$ *if and only if* $\frac{Pr}{e^{r(t-s)}-1} = E(s)$

Clearly by definition of E_{av}, $r\Delta t = \frac{Pr}{E_{av}}$. We can write $\frac{Pr}{e^{r\Delta t}-1}$ equivalently as $\frac{Pr}{e^{Pr/E_{av}}-1}$ in the above. *Theorem 1* above can therefore be restated as,

Theorem 1a: $E(t) = E_0 e^{rt}$ *if and only if* $\frac{Pr}{e^{Pr/E_{av}}-1}$ *is invariant with t*

The above *Characterization 2* can then be restated as

Characterization 2a: $E(t) = E_0 e^{rt}$ *if and only if* $\frac{Pr}{e^{Pr/E_{av}}-1} = E(s)$.

But if $\frac{Pr}{e^{Pr/E_{av}}-1} = E(s)$, then by *Characterization 2a*, $E(t) = E_0 e^{rt}$. Then, by *Characterization 1*, we must have that $\Delta E = Pr$. And so we can write equivalently $\frac{\Delta E}{e^{\Delta E/E_{av}}-1} = E(s)$. We have the following equivalence,

Characterization 3: $E(t) = E_0 e^{rt}$ *if and only if* $\frac{\Delta E}{e^{\Delta E/E_{av}}-1} = E(s)$

As we've seen above, it is always true that $\frac{Pr}{E_{av}} = r\Delta t$. But for exponential functions $E(t)$ we also have that $\Delta E = Pr$. So, for exponential functions we have the following.

Characterization 4: $E(t) = E_0 e^{rt}$ *if and only if* $\frac{\Delta E}{E_{av}} = r\Delta t$

14.2 Part II: Integrable functions

We next consider that $E(t)$ is any function. In this case, we have the following.

Theorem 2: a) For any differentiable function $E(t)$, $\lim_{t\to s}\frac{\Delta E}{e^{\Delta E/E_{av}}-1}=E(s)$

b) For any integrable function $E(t)$, $\lim_{t\to s}\frac{Pr}{e^{r\Delta t}-1}=E(s)$

Proof: Since $\frac{\Delta E}{e^{\Delta E/E_{av}}-1}\to\frac{0}{0}$ and $\frac{Pr}{e^{r\Delta t}-1}\to\frac{0}{0}$ as $t\to s$, we apply L'Hopital's Rule.

$$\lim_{t\to s}\frac{\Delta E}{e^{\Delta E/\overline{E}}-1}=\lim_{t\to s}\frac{D_tE(t)}{e^{\Delta E/\overline{E}}\cdot\left[\frac{D_tE(t)\cdot\overline{E}-D_t\overline{E}\cdot\Delta E}{\overline{E}^2}\right]}$$

$$=\lim_{t\to s}\frac{\overline{E}^2\cdot D_tE(t)}{e^{\Delta E/\overline{E}}\cdot\left[D_tE(t)\cdot\overline{E}-D_t\overline{E}\cdot\Delta E\right]}=E(s)$$

since $\Delta E\to 0$ and $\overline{E}\to E(s)$ as $t\to s$.

Likewise, we have $\lim_{t\to s}\frac{Pr}{e^{r\Delta t}-1}=\lim_{t\to s}\frac{E(s)r}{e^{r\Delta t}\cdot r}=E(s)$. *q.e.d.*

Corollary A: $\frac{\Delta E}{e^{\Delta E/\overline{E}}-1}$ *is invariant with* t *if and only if* $E(s)=\frac{\Delta E}{e^{\Delta E/\overline{E}}-1}$

Proof: Using *Theorem 2* we have $\lim_{t\to s}\frac{\Delta E}{e^{\Delta E/E_{av}}-1}=E(s)$. Since $\frac{\Delta E}{e^{\Delta E/E_{av}}-1}$ is constant with respect to *t*, we have $E(s)=\frac{\Delta E}{e^{\Delta E/E_{av}}-1}$. *Conversely,* if $E(s)=\frac{\Delta E}{e^{\Delta E/E_{av}}-1}$, then by *Characterization 3,* $E(s)=E_0e^{rs}$. Since $E(s)$ is a constant, $\frac{\Delta E}{e^{\Delta E/E_{av}}-1}$ is invariant with respect to *t*. *q.e.d*

Since it is always true by definitions that $r\Delta t=\frac{Pr}{E_{av}}$, *Theorem 2* can also be written as,

Theorem 2a: For any integrable function $E(t)$, $\lim_{t\to s}\frac{Pr}{e^{Pr/E_{av}}-1}=E(s)$

As a direct consequence of the above, we have the following interesting and important result:

Corollary B: $E(s)=\frac{\Delta E}{e^{\Delta E/E_{av}}-1}$ and $E(s)=\frac{Pr}{e^{Pr/E_{av}}-1}$ are independent of Δt, ΔE.

14.3 Part III: Independent proof of *Characterization 3*

In the following we provide a direct and independent proof of *Characterization 3*.
We first prove the following,

Lemma: For any *E*, $D_t\overline{E}(t)=\frac{E(t)-\overline{E}}{t-s}$ and $D_s\overline{E}(s)=\frac{\overline{E}-E(s)}{t-s}$

Proof: We let $\Delta t=t-s$ and $\overline{E}=\frac{1}{t-s}\int_s^t E(u)du$.

Differentiating with respect to *t* we have $(t-s)\cdot D_t\overline{E}(t)+\overline{E}=E(t)$.

Rewriting, we have $D_t\overline{E}(t)=\dfrac{E(t)-\overline{E}}{t-s}$. Differentiating with respect to s we have $(t-s)\cdot D_s\overline{E}(s)-\overline{E}=-E(s)$. Rewriting, we have $D_s\overline{E}(s)=\dfrac{\overline{E}-E(s)}{t-s}$. *q.e.d.*

Characterization 3: $E(t)=E_0e^{rt}$ *if and only if* $\dfrac{\Delta E}{e^{\Delta E/E_{av}}-1}=E(s)$

Proof: Assume that $E(t)=E_0e^{rt}$. From,

$$P=\int_s^t E_0e^{ru}du=\frac{E_0}{r}\left[e^{rt}-e^{rs}\right]=\frac{E_0e^{rs}}{r}\left[e^{r\Delta t}-1\right]=\frac{E(s)}{r}\left[e^{r\Delta t}-1\right]$$

we get, $E(s)=\dfrac{Pr}{e^{r\Delta t}-1}$. This can be rewritten as, $E(s)=\dfrac{Pr}{e^{Pr/E_{av}}-1}$. Since $\Delta E=Pr$, this can further be written as $E(s)=\dfrac{\Delta E}{e^{\Delta E/E_{av}}-1}$.

Conversely, consider next a function $E(s)$ satisfying

$$E(s)=\frac{\Delta E}{e^{\xi}-1},\ \text{where}\ \begin{cases}\Delta E=E(t)-E(s)\\ \Delta t=t-s\\ \xi=\dfrac{\Delta E}{\overline{E}}\\ \overline{E}=\dfrac{1}{\Delta t}\displaystyle\int_s^t E(u)du\end{cases}\quad \text{and } t \text{ can be any real value.}$$

From the above, we have that $e^{\xi}=\dfrac{\Delta E}{E(s)}+1=\dfrac{E(t)-E(s)+E(s)}{E(s)}=\dfrac{E(t)}{E(s)}$.

Differentiating with respect to s, we get $e^{\xi}\cdot D_s\xi=\dfrac{-E(t)\cdot D_sE(s)}{E(s)^2}=-e^{\xi}\cdot\dfrac{D_sE(s)}{E(s)}$

and so, $D_s\xi=-\dfrac{D_sE(s)}{E(s)}$ (A1)

From the above *Lemma* we have

$$D_s\overline{E}(s)=\frac{\overline{E}-E(s)}{t-s} \tag{A2}$$

Differentiating $\xi=\dfrac{\Delta E}{\overline{E}}$ with respect to s we get,

$$D_s\xi=\frac{-D_sE(s)\cdot\overline{E}-\Delta E\cdot D_s\overline{E}(s)}{\overline{E}^2} \tag{A3}$$

and combining (A1), (A2), and (A3) we have

$$-\frac{D_sE(s)}{E(s)}=\frac{-D_sE(s)\cdot\overline{E}-\dfrac{\Delta E}{\Delta t}\left(\overline{E}-E(s)\right)}{\overline{E}^2}=-\frac{D_sE(s)}{\overline{E}}-\frac{\Delta E}{\Delta t}\cdot\frac{\left(\overline{E}-E(s)\right)}{\overline{E}^2}$$

We can rewrite the above as follows,

$$\frac{D_s E(s)}{E(s)} - \frac{D_s E(s)}{\overline{E}} = D_s E(s)\left(\frac{\overline{E} - E(s)}{E(s)\cdot \overline{E}}\right) = \frac{\Delta E}{\Delta t}\cdot\frac{\left(\overline{E} - E(s)\right)}{\overline{E}^2}$$

and so, $\dfrac{D_s E(s)}{E(s)} = \dfrac{\Delta E}{\Delta t}\cdot\dfrac{1}{\overline{E}}$.

Using (A1), this can be written as

$$-D_s\xi = \frac{\xi}{\Delta t}\text{ , or as }\xi = -D_s\xi\cdot\Delta t\,. \tag{A4}$$

Differentiating (A4) above with respect to s, we get $D_s\xi = -D_s^2\xi\cdot\Delta t + D_s\xi$.
Therefore, $D_s^2\xi = 0$. Working backward, this gives $D_s\xi = -r$ = constant.
From (A1), we then have that $\dfrac{D_s E(s)}{E(s)} = r$ and therefore $E(s) = E_0 e^{rs}$. *q.e.d.*

15. Acknowledgement

I am indebted to Segun Chanillo, Prof. of Mathematics, Rutgers University for his encouragement, when all others thought my efforts were futile. Also, I am deeply grateful to Hayrani Oz, Prof. of Aerospace Engineering, Ohio State University, who discovered my posts on the web and was the first to recognize the significance of my results in Physics. Special thanks also to Miguel Bayona of The Lawrenceville School for his friendship and help with the graphics in this chapter. And Alexander Morisse who is my best and severest critic of the Physics in these results.

16. References

Frank, Adam (2010), *Who Wrote the Book of Physics?* Discover Magazine (April 2010)

Keesing, Richard (2001). *Einstein, Millikan and the Photoelectric Effect,* Open University Physics Society Newsletter, Winter 2001/2002 Vol 1 Issue 4 http://www.oufusion.org.uk/pdf/FusionNewsWinter01.pdf

Öz, H., Algebraic Evolutionary Energy Method for Dynamics and Control, in: *Computational Nonlinear Aeroelasticity for Multidisciplinary Analysis and Design,* AFRL, VA-WP-TR-2002 -XXXX, 2002, pp. 96-162.

Öz, H., Evolutionary Energy Method (EEM): An Aerothermoservoelectroelastic Application,: *Variational and Extremum Principles in Macroscopic Systems,* Elsevier, 2005, pp. 641-670.

Öz , H., *The Law Of Evolutionary Enerxaction and Evolutionary Enerxaction Dynamics* , Seminar presented at Cambridge University, England, March 27, 2008, http://talks.cam.ac.uk/show/archive/12743

Öz , Hayrani; John K. Ramsey, *Time modes and nonlinear systems,* Journal of Sound and Vibration, 329 (2010) 2565–2602, doi:10.1016/j.jsv.2009.12.021

Planck, Max (1901) *On the Energy Distribution in the Blackbody Spectrum,* Ann. Phys. 4, 553, 1901

Ragazas, C. (2010) *A Planck-like Characterization of Exponential Functions,* knol

http://knol.google.com/k/constantinos-ragazas/a-planck-like-characterization-of/ql47o1qdr604/7#
Ragazas, C. (2010) *Entropy and 'The Arrow of Time'*, knol http://knol.google.com/k/constantinos-ragazas/entropy-and-the-arrow-of-time/ql47o1qdr604/17#
Ragazas, C. (2010) *"Let there be h": An Existance Argument for Planck's Constant*, knol http://knol.google.com/k/constantinos-ragazas/let-there-be-h-an-existence-argument/ql47o1qdr604/12#
Ragazas, C. (2010) *Prime 'physis' and the Mathematical Derivation of Basic Law*, knol http://knol.google.com/k/constantinos-ragazas/prime-physis-and-the-mathematical/ql47o1qdr604/10#
Ragazas, C. (2010) *"The meaning of ψ": An Interpretation of Schroedinger's Equations*, knol http://knol.google.com/k/constantinos-ragazas/the-meaning-of-psi-an-interpretation-of/ql47o1qdr604/14#
Ragazas, C. (2010) *Planck's Law is an Exact Mathematical Identity*, knol http://knol.google.com/k/constantinos-ragazas/planck-s-law-is-an-exact-mathematical/ql47o1qdr604/3#
Ragazas, C. (2010) *The Temperature of Radiation*, knol http://knol.google.com/k/constantinos-ragazas/the-temperature-of-radiation/ql47o1qdr604/6#
Ragazas, C. (2010) *The Interaction of Measurement*, knol http://knol.google.com/k/constantinos-ragazas/the-interaction-of-measurement/ql47o1qdr604/11#
Ragazas, C. (2010) *A Time-dependent Local Representation of Energy*, knol http://knol.google.com/k/constantinos-ragazas/a-time-dependent-local-representation/ql47o1qdr604/9#
Ragazas, C. (2010)*A Plausable Explanation of the Double-slit Experiment in Physics*, knol http://knol.google.com/k/constantinos-ragazas/a-plausible-explanation-of-the-double/ql47o1qdr604/4#
Ragazas, C. (2010) *The Photoelectric Effect Without Photons*, knol http://knol.google.com/k/constantinos-ragazas/the-photoelectric-effect-without-photons/ql47o1qdr604/8#
Ragazas, C. (2010) *Stocks and Planck's Law*, knol http://knol.google.com/k/constantinos-ragazas/stocks-and-planck-s-law/ql47o1qdr604/2#
Ragazas, C. (2011) *What is The Matter With de Broglie Waves?* knol http://knol.google.com/k/constantinos-ragazas/what-is-the-matter-with-de-broglie-waves/ql47o1qdr604/18#
Ragazas, C. (2011) *"If the Speed of Light is a Constant, Then Light is a Wave"*, knol http://knol.google.com/k/constantinos-ragazas/if-the-speed-of-light-is-a-constant/ql47o1qdr604/19#
Tonomura (1989) http://www.hitachi.com/rd/research/em/doubleslit.html
Wikipedia, (n.d.) http://en.wikipedia.org/wiki/File:Firas_spectrum.jpg

4

Thermodynamics as a Tool for the Optimization of Drug Binding

Ruth Matesanz, Benet Pera and J. Fernando Díaz
Centro de Investigaciones Biológicas (C.S.I.C.)
Spain

1. Introduction

A non-covalent interaction is a kind of chemical bond, typically between macromolecules, that involves dispersed variations of electromagnetic interactions (Alberts *et al.* 1994; Connors & Mecozzi 2010). Non-covalent interactions are individually weak as compared with covalent bonds, but their net strength is higher than the sum of that of the individual interactions. There are few drugs that bind irreversibly to their targets, in pharmacology, most drugs establish non-covalent interactions with their target molecules (usually proteins).

From a chemical point of view, the affinity constant (K_a) is a very useful measurement for the study of binding reactions as it provides much information about the mechanism. In many cases some chemical or physical properties of ligand or target change with the interaction between them, these changes might help to measure binding constants. It is important to establish the stoichiometry of the complex to be sure that the constants are accurately calculated. From the affinity constants measured it is possible to calculate the standard thermodynamic quantities for the binding reaction: free-energy (ΔG), enthalpy (ΔH) and entropy (ΔS).

Our group has already demonstrated that, in some cases, binding affinity measurements are very helpful for the optimization of ligand binding as it can be determined the contribution of every single chemical modification of the ligand to the binding affinity (Buey *et al.* 2004; Matesanz *et al.* 2008)

One of the objectives of drug development is the search of new or modified compounds with improved properties such as better potency, higher selectivity, better pharmacokinetics or superior drug resistance profiles. An important goal in this objective is the optimization of drugs binding affinity towards their targets, as binding affinity is directly related to potency (Ruben *et al.* 2006). Moreover, it has been shown that extremely high affinity drugs reflect as well changes in other properties like selectivity (Ohtaka *et al.* 2004; Ohtaka & Freire 2005) or resistance overcoming ability (Matesanz *et al.* 2008).

Examples of the importance of ligand affinity in drug optimization can be observed in the development of HIV-1 protease inhibitors and statins (cholesterol lowering drugs) over the years as remarked in (Freire 2008).

In this chapter we will study the nature of non-covalent interations and the concept of binding constant for these interactions. Examples of methodologies to measure binding constants of small ligands to macromolecules will be introduced and we will emphasize the need to determine the stoichiometry of the studied system to calculate accurately the constants. Once the thermodynamic concepts were introduced, we will show the use of these kind of studies for the optimization of drug binding to its target. We will detail the role of single chemical modifications in the molecule of study to modulate its binding affinity, and the way to quantify these changes. We will finally further discuss how the selection of the best sustituents can result in the optimization of binding.

2. Non-covalent interactions

Non-covalent interactions are chemical bonds that do not involve sharing of electron pairs between orbitals of different atoms, there are no orbital overlapping in these interactions which have an electrostatic nature and are not highly directional. Covalent bonds are generally shorter than 2Å while the non-covalent ones are within the range of several angstroms. Another difference between these two types of bonds is the energy released in its formation, non-covalent interactions are weaker, with energies below 40 kJ/mol whereas covalent bonds energies range 80-800 kJ/mol.

These weak interactions have important roles in the binding of macromolecules with each other and with other molecules in the cell, in the mainteinance of the three dimensional structure of large macromolecules such as proteins or nucleic acids (e.g. DNA double helix) and they are the forces found in the majority of the drug-proteins interactions in pharmacology.

2.1 Types of non-covalent interactions

There are four commonly mentioned fundamental non-covalent interaction types including ionic interactions, hydrogen bonds, hydrophobic interactions and van der Waals forces (dispersion attractions, dipole-dipole and dipole-induced dipole interactions). All these weak interactions must work together to have significant effects. Their combined bond effect is greater than the sum of the individual ones. The free energy of multiple bonds between two molecules is different than the sum of the enthalpies of each bond due to entropic effects.

2.1.1 Ionic interactions

Ionic bonds result from the electrostatic attraction between two ionized groups of opposite charge such as carboxyl ($-COO^-$) and amino ($-NH_3^+$). These ionic interactions are directly proportional to the product of the interacting charges and inversely proportional to the dielectric constant of the medium and the distance separating the charges. This relationship is defined by Coulomb's law:

$$E = \frac{kq_1q_2}{Dr} \tag{1}$$

where E is the energy, q_1 and q_2 are the charges of two atoms, r is the distance between them, D is the dielectric constant, and k is a proportionality constant. A charged group on a molecule can attract an oppositely charged group from another molecule. By contract, an attractive interaction has a negative energy. The dielectric constant is important for the medium. In water, these bonds are very weak as the dielectric constant is much higher (D=80) than in vacuum (D=1). As an example, the electrostatic interaction between two atoms bearing single opposite charges separated by 3 Å in water has an energy of 5.9 kJ/mol (k=1389 kJ/mol).

2.1.2 Van der Waals forces

Van der Waals forces are short range attractive forces between chemical groups in contact. The forces are caused by slight charge displacements. The distribution of electronic charge around an atom changes with time. At any moment, the charge distribution is not perfectly symmetric. This transient asymmetry in the electronic charge around an atom induces a complementary asymmetry in the electron distribution around its neighboring atoms. These induced dipole effects give rise to the so called van der Waals interactions, also known as dispersion forces. The attraction between two atoms increases as they come closer to each other, until they are separated by the so called van der Waals contact distance. At a shorter distance, very strong repulsive forces become dominant because the outer electron clouds overlap. The van der Waals radius of an atom is defined where the net force between two atoms is zero. The van der Waals potential is then best described as a balance between attraction and repulsion.

Van der Waals forces are non-directional. Energies associated with them are quite small; typical interactions contribute from 2 to 4 kJ/mol per atom pair. However, when the surfaces of two large molecules come together, a large number of atoms are in van der Waals contact, and the net effect, summed over many atom pairs, can be substantial.

2.1.3 Hydrogen bonds

A hydrogen bond is an interaction between a proton donor group (a hydrogen atom covalently bound to an electronegative atom -e.g. F, O, N, S-) and a proton acceptor atom (another electronegative atom). It is a very important interaction responsible for the structure and properties of water, as well as the structure and properties of biological macromolecules (e.g. hydrogen bonds are responsible of specific base-pair formation in the DNA double helix).

Hydrogen bonds are fundamentally electrostatic interactions. The relatively electronegative atom to which the hydrogen atom is covalently bonded pulls electron density away from the hydrogen atom so that it develops a partial positive charge (δ^+). Thus, it can interact with an atom having a partial negative charge (δ^-) through an electrostatic interaction. However, this interaction is more than just an ionic or dipole-dipole interaction between the donor and the acceptor groups. Here, the distance between the hydrogen and acceptor atoms is less than the sum of their respective van der Waals radii.

Hydrogen bonds are directional toward the electronegative atom. The strongest hydrogen bonds have a tendency to be approximately straight, such that the proton donor group, the hydrogen atom, and the acceptor atom lie along a straight line, with significant weakening of the interaction if they are not colinear. They are somewhat longer than are covalent bonds. Hydrogen bonds are constantly being made and remade. Their half-life is about 10 seconds. These bonds have only 5% or so of the strength of covalent bonds. They have energies of 5-15 kJ/mol compared with approximately 420 kJ/mol for a carbon-hydrogen covalent bond. However, when many hydrogen bonds can form between two molecules (or parts of the same molecule), the resulting union can be sufficiently strong as to be quite stable. Examples of multiple hydrogen bonds are widely found in biological systems, they hold secondary structures of polypeptides, help in binding of enzymes to their substrate or antibodies to their antigen, help also transcription factors bind to each other or to DNA.

2.1.4 Hydrophobic interactions

Hydrophobic interactions result when non-polar molecules are in a polar solvent (e.g. water). The non-polar molecules group together to exclude water so that they minimize the

surface area in contact with the polar solvent. Unlike the non-covalent interactions mentioned above, which are pairwise interactions between atoms or parts of molecules, the nature of the hydrophobic interaction is very different. It involves a considerable number of (water) molecules, and does not arise from a direct force between the non-polar molecules.
Nonpolar molecules are not good acceptors of hydrogen bonds. When a non-polar molecule is placed in water, the hydrogen bonding network of water is disrupted. Water molecules must reorganize around the solute and make a kind of cage, similar to the structure of water in ice, in order to gain back the broken hydrogen bonds. This reorganization results in a considerable loss in the configurational entropy of water and therefore, in an increase in the free energy. If there are more than one such non-polar molecules, the configuration in which they are clustered together is preferred because now the hydrogen bonding network of water is disrupted in just one (albeit bigger) pocket, rather than in several small pockets. Therefore, the entropy of water is larger when the non-polar molecules are clustered together, leading to a decrease in the free energy.
Hydrophobic interactions have strengths comparable in energy to hydrogen bonds.

3. Binding constants

Most drugs have a non-covalent binding to their targets, thus these interactions are of great importance for our studies. Measurements of equilibrium constants, their dependence with temperature, the determination of stoichiometry, provide main information on the mechanism of the chemical process involved. The basic process can be taken out of the association of ligand (or ligands) to its target. The binding reaction can be writen as follows:

$$mP + nL \leftrightarrow P_m L_n \tag{2}$$

Regardless of mechanism, every reversible reaction reaches an equilibrium distribution of reactants and products. At some point the rates of the opposing reactions (association and dissociation in our case) become equal and there would no longer be any change in the concentration of the molecules implied.

$$v_{ass} = k_{ass}[P]^m[L]^n \tag{3}$$

$$v_{diss} = k_{diss}[P_m L_n] \tag{4}$$

Under these conditions ($v_{ass} = v_{diss}$):

$$\frac{k_{ass}}{k_{diss}} = \frac{[P_m L_n]}{[P]^m[L]^n} = K_a \tag{5}$$

that will be the equilibrium association constant assuming that activities are equal to concentrations.
In this section we will discuss the cases for one single site in the target, multiple sites with same affinities and multiple sites with different affinities.

3.1 One-site binding

In the simplest case, where there is only one site per target molecule, n and m are 1. It is possible to define the fraction of occupied binding sites (υ) as:

$$\upsilon = \frac{L_{bound}}{P_{total}} = \frac{[PL]}{[PL] + [P]} \quad (6)$$

Determining υ is often easy in spectrophotometric manipulation as will be discussed later. Given υ, equation 5 can then be solved for [PL] and the answer substituted into equation 6 to obtain the quantitative 1:1 stoichiometric model:

$$\upsilon = \frac{K_a[L]}{1 + K_a[L]} \quad (7)$$

This equation is the 1:1 binding isotherm also known as the Langmuir isotherm or the "direct" plot. Its functional form is a rectangular hyperbola whose midpoint will yield K_a. Chemical interpretation of 1:1 binding is that the target P has a single "binding site", as has the ligand L; and when the complex PL forms, no further sites are available for the binding of any additional ligand. To test the 1:1 stoichoimetry equation 7 may be rearrange into a linear plotting form. Since υ is the bound fraction, then 1-υ is the free one: (1- υ) = $1/(1+K_a[L])$. Thus $\upsilon/(1-\upsilon) = K_a[L]$, and:

$$\log \frac{\upsilon}{1-\upsilon} = \log\ [L]\ +\ \log K_a \quad (8)$$

This log-log plot should be linear with a slope of one if the stoichoimetry is really 1:1. This is called a Hill plot. Equation 7 can be also rearranged to three different non-logarithmic linear plotting forms. Taking simply the reciprocal of the equation yields the double-reciprocal plot (used by plotting 1/υ against 1/[L]):

$$\frac{1}{\upsilon} = \frac{1}{K_a[L]} + 1 \quad (9)$$

In spectroscopic studies this plot is commonly known as the Benesi-Hildebrand plot (Benesi & Hildebrand 1949).

Another plot is that of [L]/υ against [L] which is expected to be linear:

$$\frac{[L]}{\upsilon} = [L] + \frac{1}{K_a} \quad (10)$$

And the third plotting of υ/[L] agains υ, sometimes called Scatchard plot (Scatchard 1949):

$$\frac{\upsilon}{[L]} = \upsilon K_a + K_a \quad (11)$$

Linearity in all of these plots is a necessary condition if the 1:1 model is valid; and from the parameters of equations K_a can be evaluated. Usually υ is not measured directly but rather some experimental quantity related to it, so that the interpretation of the plots depends on the particular experimental methodology.

3.2 Multi-site binding

Most biological systems tend to have more than one binding site, that is the case of many systems of small molecules binding to proteins. In these cases we may consider that n ligands may bind to a single target molecule. The average number of ligand molecules bound per target molecule (b) is defined as:

$$b = \frac{L_{total} - [L]}{P_{total}} \quad (12)$$

Assuming that all n binding sites in the target molecule are identical and independent, it is possible to establish:

$$b = \frac{nk[L]}{1 + k[L]} \tag{13}$$

where k is the constant for binding to a single site. According to this equation this system follows the hyperbolic function characteristic for the one-site binding model. To define the model n and k can be evaluated from a Scatchard plot. The affinity constant k is an average over all binding sites, it is in fact constant if all sites are truly identical and independent. A stepwise binding constant (K_{st}) can be defined which would vary statistically depending on the number of target sites previously occupied. It means that for a target with n sites will be much easier for the first ligand added to find a binding site than it will be for each succesive ligand added. The first ligand would have n sites to choose while the nth one would have just one site to bind. The stepwise binding constant can be defined as:

$$K_{st} = \frac{\text{number of free target sites}}{\text{number of bound sites}} k = \frac{n - b + 1}{b} k \tag{14}$$

It is interesting to notice that a deviation from linearity in the Scatchard plot (and to a lesser extent in the Benesi-Hildebrand) gives information on the nature of binding sites. A curved plot denotes that the binding sites are not identical and independent.

3.3 Allosteric interactions

Another common situation in biological systems is the cooperative effect, in that case several identical but dependent binding sites are found in the target molecule. It is important to define the effect of the binding of succesive ligands to the target to describe the system. An useful model for that issue is the Hill plot (Hill 1910). In this case the number of ligands bound per target molecule will be (take into account that the situation in this system for equation 2 is m=1 and n≠1):

$$b = \frac{n[PL_n]}{[PL_n] + [P]} \tag{15}$$

if equation 5 is solved for $[PL_n]$ and substitute into equation 15, then:

$$b = \frac{nK_a[L]^n}{K_a[L]^n + 1} \tag{16}$$

This expression can be rewritten as:

$$\frac{b}{n - b} = K_a[L]^n \tag{17}$$

Note that the fraction of sites bound, υ (see equation 6), is the number of sites occupied, b, divided by the number of sites available, n. Then equation 17 becomes:

$$\frac{\upsilon}{1 - \upsilon} = K_a[L]^n \tag{18}$$

Equation 18 is known as the Hill equation. From the Hill equation we arrive at the Hill plot by taking logarithms at both sides:

$$\log \frac{\upsilon}{1 - \upsilon} = n_H \log [L] + \log K_a \tag{19}$$

Plotting log(υ/(1-υ)) against log[L] will yield a straight line with slope n_H (called the Hill coefficient). The Hill coefficient is a qualitative measure of the degree of cooperativity and it is experimentally less than the actual number of binding sites in the target molecule. When $n_H > 1$, the system is said to be positively cooperative, while if $n_H < 1$, it is said to be anti-cooperative. Positively cooperative binding means that once the first ligand is bound to its target molecule the affinity for the next ligand increases, on the other hand the affinity for subsequent ligand binding decreases in negatively cooperative (anti-cooperative) systems. In the case of $n_H = 1$ a non-cooperative binding occurs, here ligand affinity is independent of whether another ligand is already bound or not.

Since equation 19 assumes that $n_H = n$, it does not described exactly the real situation. When a Hill plot is constructed over a wide range of ligand concentrations, the continuity of the plot is broken at the extremes concentrations. In fact, the slope at either end is approximately one. This phenomenon can be easily explained: when ligand concentration is either very low or very high, cooperativity does not exist. For low concentrations it is more probable for individual ligands to find a target molecule "empty" rather than to occupy succesive sites on a pre-bound molecule, thus single-binding is happening in this situation. At the other extreme, for high concentrations, every binding-site in the target molecule but one will be filled, thus we find again single-binding situation. The larger the number of sites in a single target molecule is, the wider range of concentrations the Hill plot will show cooperativity.

4. Determination of binding constants

As discuss above the binding constant provides important and interesting information about the system studied. We will present a few of the multiple experimental posibilities to measure this constant (further information could be found in the literature (Johnson *et al.* 1960; Connors 1987; Hirose 2001; Connors&Mecozzi 2010; Pollard 2010)). It is essencial to keep in mind some crucial details to be sure to calculate the constants properly: it is important to control the temperature, to be sure that the system has reached the equilibrium and to use the correct equilibrium model. One common mistake that should be avoid is confuse the total and free concentrations in the equilibrium expression.

Different techniques are commonly used to study the binding of ligands to their targets. These techniques can be classified as calorimetry, spectroscopy and hydrodynamic methods. Hydrodynamic techniques are tipically separation methodologies such as different chromatographies, ultracentrifugation or equilibrium dialysis with which free ligand, free target and complex are physically separated from each other at equilibrium, thus concentrations of each can be measured. Spectroscopic methodologies include optical spectroscopy (e.g. absorbance, fluorescence), nuclear magnetic resonance or surface plasmon resonance. Calorimetry includes isothermal titration and differential scanning. Calorimetry and spectroscopy methods allow accurately determination of thermodynamics and kinetics of the binding, as well as can give information about the structure of binding sites.

Once the bound (or free) ligand concentration is measured, the binding proportion can be calculated. Other thermodynamic parameters can be calculated by varying ligand or target concentrations or the temperature of the system.

4.1 Determination of stoichiometry. Continuous variation method.

Since correct reaction stoichiometry is crucial for correct binding constant determination we will study how can it be evaluated. There are different methods of calculating the stoichoimetry: continuous variation method, slope ratio method, mole ratio method, being the first one, the continuous variation method the most popular. In order to determine the stoichiometry by this method the concentration of the produced complex (or any property proportional to it) is plotted versus the mole fraction ligand ($[L]_{total}/([P]_{total}+[L]_{total})$) over a number of tritation steps where the sum of $[P]_{total}$ and $[L]_{total}$ is kept constant (α) changing $[L]_{total}$ from 0 to α. The maxima of this plot (known as Job's plot, (Job 1928; Ingham 1975)) indicates the stoichiometry of the binding reaction: 1:1 is indicated by a maximum at 0.5 since this value corresponds to n/(n+m). For the understanding of the theoretical background of the method, it is important to remember equations 2 and 5; notice that:

$$[P]_{total} = [P] + m[P_mL_n] \quad (20)$$

$$[L]_{total} = [L] + n[P_mL_n] \quad (21)$$

$$\alpha = [L]_{total} + [P]_{total} \quad (22)$$

$$x = \frac{[L]_{total}}{[P]_{total} + [L]_{total}} \quad (23)$$

$$y = [P_mL_n] \quad (24)$$

Substitution of $[P]_{total}$ and $[L]_{total}$ by the functions of α and x from equation 23 and 24 yields:

$$[P]_{total} = \alpha - \alpha x \quad (25)$$

$$[L]_{total} = \alpha x \quad (26)$$

from equations 2, 5, 20, 21, 24, 25, 26:

$$y = K_a(\alpha - my - \alpha x)^m(\alpha x - ny)^n \quad (27)$$

Equation 27 is differentiated, and the dy/dx substituted by zero to obtain the x-coordinate at the maximum:

$$x = \frac{n}{n+m} \quad (28)$$

This equation shows the correlation between stoichiometry and the x-coordinate at the maximum in Job's plot. That's why a maximum at x = 0.5 means a 1:1 stoichiometry (n = m = 1). In the case of 1:2 the maximum would be at x = 1/3.

4.2 Calorimetry

Isothermal titration calorimetry (ITC) is a useful tool for the characterization of thermodynamics and kinetics of ligands binding to macromolecules. With this method the rate of heat flow induced by the change in the composition of the target solution by tritation of a ligand (or vice versa) is measured. This heat is proportional to the total amount of binding. Since the technique measures heat directly, it allows simultaneous determination of the stoichiometry (n), the binding constant (K_a) and the enthalpy (ΔH^0) of binding. The free energy (ΔG^0) and the entropy (ΔS^0) are easily calculated from ΔH^0 and K_a. Note that the binding constant is related to the free energy by:

$$\Delta G^0 = -RT \ln K_a \quad (29)$$

where R is the gas constant and T the absolute temperature. The free energy can be dissected into enthalpic and entropic components by:

$$\Delta G^0 = \Delta H^0 - T\Delta S^0 \tag{30}$$

On the other hand, the heat capacity (ΔC_p -p subscript indicates that the system is at constant pressure-) of a reaction predicts the change of ΔH^0 and ΔS^0 with temperature and can be expressed as:

$$\Delta C_p = \frac{\Delta H^0{}_{T2} - \Delta H^0{}_{T1}}{T_2 - T_1} \tag{31}$$

or

$$\Delta C_p = \frac{\Delta S^0{}_{T2} - \Delta S^0{}_{T1}}{\ln \frac{T_2}{T_1}} \tag{32}$$

In an ITC experiment a constant temperature is set, a precise amount of ligand is added to a known target molecule concentration and the heat difference is measured between reference and sample cells. To eliminate heats of mixing effects, the ligand and target as well as the reference cell contain identical buffer composition. Subsequent injections of ligand are done until no further heat of binding is observed (all sites are then bound with ligand molecules). The remaining heat generated now comes from dilution of ligand into the target solution. Data should be corrected for the heat of dilution. The heat of binding calculated for every injection is plotted versus the molar ratio of ligand to protein. K_a is related to the curve shape and binding capacity (n) determined from the ratio of ligand to target at the equivalence point of the curve. Data must be fitted to a binding model. The type of binding must be known from other experimental techniques. Here, we will study the simplest model with a single site. Equations 6 and 7 can be rearranged to find the following relation between υ and K_a:

$$K_a = \frac{\upsilon}{(1-\upsilon)[L]} \tag{33}$$

Total ligand concentration is known and can be represented as (remember that we are assuming m=n=1):

$$[L]_{total} = [L] + \upsilon[P]_{total} \tag{34}$$

Combining equations 33 and 34 gives:

$$\upsilon^2 - \left(\frac{[L]_{total}}{[P]_{total}} + \frac{1}{K_a[P]_{total}} + 1\right)\upsilon + \frac{[L]_{total}}{[P]_{total}} = 0 \tag{35}$$

Solving for υ:

$$\upsilon = \frac{1}{2}\left[\left(\frac{[L]_{total}}{[P]_{total}} + \frac{1}{K_a[P]_{total}} + 1\right) - \sqrt{\left(\frac{[L]_{total}}{[P]_{total}} + \frac{1}{K_a[P]_{total}} + 1\right)^2 - \frac{4\,[L]_{total}}{[P]_{total}}}\right] \tag{36}$$

The total heat content (Q) in the sample cell at volume (V) can be defined as:

$$Q = [PL]\Delta H^0 V = \upsilon[P]_{total}\Delta H^0 V \tag{37}$$

where ΔH^0 is the heat of binding of the ligand to its target. Substituing equation 36 into 37 yields:

$$Q = \frac{[P]_{total}\Delta H^0 V}{2}\left[\left(\frac{[L]_{total}}{[P]_{total}} + \frac{1}{K_a[P]_{total}} + 1\right) - \sqrt{\left(\frac{[L]_{total}}{[P]_{total}} + \frac{1}{K_a[P]_{total}} + 1\right)^2 - \frac{4\,[L]_{total}}{[P]_{total}}}\,\right] \tag{38}$$

Therefore Q is a function of K_a and ΔH^0 (and n, but here we considered it as 1 for simplicity) since $[P]_{total}$, $[L]_{total}$ and V are known for each experiment.

4.3 Optical spectroscopy

The goal to be able to determine binding affinity is to measure the equilibrium concentration of the species implied over a range of concentrations of one of the reactants (P or L). Measuring one of them should be sufficient as total concentrations are known and therefore the others can be calculated by difference from total concentrations and measured equilibrium concentration of one of the species. Plotting the concentration of the complex (PL) against the free concentration of the varying reactant, the binding constant could be calculated.

4.3.1 Absorbance

As an example a 1:1 stoichiometry model will be shown, wherein the Lambert-Beer law is obeyed by all the reactants implied. To use this technique we should ensured that the complex (PL) has a significantly different absorption spectrum than the target molecule (P) and a wavelenght at which both molar extinction coefficients are different should be selected. At these conditions the absorbance of the target molecule in the absence of ligand will be:

$$Abs_0 = \varepsilon_P\, l\, [P]_{total} \tag{39}$$

If ligand is added to a fixed total target concentration, the absorbance of the mix can be written as:

$$Abs_{mix} = \varepsilon_P\, l\, [P] + \varepsilon_L\, l\, [L] + \varepsilon_{PL}\, l\, [PL] \tag{40}$$

Since $[P]_{total} = [P] + [PL]$ and $[L]_{total} = [L] + [PL]$, equation 40 can be rewritten as:

$$Abs_{mix} = \varepsilon_P\, l\, [P]_{total} + \varepsilon_L\, l\, [L]_{total} + \Delta\varepsilon\, l\, [PL] \tag{41}$$

where $\Delta\varepsilon = \varepsilon_{PL}-\varepsilon_P-\varepsilon_L$. If the blank solution against which samples are measured contains $[L]_{total}$, then the observed absorbance would be:

$$Abs_{obs} = \varepsilon_P\, l\, [P]_{total} + \Delta\varepsilon\, l\, [PL] \tag{42}$$

Substracting equation 39 from 42 and incorporating K_a (equation 5):

$$\Delta Abs = K_a\, \Delta\varepsilon\, l\, [P]\, [L] \tag{43}$$

$[P]_{total}$ can be written as $[P]_{total} = [P](1+K_a[L])$ which included in equation 43 yields:

$$\frac{\Delta Abs}{l} = \frac{[P]_{total}\, K_a\, \Delta\varepsilon\, [L]}{1 + K_a\, [L]} \tag{44}$$

which is the direct plot expressed in terms of spectrophotometric observation. Note that the dependence of $\Delta Abs/l$ on [L] is the same as the one shown in equation 7.
The free ligand concentration is actually unknown. The known concentrations are $[P]_{total}$ to which a known $[L]_{total}$ is added. In a similar way as shown above for $[P]_{total}$, $[L]_{total}$ can be written as:

$$[L]_{total} = [L] \; \frac{[P]_{total} \, K_a \, [L]}{1 + K_a \, [L]} \tag{45}$$

From equations 44 and 45 a complete description of the system is obtained. If $[L]_{total}$ >>$[P]_{total}$ we will have that $[L]_{total} \approx [L]$ from equation 45, equation 44 can be then analysed with this approximation. With this first rough estimate of K_a, equation 45 can be solved for the [L] value for each $[L]_{total}$. These values can be used in equation 44 to obtain an improved estimation of K_a, and this process should be repeated until the solution for K_a reaches a constant value. Equation 44 can be solved graphically using any of the plots presented in section 3.1.

4.3.2 Fluorescence

Fluorescence spectroscopy is a widely used tool in biochemistry due to its ease, sensitivity to local environmental changes and ability to describe target-ligand interactions qualitatively and quantitatively in equilibrium conditions. In this technique the fluorophore molecule senses changes in its local environment. To analyse ligand-target interactions it is possible to take advantage of the nature of ligands, excepcionally we can find molecules which are essentially non or weakely fluorescent in solution but show intense fluorescence upon binding to their targets (that is the case, for example, of colchicines and some of its analogues). Fluorescence moieties such as fluorescein can be also attached to naturally non-fluorescent ligands to make used of these methods. The fluorescent dye may influence the binding, so an essential control with any tagged molecule is a competition experiment with the untagged molecule. Finally, in a few favourable cases the intrinsic tryptophan fluorescence of a protein changes when a ligand binds, usually decreasing (fluorescence quenching). Again, increasing concentrations of ligand to a fixed concentration of target (or vice versa) are incubated at controlled temperature and fluorescence changes measured until saturation is reached. Binding constant can be determined by fitting data according to equation 11 (Scatchard plot). From fluorescence data (F), υ can be calculated from the relantionship:

$$\upsilon = \frac{F_{max} - F}{F_{max}} \tag{46}$$

If free ligand has an appreciable fluorescence as compared to ligand bound to its target, then the fluorescence enhancement factor (Q) should be determined. Q is defined as (Mas & Colman 1985):

$$Q = \frac{F_{bound}}{F_{free}} - 1 \tag{47}$$

To determine it, a reverse titration should be done. The enhancement factor can be obtained from the intercept of linear plot of $1/((F/F_0)-1)$ against $1/P$, where F and F_0 are the observed fluorescence in the presence and absence of target, respectively. Once it is known, the concentration of complex can be determine from a fluorescence titration experiment using:

$$[PL] = [L]_{total} \frac{(F/F_0) - 1}{Q - 1} \tag{48}$$

Thus the binding constant can be determined from the Scatchard plot as described above.

4.3.3 Fluorescence anisotropy

Fluorescence anisotropy measures the rotational diffusion of a molecule. The effective size of a ligand bound to its target usually increases enormously, thus restricting its motion considerably. Changes in anisotropy are proportional to the fraction of ligand bound to its target. Using suitable polarizers at both sides of the sample cuvette, this property can be measured. In a tritation experiment similar to the ones described above, the fraction of ligand bound ($X_L=[PL]/[L]_{total}$) is determined from:

$$X_L = \frac{r - r_0}{r_{max} - r_0} \tag{49}$$

where r is the anisotropy of ligand in the presence of the target molecule, r_0 is the anisotropy of ligand in the absence of target and r_{max} is the anisotropy of ligand fully bound to its target (note that equation 49 can be used only in the case where ligand fluorescence intensity does not change, otherwise appropriate corrections should be done, see (Lakowicz 1999)). [P] can be calculated from:

$$[P] = [P]_{total} - X_L[L]_{total} \tag{50}$$

The binding constant can be determined from the hyperbola:

$$X_L = \frac{K_a[P]}{1 + K_a[P]} \tag{51}$$

4.4 Competition methods

The characterization of a ligand binding let us determine the binding constant of any other ligand competing for the same binding site. Measurements of ligand (L), target (P), reference ligand (R) and both complexes (PR and PL) concentrations in the equilibrium permit the calculation of the binding constant (K_L) from equation 53 (see below) as the binding constant of the reference ligand (K_R) is already known.

$$L + R + P \leftrightarrow PL + PR \tag{52}$$

$$K_L = K_R \frac{[PL][R]}{[L][PR]} \tag{53}$$

In the case that the reference ligand has been characterized due to the change of a ligand physical property (i.e. fluorescence, absorbance, anisotropy) upon binding, would permit us also following the displacement of this reference ligand from its site by competition with a ligand „blind" to this signal (Diaz & Buey 2007). In this kind of experiment equimolar concentrations of the reference ligand and the target molecule are incubated, increasing concentrations of the problem ligand added and the appropiate signal measured. It is possible then to determine the concentration of ligand at which half the reference ligand is bound to its site (EC_{50}). Thus K_L is calculated from:

$$K_L = \frac{1 + [R]K_R}{EC_{50}} \tag{54}$$

5. Drug optimization

Microtubule stabilizing agents (MSA) comprise a class of drugs that bind to microtubules and stabilize them against disassembly. During the last years, several of these compounds have been approved as anticancer agents or submitted to clinical trials. That is the case of taxanes (paclitaxel, docetaxel) or epothilones (ixabepilone) as well as discodermolide (reviewed in (Zhao *et al.* 2009)). Nevertheless, anticancer chemotherapy has still unsatisfactory clinical results, being one of the major reasons for it the development of drug resistance in treated patients (Kavallaris 2010). Thus one interesting issue in this field is drug optimization with the aim of improving the potential for their use in clinics: minimizing side-effects, overcoming resistances or enhancing their potency.

Our group has studied the influence of different chemical modifications on taxane and epothilone scaffolds in their binding affinities and the consequently modifications in ligand properties like citotoxicity. The results from these studies firmly suggest thermodynamic parameters as key clues for drug optimization.

5.1 Epothilones

Epothilones are one of the most promising natural products discovered with paclitaxel-like activity. Their advantages come from the fact that they can be produced in large amounts by fermentation (epothilones are secondary metabolites from the myxobacterium *Sorangiun celulosum*), their higher solubility in water, their simplicity in molecular architecture which makes possible their total synthesis and production of many analogs, and their effectiveness against multi-drug resistant cells due to they are worse substrates for P-glycoprotein.

The structure affinity-relationship of a group of chemically modified epothilones was studied. Epothilones derivatives with several modifications in positions C12 and C13 and the side chain in C15 were used in this work.

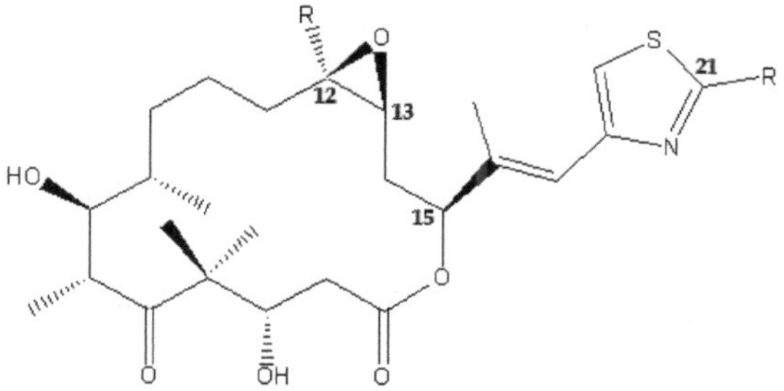

Fig. 1. Epothilone atom numbering.

Epothilone binding affinities to microtubules were measured by displacement of Flutax-2, a fluorescent taxoid probe (fluorescein tagged paclitaxel). Both epothilones A and B binding constants were determined by direct sedimentation which further validates Flutax-2 displacement method.

All compounds studied are related by a series of single group modifications. The measurement of the binding affinity of such a series can be a good approximation of the incremental binding energy provided by each group. Binding free energies are easily calculated from binding constants applying equation 29. The incremental free energies (ΔG^0) change associated with the modification of ligand L into ligand S is defined as:

$$\Delta\Delta G^0(L\rightarrow S) = \Delta G^0(L) - \Delta G^0(S) \quad (55)$$

These incremental binding energies were calculated for a collection of 20 different epothilones as reported in (Buey *et al.* 2004).

Site	Modification	Compounds	ΔΔG
C15	*S* → *R*	4 → 17	~ 27
		7 → 18	~ 27
		14 → 16	17.8 ± 0.3
	Thiazole → Pyridine	5 → 7	-2.9 ± 0.2
		6 → 8	-2.1 ± 0.3
		14 → 4	-0.2 ± 0.4
		16 → 17	~ 9.4
C21	Methyl → Thiomethyl	2 → 3	-2.8 ± 0.8
		5 → 10	-5.9 ± 0.6
		6 → 11	-3.6 ± 0.3
		8 → 12	2.6 ± 0.3
	Methyl → Hydroxymethyl	8 → 9	1.4 ± 0.3
	5-Thiomethyl-pyridine → 6-Thiomethyl-pyridine	12 → 13	4.1 ± 0.5
C12	*S* → *R*	4 → 7	-2.1 ± 0.3
		14 → 5	0.6 ± 0.3
		17 → 18	~ -2
		19 → 11	9.0 ± 0.6
		20 → 8	1.9 ± 0.4
	Epoxide → Cyclopropyl	1 → 14	-4.7 ± 0.4
		3 → 19	-5.4 ± 0.8
	Cyclopropyl → Cyclobutyl	5 → 15	4.1 ± 0.2
	S H → Methyl	1 → 2	-8.1 ± 0.6
		4 → 20	-1.8 ± 0.5
	R H → Methyl	5 → 6	0.4 ± 0.3
		7 → 8	1.2 ± 0.2
		10 → 11	2.7 ± 0.7

Table 1. Incremental binding energies of epothilone analogs to microtubules. (ΔΔG in kJ/mol at 35°C). Data from (Buey *et al.* 2004).

The data in table 1 show that the incremental binding free energy changes of single modifications give a good estimation of the binding energy provided by each group. Moreover, the effect of the modifications is accumulative, resulting the epothilone derivative with the most favourable modifications (a thiomethyl group at C21 of the thiazole side chain, a methyl group at C12 in the *S* configuration, a pyridine side chain with C15 in the *S* configuration and a cyclopropyl moiety between C12 and C13) the one with the highest affinity of all the compounds studied (K_a 2.1±0.4 x 10^{10} M^{-1} at 35°C).

The study of these compounds showed also a correlation between their citotoxic potencial and their affinities to microtubules. The plot of log IC_{50} in human ovarian carcinoma cells versus log K_a shows a good correlation (figure 2), suggesting binding affinity as an important parameter affecting citotoxicity.

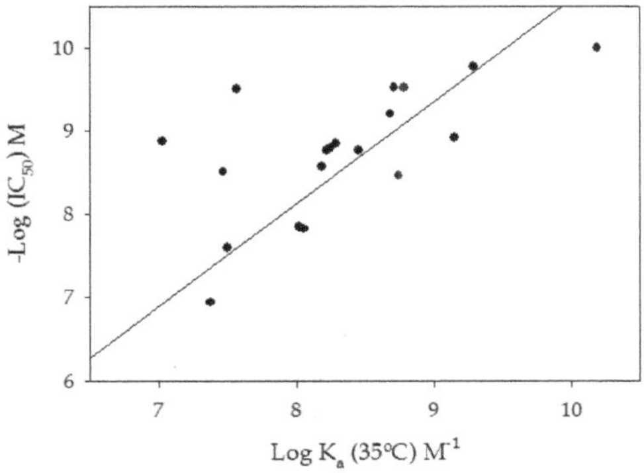

Fig. 2. Dependence of the IC_{50} of epothilone analogs against 1A9 cells on their K_a to microtubules. Data from (Buey *et al.* 2004).

5.2 Taxanes

Paclitaxel and docetaxel are widely used in the clinics for the treatment of several carcinoma and Kaposi's sarcoma. Nevertheless, their effectiveness is limited due to the development of resistance, beeing its main cause the overexpression and drug efflux activity of transmembrane proteins like P-glycoprotein (Shabbits *et al.* 2001).
We have studied the thermodynamics of binding of a set of nearly 50 taxanes to crosslinked stabilized microtubules with the aim to quantify the contributions of single modifications at four different locations of the taxane scaffold (C2, C13, C7 and C10).

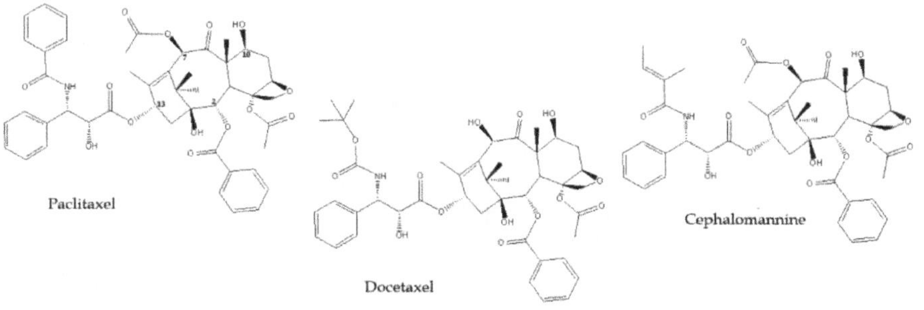

Fig. 3. Taxanes head compounds. Atom numbering

Once confirmed that all the compounds were paclitaxel-like MSA, their affinities were measured using the same competition method mentioned above (section 5.1. displacement of Flutax-2). Seven of the compounds completely displaced Flutax-2 at equimolar concentrations indicating that they have very high affinities and so they are in the limit of the range to be accurately calculated by this method (Diaz&Buey 2007). The affinities of these compounds were then measured using a direct competition experiment with epothilone-B, a higher-affinity ligand (K_a 75.0 x 10^7 at 35°C compared with 3.0 x 10^7 for Flutax-2). With all the binding constants determined at a given temperature, it is possible to determine the changes in binding free energy caused by every single modification as discussed above for epothilones (table 2).

Site	Modification	Compounds	ΔΔG	Average
C2	benzoyl → benzylether	T → 25	13.2	+13.0 ± 0.2
		21 → 24	12.8	
	benzoyl → benzylsulphur	T → 27	13.6	+15.9 ± 2.3
		21 → 26	18.1	
	benzoyl → benzylamine	T → 38	18.6	+20.1 ± 1.5
		21 → 39	21.6	
	benzoyl → thiobenzoyl	T → 23	19.6	+15.9 ± 3.8
		21 → 22	12.1	
	benzoyl → benzamide	21 → 42	19.2	
	benzamide → 3-methoxy-benzamide	42 → 43	-3.4	
	benzamide → 3-Cl-benzamide	42 → 44	5.3	
	benzoyl → 3 methyl- 2 butenoyl	1 → 2	6.2	
	benzoyl → 3 methyl- 3 butenoyl	1 → 3	4.9	
	benzoyl → 2(E)-butenoyl	1 → 9	7.3	
	benzoyl → 3 methyl- butanoyl	1 → 10	6.3	
	benzoyl → 2-debenzoyl-1,2-carbonate	C → 16	5.8	
	benzoyl → 3-azido-benzoyl	1 → 4	-8	-11.2 ± 1.3
		T → 12	-13.9	
		C → 14	-12.2	
		18 → 20	-10.6	
	benzoyl → 3-methoxy-benzoyl	1 → 5	-6.2	-7.2 ± 0.6
		T → 11	-8.3	
		C → 13	-8.1	
		18 → 19	-6.3	
	benzoyl → 3-Cl-benzoyl	1 → 6	-3.1	
	benzoyl → 3-Br-benzoyl	1 → 34	-2.3	
	benzoyl → 3-I-benzoyl	1 → 30	-3.3	
	benzoyl → 3-ciano-benzoyl	1 → 7	0.6	
	benzoyl → 3-methyl-benzoyl	1 → 8	0	
	benzoyl → 3-hydroxymethyl-benzoyl	1 → 36	7.2	
	benzoyl → 3-hydroxy-benzoyl	18 → 37	9.2	
	3-Cl-benzoyl → 2,4-di-Cl-benzoyl	6 → 29	4.8	
	benzoyl → 2,4-di-F-benzoyl	1 → 28	2.7	
	3-methoxy-benzoyl → 2,5-di-methoxy-benzoyl	5 → 35	4.6	
	benzoyl → 2-thienoyl	1 → 31	4.1	
	benzoyl → 3-thienoyl	1 → 32	1.8	
	benzoyl → 6-carboxy-pyran-2-one	1 → 41	8.1	
C13	paclitaxel → cephalomannine	T → C	1.9	+2.0 ± 0.2
		11 → 13	1.9	
		12 → 14	1.6	
		15 → 17	2.4	
	paclitaxel → docetaxel	23 → 22	-1.7	-3.2 ± 0.9
		25 → 24	-6.2	

		27 → 26	-1.3	
		38 → 39	-2.8	
		T → 21	-4.2	
	cephalomannine → docetaxel	C → 21	-3.8	-5.6 ± 1.1
		17 → D	-7.7	
		20 → 40	-5.2	
C10	acetyl → hydroxyl	T → 15	-1.3	-1.7 ± 0.8
		C → 17	-0.7	
		21 → D	-3.2	
	propionyl → hydroxyl	18 → 17	0.9	
	acetyl → propionyl	C → 18	-1.6	-0.5 ± 0.4
		13 → 19	0.2	
		14 → 20	0	
C7	propionyl → hydroxyl	17 → 1	-1.6	

Table 2. Incremental binding energies of taxane analogs to microtubules. (ΔΔG in kJ/mol at 35°C). Data from (Matesanz *et al.* 2008).

In this way, it is possible to select the most favourable substituents at the positions studied and design optimized taxanes. According to the data obtained, the optimal taxane should have the docetaxel side chain at C13, a 3-N_3-benzoyl at C2, a propionyl at C10, and a hydroxyl at C7. From compound 1 with a binding energy of -39.4 kJ/mol, the modifications selected would increase the binding affinity in -5.6 kJ/mol from the change of the cephalomannine side chain at C13 to the docetaxel one, -11.2 kJ/mol from the introduction of 3-N_3-benzoyl instead of benzoyl at C2, -1.6 kJ/mol from the substitution of a propionyl at C7 with a hydroxyl, and -0.9 kJ/mol from the change of a hydroxyl at C10 to a propionyl. Thus, this optimal taxane would have a predicted ΔG at 35°C of -58.7 kJ/mol. This molecule was synthesized (compound 40) and its binding affinity measured using the epothilone-B displacement method and the value obtained is in good corespondence with the predicted one: K_a = 6.28±0.15 x 10^9 M^{-1}; ΔG = -57.7±0.1 kJ/mol (Matesanz *et al.* 2008). This value means a 500-fold increment over the paclitaxel affinity.

It is also possible to check the influence of the modifications on the cytotoxic activity determining the IC_{50} of each compound in the human ovarian carcinoma cells A2780 and their MDR counterparts (A2780AD). The plots of log IC_{50} versus log K_a (figure 4) indicate that, as in the case of epothilones, both magnitudes are related, and the binding affinity acts as a good predictor of citotoxicity. In this type of MDR cells the high-affinity drugs are circa 100-fold more cytotoxic than the clinically used taxanes (paclitaxel and docetaxel) and exhibit very low resistance indexes.

The plot of log resistance index against log K_a shows a bell-shaped curve (figure 5). Resistance index present a maximum for taxanes with similar affinities for microtubules and P-glycoprotein, then rapidly decreases when the affinity for microtubules either increases or decreases. To find an explanation for this behaviour we should note that the intracellular free concentration of the high-affinity compounds will be low. To be pumped out by P-glycoprotein ligands must first bind it, so ligand outflow will decrease with lower free ligand concentrations (discussed in (Matesanz *et al.* 2008)). In the case of the low-affinity drugs, the concentrations needed to exert their citotoxicity are so high that the pump gets saturated and cannot effectively reduced the intracellular free ligand concentration.

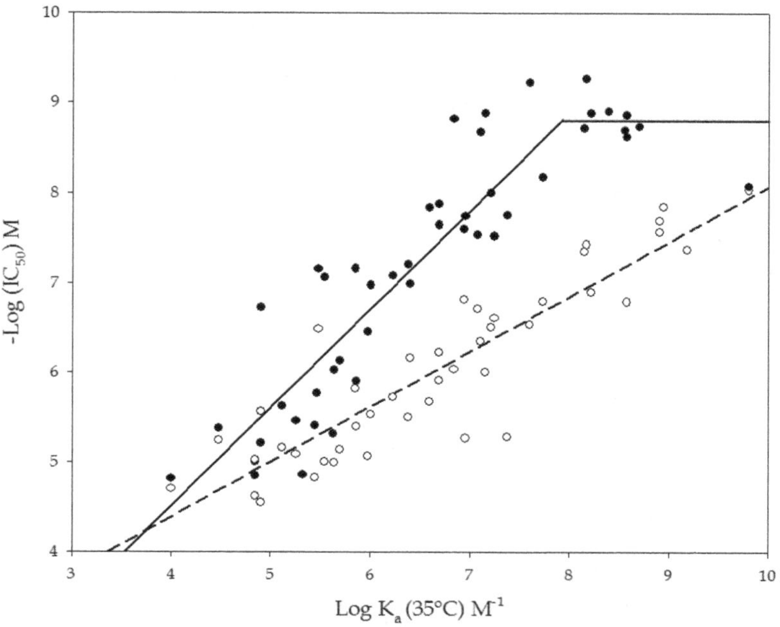

Fig. 4. Dependence of the IC_{50} of taxane analogs against A2780 non-resistant cells (black circles, solid line) and A2780AD resistant cells (white circles, dashed line) on their K_a to microtubules. Data from (Matesanz *et al.* 2008).

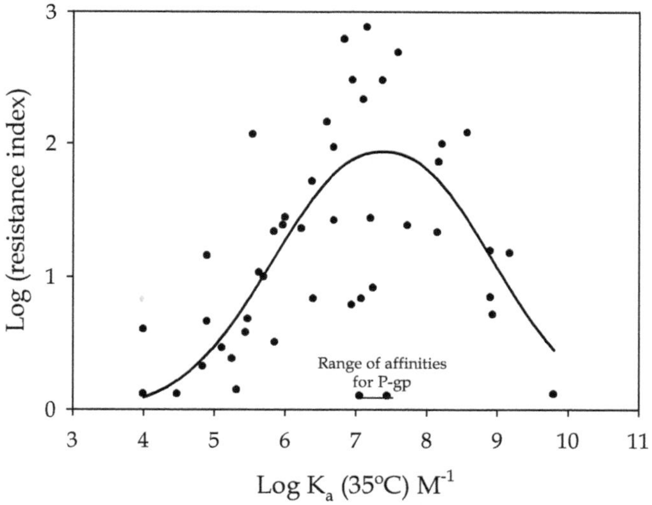

Fig. 5. Dependence of the resistance index of the A2780AD MDR cells on the K_a of the taxanes to microtubules. Data from (Yang et al. 2007; Matesanz et al. 2008).

6. Conclusion

We found a correlation between binding affinities of paclitaxel-like MSA to microtubules and their citotoxicities in tumoral cells both MDR and non-resistant. The results with taxanes further validate the binding affinity approach as a tool to be used in drug optimization as it was previously discuss for the case of epothilones. Moreover, from the thermodynamic data we could design novel high-affinity taxanes with the ability to overcome resistance in P-glycoprotein overexpressing cells. Anyway, there is a limit concentration below which MSA are not able to kill cells (discussed in (Matesanz *et al.* 2008)), the highest-affinity compounds studied have no dramatically better citotoxicities than paclitaxel or docetaxel have. Thus, the goal is not to find the drug with the highest cytotoxicity possible but rather to find one able to overcome resistances. The study of taxanes indicates that increased drug affinity could be an improvement in this direction. The extreme example of that come from the covalent binding of cyclostreptin (Buey *et al.* 2007) (that might be consider as infinite affinity) having a resistance index close to one.

However, in the case of chemically diverse paclitaxel-like MSA, the inhibition of cell proliferation correlates better with enthalpy change than with binding constants (Buey *et al.* 2005) suggesting that favourable enthalpic contributions to the binding are important to improve drug activity as it has been shown for statins and HIV protease inhibitors (Freire 2008).

7. References

Alberts, B., D. Bray, J. Lewis, M. Raff, K. Roberts&J. D. Watson, Eds. (1994). Molecular Biology of the Cell. New York, Garland Science.

Benesi, H. A.&J. H. Hildebrand (1949). "A Spectrophotometric Investigation of the Interaction of Iodine with Aromatic Hydrocarbons." journal of the american chemical society 71(8): 2703-2707.

Buey, R. M., I. Barasoain, E. Jackson, A. Meyer, P. Giannakakou, I. Paterson, S. Mooberry, J. M. Andreu&J. F. Diaz (2005). "Microtubule interactions with chemically diverse stabilizing agents: thermodynamics of binding to the paclitaxel site predicts cytotoxicity." Chem Biol 12(12): 1269-1279.

Buey, R. M., E. Calvo, I. Barasoain, O. Pineda, M. C. Edler, R. Matesanz, G. Cerezo, C. D. Vanderwal, B. W. Day, E. J. Sorensen, J. A. Lopez, J. M. Andreu, E. Hamel&J. F. Diaz (2007). "Cyclostreptin binds covalently to microtubule pores and lumenal taxoid binding sites." Nat Chem Biol 3(2): 117-125.

Buey, R. M., J. F. Diaz, J. M. Andreu, A. O'Brate, P. Giannakakou, K. C. Nicolaou, P. K. Sasmal, A. Ritzen&K. Namoto (2004). "Interaction of epothilone analogs with the paclitaxel binding site: relationship between binding affinity, microtubule stabilization, and cytotoxicity." Chem Biol 11(2): 225-236.

Connors, K. A., Ed. (1987). Binding Constants: The Measurement of Molecular Complex Stability. New York, wiley-interscience.

Connors, K. A.&S. Mecozzi, Eds. (2010). Thermodynamics of Pharmaceutical Systems. An Introduction to Theory and Applications. new york, wiley-intersciences.

Diaz, J. F.&R. M. Buey (2007). "Characterizing ligand-microtubule binding by competition methods." Methods Mol Med 137: 245-260.

Freire, E. (2008). "Do enthalpy and entropy distinguish first in class from best in class?" Drug Discovery Today 13(19-20): 869-874.

Hill, A. V. (1910). "The possible effects of the aggregation of the molecules of haemoglobin on its dissociation curves." The Journal of Physiology 40(Suppl): iv-vii.

Hirose, K. (2001). "A Practical Guide for the Determination of Binding Constants." Journal of Inclusion Phenomena and Macrocyclic Chemistry 39(3): 193-209.

Ingham, K. C. (1975). "On the application of Job's method of continuous variation to the stoichiometry of protein-ligand complexes." Analytical Biochemistry 68(2): 660-663.

Job, P. (1928). "Formation and stability of inorganic complexes in solution." Annali di Chimica 9: 113-203.

Johnson, I. S., H. F. Wright, G. H. Svoboda&J. Vlantis (1960). "Antitumor principles derived from Vinca rosea Linn. I. Vincaleukoblastine and leurosine." Cancer Res 20: 1016-1022.

Kavallaris, M. (2010). "Microtubules and resistance to tubulin-binding agents." Nat Rev Cancer 10(3): 194-204.

Lakowicz, J. R. (1999). Principles of fluorescence spectroscopy. New York, Kluwer Academic/ Plenum Publishers.

Mas, M. T.&R. F. Colman (1985). "Spectroscopic studies of the interactions of coenzymes and coenzyme fragments with pig heart oxidized triphosphopyridine nucleotide specific isocitrate dehydrogenase." Biochemistry 24(7): 1634-1646.

Matesanz, R., I. Barasoain, C. G. Yang, L. Wang, X. Li, C. de Ines, C. Coderch, F. Gago, J. J. Barbero, J. M. Andreu, W. S. Fang&J. F. Diaz (2008). "Optimization of taxane binding to microtubules: binding affinity dissection and incremental construction of a high-affinity analog of paclitaxel." Chem Biol 15(6): 573-585.

Ohtaka, H.&E. Freire (2005). "Adaptive inhibitors of the HIV-1 protease." Progress in Biophysics and Molecular Biology 88(2): 193-208.

Ohtaka, H., S. Muzammil, A. Schön, A. Velazquez-Campoy, S. Vega&E. Freire (2004). "Thermodynamic rules for the design of high affinity HIV-1 protease inhibitors with adaptability to mutations and high selectivity towards unwanted targets." The International Journal of Biochemistry & Cell Biology 36(9): 1787-1799.

Pollard, T. D. (2010). "A Guide to Simple and Informative Binding Assays." Mol. Biol. Cell 21(23): 4061-4067.

Ruben, A. J., Y. Kiso&E. Freire (2006). "Overcoming Roadblocks in Lead Optimization: A Thermodynamic Perspective." Chemical Biology & Drug Design 67(1): 2-4.

Scatchard, G. (1949). "The attractions of proteins for small molecules and ions." Annals of the New York Academy of Sciences 51(4): 660-672.

Shabbits, J. A., R. Krishna&L. D. Mayer (2001). "Molecular and pharmacological strategies to overcome multidrug resistance." Expert Rev Anticancer Ther 1(4): 585-594.

Yang, C. G., I. Barasoain, X. Li, R. Matesanz, R. Liu, F. J. Sharom, D. L. Yin, J. F. Diaz&W. S. Fang (2007). "Overcoming Tumor Drug Resistance with High-Affinity Taxanes: A SAR Study of C2-Modified 7-Acyl-10-Deacetyl Cephalomannines." ChemMedChem 2(5): 691-701.

Zhao, Y., W.-S. Fang&K. Pors (2009). "Microtubule stabilising agents for cancer chemotherapy." Expert Opinion on Therapeutic Patents 19(5): 607-622.

5

Statistical Thermodynamics

Anatol Malijevský
Department of Physical Chemistry, Institute of Chemical Technology, Prague
Czech Republic

1. Introduction

This chapter deals with the statistical thermodynamics (statistical mechanics) a modern alternative of the classical (phenomenological) thermodynamics. Its aim is to determine thermodynamic properties of matter from forces acting among molecules. Roots of the discipline are in kinetic theory of gases and are connected with the names Maxwelland Boltzmann. Father of the statistical thermodynamics is Gibbs who introduced its concepts such as the statistical ensemble and others, that have been used up to present.

Nothing can express an importance of the statistical thermodynamics better than the words of Richard Feynman Feynman et al. (2006), the Nobel Prize winner in physics: *If, in some cataclysm, all of scientific knowledge were to be destroyed, and only one sentence passed on to the next generations of creatures, what statement would contain the most information in the fewest words? I believe it is the atomic hypothesis (or the atomic fact, or whatever you wish to call it) that* **All things are made of atoms – little particles that move around in perpetual motion, attracting each other when they are a little distance apart, but repelling upon being squeezed into one another.**

In that one sentence, you will see, there is an enormous amount of information about the world, if just a little imagination and thinking are applied.

The chapter is organized as follows. Next section contains axioms of the phenomenological thermodynamics. Basic concepts and axioms of the statistical thermodynamics and relations between the partition function and thermodynamic quantities are in Section 3. Section 4 deals with the ideal gas and Section 5 with the ideal crystal. Intermolecular forces are discussed in Section 6. Section 7 is devoted to the virial expansion and Section 8 to the theories of dense gases and liquids. The final section comments axioms of phenomenological thermodynamics in the light of the statistical thermodynamics.

2. Principles of phenomenological thermodynamics

The phenomenological thermodynamics or simply thermodynamics is a discipline that deals with the thermodynamic system, a macroscopic part of the world. The thermodynamic state of system is given by a limited number of thermodynamic variables. In the simplest case of one-component, one-phase system it is for example volume of the system, amount of substance (*e.g.* in moles) and temperature. Thermodynamics studies changes of thermodynamic quantities such as pressure, internal energy, entropy, *e.t.c.* with thermodynamic variables.

The phenomenological thermodynamics is based on six axioms (or postulates if you wish to call them), four of them are called the laws of thermodynamics:

- **Axiom of existence of the thermodynamic equilibrium**
  For thermodynamic system at unchained external conditions there exists a state of the thermodynamic equilibrium in which its macroscopic parameters remain constant in time. The thermodynamic system at unchained external conditions always reaches the state of the thermodynamic equilibrium.
- **Axiom of additivity**
  Energy of the thermodynamic system is a sum of energies of its macroscopic parts. This axiom allows to define extensive and intensive thermodynamic quantities.
- **The zeroth law of thermodynamics**
  When two systems are in the thermal equilibrium, *i.e.* no heat flows from one system to the other during their thermal contact, then both systems have the same temperature as an intensive thermodynamic parameter. If system A has the same temperature as system B and system B has the same temperature as system C, then system A also has the same temperature as system C (temperature is transitive).
- **The first law of thermodynamics**
  There is a function of state called internal energy U. For its total differential $\mathrm{d}U$ we write

$$\mathrm{d}U = đW + đQ\,, \tag{1}$$

  where the symbols $đQ$ and $đW$ are not total differentials but represent infinitesimal values of heat Q and work W supplied to the system.
- **The second law of thermodynamics**
  There is a function of state called entropy S. For its total differential $\mathrm{d}S$ we write

$$\mathrm{d}S = \frac{đQ}{T}\,, \qquad \text{[reversible process]}\,, \tag{2}$$

$$\mathrm{d}S > \frac{đQ}{T}\,, \qquad \text{[irreversible process]}\,. \tag{3}$$

- **The third law of thermodynamics**
  At temperature of 0 K, entropy of a pure substance in its most stable crystalline form is zero

$$\lim_{T\to 0} S = 0\,. \tag{4}$$

  This postulate supplements the second law of thermodynamics by defining a natural referential value of entropy. The third law of thermodynamics implies that temperature of 0 K cannot be attained by any process with a finite number of steps.

Phenomenological thermodynamics using its axioms radically reduces an amount of experimental effort necessary for a determination of the values of thermodynamic quantities. For example enthalpy or entropy of a pure fluid need not be measured at each temperature and pressure but they can be calculated from an equation of state and a temperature dependence of the isobaric heat capacity of ideal gas. However, empirical constants in an equation of state and in the heat capacity must be obtained experimentally.

3. Principles of statistical thermodynamics

3.1 Basic concepts

The statistical thermodynamics considers thermodynamic system as an assembly of a very large number (of the order of 10^{23}) of mutually interacting particles (usually molecules). It uses the following concepts:

- **Microscopic state of system**
  The microscopic state of thermodynamic system is given by positions and velocities of all particles in the language of the Newton mechanics, or by the quantum states of the system in the language of quantum mechanics. There is a huge number of microscopic states that correspond to a given thermodynamic (macroscopic) state of the system.
- **Statistical ensemble**
  Statistical ensemble is a collection of all systems that are in the same thermodynamic state but in the different microscopic states.
- **Microcanonical ensemble** or **NVE** ensemble is a collection of all systems at a given number of particles N, volume V and energy E.
- **Canonical ensemble** or **NVT** ensemble is a collection of all systems at a given number of particles N, volume V and temperature T.

  There is a number of ensembles, *e.g.* the grandcanonical (μ**VT**) or isothermal isobaric (**NPT**) that will not be considered in this work.
- **Time average of thermodynamic quantity**
  The time average $\overline{X}_\tau$ of a thermodynamic quantity X is given by

$$\overline{X}_\tau = \frac{1}{\tau} \int_0^\tau X(t)\,\mathrm{d}t\,, \tag{5}$$

  where $X(t)$ is a value of X at time t and, τ is a time interval of a measurement.
- **Ensemble average of thermodynamic quantity**
  The ensemble average $\overline{X}_s$ of a thermodynamic quantity X is given by

$$\overline{X}_s = \sum_i P_i X_i\,, \tag{6}$$

  where X_i is a value in the quantum state i, and P_i is the probability of the quantum state.

3.2 Axioms of the statistical thermodynamics

The statistical thermodynamics is bases on two axioms:

Axiom on equivalence of average values

It is postulated that the time average of thermodynamic quantity X is equivalent to its ensemble average

$$\overline{X}_\tau = \overline{X}_s\,. \tag{7}$$

Axiom on probability

Probability P_i of a quantum state i is only a function of energy of the quantum state, E_i,

$$P_i = f(E_i)\,. \tag{8}$$

3.3 Probability in the microcanonical and canonical ensemble

From Eq.(8) relations between the probability and energy can be derived:

Probability in the microcanonical ensemble

All the microscopic states in the microcanonical ensemble have the same energy. Therefore,

$$P_i = \frac{1}{W} \quad \text{for} \quad i = 1, 2, \ldots, W, \tag{9}$$

where W is a number of microscopical states (the statistical weight) of the microcanonical ensemble.

Probability in the canonical ensemble

In the canonical ensemble it holds

$$P_i = \frac{\exp(-\beta E_i)}{Q}, \tag{10}$$

where $\beta = \dfrac{1}{k_B T}$, k_B is the Boltzmann constant, T temperature and Q is the *partition function*

$$Q = \sum_i \exp(-\beta E_i), \tag{11}$$

where the sum is over the microscopic states of the canonical ensemble.

3.4 The partition function and thermodynamic quantities

If the partition function is known thermodynamic quantities may be determined. The following relations between the partition function in the canonical ensemble and thermodynamic quantities can be derived

$$A = -k_B T \ln Q \tag{12}$$

$$U = k_B T^2 \left(\frac{\partial \ln Q}{\partial T}\right)_V \tag{13}$$

$$S = k_B \ln Q + k_B T \left(\frac{\partial \ln Q}{\partial T}\right)_V . \tag{14}$$

$$C_V = \left(\frac{\partial U}{\partial T}\right)_V = k_B T^2 \frac{\partial^2 \ln Q}{\partial T^2} + 2k_B T \left(\frac{\partial \ln Q}{\partial T}\right)_V, \tag{15}$$

$$p = -\left(\frac{\partial A}{\partial V}\right)_T = k_B T \left(\frac{\partial \ln Q}{\partial V}\right)_T, \tag{16}$$

$$H = U + pV = k_B T^2 \left(\frac{\partial \ln Q}{\partial T}\right)_V + V k_B T \left(\frac{\partial \ln Q}{\partial V}\right)_T, \tag{17}$$

$$G = A + pV = -k_B T \ln Q + V k_B T \left(\frac{\partial \ln Q}{\partial V}\right)_T, \tag{18}$$

$$C_p = \left(\frac{\partial H}{\partial T}\right)_V = C_V + V k_B \frac{\partial^2 \ln Q}{\partial V \partial T} . \tag{19}$$

A is Helmholtz free energy, U internal energy, S entropy, C_V isochoric heat capacity, p pressure, H enthalpy, G Gibbs free energy and C_p isobaric heat capacity.

Unfortunately, the partition function is known only for the simplest cases such as the ideal gas (Section 4) or the ideal crystal (Section 5). In all the other cases, real gases and liquids considered here, it can be determined only approximatively.

3.5 Probability and entropy

A relation between entropy S and probabilities P_i of quantum states of a system can be proved in the canonical ensemble

$$S = -k_B \sum_i P_i \ln P_i \,. \tag{20}$$

For the microcanonical ensemble a similar relation holds

$$S = k_B \ln W \,, \tag{21}$$

where W is a number of accessible states. This equation (with log instead of ln) is written in the grave of Ludwig Boltzmann in Central Cemetery in Vienna, Austria.

4. Ideal gas

The ideal gas is in statistical thermodynamics modelled by a assembly of particles that do not mutually interact. Then the energy of i-th quantum state of system, E_i, is a sum of energies of individual particles

$$E_i = \sum_{i=1}^{N} \epsilon_{i,j} \,. \tag{22}$$

In this way a problem of a determination of the partition function of system is dramatically simplified. For one-component system of N molecules it holds

$$Q = \frac{q^N}{N!} \,, \tag{23}$$

where

$$q = \sum_j \exp(-\beta \epsilon_j) \tag{24}$$

is the partition function of molecule.

The partition function of molecule may be further simplified. The energy of molecule can be approximated by a sum of the translational ϵ_{trans}, the rotational ϵ_{rot}, the vibrational ϵ_{vib}, and the electronic ϵ_{el} contributions (subscript j in ϵ_j is omitted for simplicity of notation)

$$\epsilon = \epsilon_0 + \epsilon_{\text{trans}} + \epsilon_{\text{rot}} + \epsilon_{\text{vib}} + \epsilon_{el} \,, \tag{25}$$

where ϵ_0 is the zero point energy. The partition function of system then becomes a product

$$Q = \frac{\exp(-N\beta\epsilon_0)}{N!} q_{\text{trans}} q_{\text{rot}} q_{\text{vib}} q_{\text{el}} \,. \tag{26}$$

Consequently all thermodynamic quantities of the ideal gas become sums of the corresponding contributions. For example the Helmholtz free energy is

$$\begin{aligned} A &= -k_B T \ln Q \\ &= k_B T \ln N! + U_0 - N k_B T \ln q_{\text{tr}} - N k_B T \ln q_{\text{rot}} - N k_B T \ln q_{\text{vib}} - N k_B T \ln q_{\text{el}} \\ &= k_B T \ln N! + U_0 + A_{\text{tr}} + A_{\text{rot}} + A_{\text{vib}} + A_{\text{el}} \,, \end{aligned} \tag{27}$$

where $U_0 = N\epsilon_0$ and A_{tr}, A_{rot}, A_{vib}, A_{el} are the translational, rotational, vibrational, electronic contributions to the Helmholtz free energy, respectively.

4.1 Translational contributions

Translational motions of a molecule are modelled by a particle in a box. For its energy a solution of the Schrödinger equation gives

$$\epsilon_{\mathrm{tr}} = \frac{h^2}{8m}\left(\frac{n_x^2}{a^2}+\frac{n_y^2}{b^2}+\frac{n_z^2}{c^2}\right), \tag{28}$$

where h is the Planck constant, m mass of molecule, and $a\,b\,c = V$ where V is volume of system. Quantities n_x, n_y, n_z are the quantum numbers of translation. The partition function of translation is

$$q_{\mathrm{tr}} = \left(\frac{2\pi m k_B T}{h^2}\right)^{3/2} V. \tag{29}$$

Translational contribution to the Helmholtz energy is

$$A_{\mathrm{tr}} = -RT\ln q_{\mathrm{tr}} = -RT\ln\left(\lambda^{-3}V\right), \tag{30}$$

where $R = Nk_B$ is the gas constant and $\lambda = h/\sqrt{2\pi m k_B T}$ is the Broglie wavelength. The remaining thermodynamic functions are as follows

$$S_{\mathrm{tr}} = -\left(\frac{\partial A_{\mathrm{tr}}}{\partial T}\right)_V = R\ln\left(\lambda^{-3}V\right)+\frac{3}{2}R, \tag{31}$$

$$p_{\mathrm{tr}} = -\left(\frac{\partial A_{\mathrm{tr}}}{\partial V}\right)_T = \frac{RT}{V}, \tag{32}$$

$$U_{\mathrm{tr}} = A_{\mathrm{tr}} + TS_{\mathrm{tr}} = \frac{3}{2}RT, \tag{33}$$

$$H_{\mathrm{tr}} = U_{\mathrm{tr}} + p_{\mathrm{tr}}V = \frac{5}{2}RT, \tag{34}$$

$$G_{\mathrm{tr}} = A_{\mathrm{tr}} + p_{\mathrm{tr}}V = -RT\ln\left(\lambda^{-3}V\right)+RT, \tag{35}$$

$$C_{V,\mathrm{tr}} = \left(\frac{\partial U_{\mathrm{tr}}}{\partial T}\right)_V = \frac{3}{2}R, \tag{36}$$

$$C_{p,\mathrm{tr}} = \left(\frac{\partial H_{\mathrm{tr}}}{\partial T}\right)_p = \frac{5}{2}R. \tag{37}$$

4.2 Rotational contributions

Rotations of molecule are modelled by the rigid rotator. For linear molecules there are two independent axes of rotation, for non-linear molecules there are three.

4.2.1 Linear molecules

For the partiton function of rotation it holds

$$q_{\mathrm{rot}} = \frac{8\pi\, I\, k_B\, T}{\sigma\, h^2}, \tag{38}$$

where σ is the symmetry number of molecule and I its moment of inertia

$$I = \sum_1^n m_i r_i^2,$$

with n a number of atoms in molecule, m_i their atomic masses and r_i their distances from the center of mass. Contributions to the thermodynamic quantities are

$$A_{\rm rot} = -RT \ln q_{\rm rot} = -RT \ln \left(\frac{8\pi^2 I k_B T}{\sigma h^2} \right) , \tag{39}$$

$$S_{\rm rot} = R \ln \left(\frac{8\pi^2 I k_B T}{\sigma h^2} \right) + R , \tag{40}$$

$$p_{\rm rot} = 0 , \tag{41}$$

$$U_{\rm rot} = RT , \tag{42}$$

$$H_{\rm rot} = U_{\rm rot} , \tag{43}$$

$$G_{\rm rot} = F_{\rm rot} , \tag{44}$$

$$C_{V,\rm rot} = R , \tag{45}$$

$$C_{p,\rm rot} = C_{V,\rm rot} . \tag{46}$$

4.2.2 Non-linear molecules

The partition function of rotation of a non-linear molecule is

$$q_{\rm rot} = \frac{1}{\sigma} \left(\frac{8\pi^2 k_B T}{h^2} \right)^{3/2} (\pi I_A I_B I_C)^{1/2} , \tag{47}$$

where I_A, I_B and I_C the principal moments of inertia. Contributions to the thermodynamic quantities are

$$A_{\rm rot} = -RT \ln q_{\rm rot} = -RT \ln \left[\frac{1}{\sigma} \left(\frac{8\pi^2 k_B T}{h^2} \right)^{3/2} (\pi I_A I_B I_C)^{1/2} \right] , \tag{48}$$

$$S_{\rm rot} = R \ln \left[\frac{1}{\sigma} \left(\frac{8\pi^2 k_B T}{h^2} \right)^{3/2} (\pi I_A I_B I_C)^{1/2} \right] + \frac{3}{2} R , \tag{49}$$

$$p_{\rm rot} = 0 , \tag{50}$$

$$U_{\rm rot} = \frac{3}{2} RT , \tag{51}$$

$$H_{\rm rot} = U_{\rm rot} , \tag{52}$$

$$G_{\rm rot} = A_{\rm rot} , \tag{53}$$

$$C_{V,\rm rot} = \frac{3}{2} R , \tag{54}$$

$$C_{p,\rm rot} = C_{V,\rm rot} . \tag{55}$$

4.3 Vibrational contributions

Vibrations of atoms in molecule around their equilibrium states may be at not very high temperatures approximated by harmonic oscillators.

4.3.1 Diatomic molecules

In a diatomic molecule there is only one vibrational motion. Its partition function is

$$q_{\rm vib} = [1 - \exp(h\nu_0 / k_B T)]^{-1} , \tag{56}$$

where ν_0 is the fundamental harmonic frequency. Vibrational contributions to thermodynamic quantities are

$$A_{\text{vib}} = -RT \ln q_{\text{vib}} = RT \ln \left(1 - e^{-x}\right) , \tag{57}$$

$$S_{\text{vib}} = R \frac{x e^{-x}}{1 - e^{-x}} - R \ln \left(1 - e^{-x}\right) , \tag{58}$$

$$p_{\text{vib}} = 0 , \tag{59}$$

$$U_{\text{vib}} = RT \frac{x e^{-x}}{1 - e^{-x}} , \tag{60}$$

$$H_{\text{vib}} = U_{\text{vib}} , \tag{61}$$

$$G_{\text{vib}} = A_{\text{vib}} , \tag{62}$$

$$C_{V,\text{vib}} = R \frac{x^2 e^{-x}}{(1 - e^{-x})^2} , \tag{63}$$

$$C_{p,\text{vib}} = C_{V,\text{vib}} , \tag{64}$$

where $x = \dfrac{h\nu_0}{k_B T}$.

4.3.2 Polyatomic molecules

In n-atomic molecule there is f fundamental harmonic frequencies ν_i where

$$f = \begin{cases} 3n - 5 & \text{linear molecule} \\ 3n - 6 & \text{non-linear molecule} \end{cases}$$

The partition function of vibration is

$$q_{\text{vib}} = \prod_{i=1}^{f} \frac{1}{1 - \exp(-h\nu_i / k_B T)} . \tag{65}$$

For the thermodynamic functions of vibration we get

$$A_{\text{vib}} = -RT \ln q_{\text{vib}} = RT \sum_{i=1}^{f} \ln \left(1 - e^{-x_i}\right) , \tag{66}$$

$$S_{\text{vib}} = R \sum_{i=1}^{f} \frac{x_i e^{-x_i}}{1 - e^{-x_i}} - R \sum_{i=1}^{f} \ln \left(1 - e^{-x_i}\right) , \tag{67}$$

$$p_{\text{vib}} = 0 , \tag{68}$$

$$U_{\text{vib}} = RT \sum_{i=1}^{f} \frac{x_i e^{-x_i}}{1 - e^{-x_i}} , \tag{69}$$

$$H_{\text{vib}} = U_{\text{vib}} , \tag{70}$$

$$G_{\text{vib}} = A_{\text{vib}} , \tag{71}$$

$$C_{V,\text{vib}} = R \sum_{i=1}^{f} \frac{x_i^2 e^{-x_i}}{(1 - e^{-x_i})^2} , \tag{72}$$

$$C_{p,\text{vib}} = C_{V,\text{vib}} , \tag{73}$$

where $x_i = \dfrac{h\nu_i}{k_B T}$.

4.4 Electronic contributions

The electronic partition function reads

$$q_{\rm el} = \sum_{\ell=0}^{\infty} g_{{\rm el},\ell} e^{-\varepsilon_{{\rm el},\ell}/k_B T}\,, \tag{74}$$

where $\varepsilon_{{\rm el},\ell}$ the energy level ℓ, and $g_{{\rm el},\ell}$ is its degeneracy. In most cases the electronic contributions to the thermodynamic functions are negligible at not very hight temperatures. Therefore they are not written here.

4.5 Ideal gas mixture

Let us consider two-component mixture of N_1 non-interacting molecules of component 1 and N_2 non-interacting molecules of component 2 (extension to the case of a multi-component mixture is straightforward). The partition function of mixture is

$$Q = \frac{q_1^{N_1}}{N_1!}\frac{q_2^{N_2}}{N_2!} \tag{75}$$

where q_1 and q_2 are the partition functions of molecules 1 and 2, respectively. Let us denote $X_{m,i}$ the molar thermodynamic quantity of pure component i, $i = 1, 2$ and $x_i = \frac{N_i}{N_1+N_2}$ its mole fraction. Then

$$A = RT\left(x_1 \ln x_1 + x_2 \ln x_2\right) + x_1 A_{m,1} + x_2 A_{m,2} \tag{76}$$

$$S = -R\left(x_1 \ln x_1 + x_2 \ln x_2\right) + x_1 S_{m,1} + x_2 S_{m,2}\,, \tag{77}$$

$$G = RT\left(x_1 \ln x_1 + x_2 \ln x_2\right) + x_1 G_{m,1} + x_2 G_{m,2}\,, \tag{78}$$

$$p = x_1 p_{m,1} + x_2 p_{m,2} = x_1 \frac{RT}{V_m} + x_2 \frac{RT}{V_m}\,, \tag{79}$$

$$U = x_1 U_{m,1} + x_2 U_{m,2}\,, \tag{80}$$

$$H = x_1 H_{m,1} + x_2 H_{m,2}\,, \tag{81}$$

$$C_V = x_1 C_{Vm,1} + x_2 C_{Vm,2}\,, \tag{82}$$

$$C_p = x_1 C_{pm,1} + x_2 C_{pm,2}\,. \tag{83}$$

5. Ideal crystal

We will call the ideal crystal an assembly of molecules displayed in a regular lattice without any impurities or lattice deformations. Distances among lattice centers will not depend on temperature and pressure. For simplicity we will consider one-atomic molecules. The partition function of crystal is

$$Q = e^{-U_0/k_B T} Q_{\rm vib}\,, \tag{84}$$

where U_0 is the lattice energy.

We will discuss here two models of the ideal crystal: the Einstein approximation and the Debye approximation.

5.1 Einstein model

An older and simpler Einstein model is based on the following postulates

1. Vibrations of molecules are independent:

$$Q_{\text{vib}} = q_{\text{vib}}^N , \tag{85}$$

   where q_{vib} is the vibrational partition function of molecule.

2. Vibrations are isotropic:

$$q_{\text{vib}} = q_x q_y q_z = q_x^3 . \tag{86}$$

3. Vibrations are harmonical

$$q_x = \sum_{v=0}^{\infty} e^{-\epsilon_v / k_B T} , \tag{87}$$

   where

$$\epsilon_v = h\nu \left(v + \frac{1}{2} \right)$$

   is the energy in quantum state v and ν is the fundamental vibrational frequency.

Combining these equations one obtains

$$Q = e^{-U_0/k_B T} \left(\frac{e^{-\Theta_E/(2T)}}{1 - e^{-\Theta_E/T}} \right)^{3N} , \tag{88}$$

where

$$\Theta_E = \frac{h\nu}{k_B}$$

is the *Einstein characteristic temperature.*
For the isochoric heat capacity it follows

$$C_V = 3Nk_B \left(\frac{\Theta_E}{T} \right)^2 \frac{e^{-\Theta_E/T}}{(1 - e^{-\Theta_E/T})^2} . \tag{89}$$

5.2 Debye model

Debye considers crystal as a huge molecule (*i.e* he replaces the postulates of independence and isotropy in the Einstein model) of an ideal gas; the postulate of harmonicity of vibrations remains. From these assumptions it can be derived for the partition function

$$\ln Q = -\frac{U_0}{k_B T} - \frac{9}{8} N \frac{\Theta_D}{T} - 9N \left(\frac{T}{\Theta_D} \right)^3 \int_0^{\Theta_D/T} x^2 \ln(1 - e^{-x}) \, \mathrm{d}x , \tag{90}$$

where

$$\Theta_D = \frac{h\nu_{\max}}{k_B}$$

is the *Debye characteristic temperature* with $\nu_{\max}$ being the highest frequency of crystal.
For the isochoric heat capacity it follows

$$C_V = 3R \left(4D(u) - \frac{3u}{e^u - 1} \right) . \tag{91}$$

where $u = \Theta_D/T$ and

$$D(u) = \frac{3}{u^3}\int_0^u \frac{x^3}{e^x - 1}\,\mathrm{d}x\,.$$

It can be proved that at low temperatures the heat capacity becomes a cubic function of temperature

$$C_V = 36R\left(\frac{T}{\Theta_D}\right)^3 \int_0^\infty \frac{x^3}{e^x - 1}\,\mathrm{d}x = a\,T^3\,,$$

while the Einstein model incorrectly gives

$$C_V = 3R\left(\frac{\Theta_E}{T}\right)^2 e^{-\Theta_E/T}\,.$$

Both models give a correct high-temperature limit (the Dulong-Petit law)

$$C_V = 3R\,.$$

The same is true for the zero temperature limit

$$\lim_{T\to 0} C_V = 0\,.$$

5.3 Beyond the Debye model

Both the Einstein and the Debye models assume harmonicity of lattice vibrations. This is not true at high temperatures near the melting point. The harmonic vibrations are not assumed in the lattice theories (the cell theory, the hole theory, ...) that used to be popular in forties and fifties of the last century for liquids. It was shown later that they are poor theories of liquids but very good theories for solids.
Thermodynamic functions cannot be obtained analytically in the lattice theories.

6. Intermolecular forces

Up to now forces acting among molecules have been ignored. In the ideal gas (Section 4) molecules are assumed to exert no forces upon each other. In the ideal crystal (Section 5) molecules are imprisoned in the lattice, and the intermolecular forces are counted indirectly in the lattice energy and in the Einstein or Debye temperature. For real gases and liquids the intermolecular force must be included explicitly.

6.1 The configurational integral and the molecular interaction energy

The partition function of the real gas or liquid may be written in a form

$$Q = \frac{1}{N!}\exp(-N\beta\epsilon_0)q_{\mathrm{int}}^N\left(\frac{2\pi m k_B T}{h^2}\right)^{\frac{3}{2}N} Z\,. \tag{92}$$

where $q_{\mathrm{int}} = q_{\mathrm{rot}}q_{\mathrm{vib}}q_{\mathrm{el}}$ is the partition function of the internal motions in molecule. Quantity Z is the configurational integral

$$Z = \int_{(V)}\int_{(V)}\cdots\int_{(V)} \exp[-\beta u_N(\vec{r}_1,\vec{r}_2,\ldots,\vec{r}_N)]\mathrm{d}\vec{r}_1\,\mathrm{d}\vec{r}_2\,\ldots\,\mathrm{d}\vec{r}_N\,, \tag{93}$$

where symbol

$$\int_{(V)} \cdots \mathrm{d}\vec{r}_i = \int_0^L \int_0^L \int_0^L \cdots \mathrm{d}x_i \mathrm{d}y_i \mathrm{d}z_i \quad \text{and} \quad L^3 = V\,.$$

The quantity $u_N(\vec{r}_1, \vec{r}_2, \ldots, \vec{r}_N)$ is the potential energy of an assembly of N molecules. Here and in Eq.(93) one-atomic molecules are assumed for simplicity. More generally, the potential energy is a function not only positions of centers of molecules $\vec{r}_i$ but also of their orientations $\vec{\omega}_i$. However, we will use the above simplified notation.
The interaction potential energy u_N of system may be written as an expansion in two-body, three-body, *e.t.c* contributions

$$u_N(\vec{r}_1, \vec{r}_2, \ldots, \vec{r}_N) = \sum_{i<j} u_2(\vec{r}_i, \vec{r}_j) + \sum_{i<j<k} u_3(\vec{r}_i, \vec{r}_j, \vec{r}_k)) + \cdots \tag{94}$$

Most often only the first term is considered. This approximation is called *the rule of pairwise additivity*

$$u_N = \sum_{i<j} u_2(\vec{r}_i, \vec{r}_j)\,, \tag{95}$$

where u_2 is the pair intermolecular potential. The three-body potential u_3 is used rarely at very accurate calculations, and u_4 and higher order contributions are omitted as a rule.

6.2 The pair intermolecular potential

The pair potential depends of a distance between centers of two molecules r and on their mutual orientation $\vec{\omega}$. For simplicity we will omit the angular dependence of the pair potential (it is true for the spherically symmetric molecules) in further text, and write

$$u_2(\vec{r}_i, \vec{r}_j) = u_2(r_{ij}, \vec{\omega}_{ij}) = u(r)$$

where subscripts 2 and ij are omitted, too.
The following model pair potentials are most often used.

6.2.1 Hard spheres

It is after the ideal gas the simplest model. It ignores attractive interaction between molecules, and approximates strong repulsive interactions at low intermolecular distances by an infinite barrier

$$u(r) = \begin{cases} \infty & r < \sigma \\ 0 & r > \sigma \end{cases} \tag{96}$$

where σ is a diameter of molecule.

6.2.2 Square well potential

Molecules behave like hard spheres surrounded by an area of attraction

$$u(r) = \begin{cases} \infty & r < \sigma \\ -\epsilon & \sigma < r < \lambda\sigma \\ 0 & r > \lambda\sigma \end{cases} \tag{97}$$

Here σ is a hard-sphere diameter, ϵ a depth of the attractive well, and the attraction region ranges from σ to $\lambda\sigma$.

6.2.3 Lennard-Jones potential

This well known pair intermolecular potential realistically describes a dependence of pair potential energy on distance

$$u(r) = 4\epsilon \left[\left(\frac{\sigma}{r}\right)^{12} - \left(\frac{\sigma}{r}\right)^{6} \right] . \tag{98}$$

ϵ is a depth of potential at minimum, and $2^{1/6}\sigma$ is its position.
More generally, the Lennard-Jones n-m potential is

$$u(r) = 4\epsilon \left[\left(\frac{\sigma}{r}\right)^{n} - \left(\frac{\sigma}{r}\right)^{m} \right] . \tag{99}$$

6.2.4 Pair potentials of non-spherical molecules

There are analogues of hard spheres for non-spherical particles: hard diatomics or dumbbells made of two fused hard spheres, hard triatomics, hard multiatomics, hard spherocylinders, hard ellipsoids, and so on.
Examples of soft pair potentials are Lennard-Jones multiatomics, molecules whose atoms interact according to the Lennard-Jones potential (98).
Another example is the Stockmayer potential, the Lennard-Jones potential with an indebted dipole moment

$$u(r,\theta_1,\theta_2,\phi) = 4\epsilon \left[\left(\frac{\sigma}{r}\right)^{12} - \left(\frac{\sigma}{r}\right)^{6} \right] - \frac{\mu^2}{r^3} \left[2\cos\theta_1\cos\theta_2 - \sin\theta_1\sin\theta_2\cos\phi\right] , \tag{100}$$

where μ is the dipole moment.

6.2.5 Pair potentials of real molecules

The above model pair potentials, especially the Lennard-Jones potential and its extensions, may be used to calculate properties of the real substances. In this case their parameters, for example ϵ and σ, are fitted to the experimental data such as the second virial coefficients, rare-gas transport properties and molecular properties.
More sophisticated approach involving a realistic dependence on the interparticle separation with a number of adjustable parameters was used by Aziz, see Aziz (1984) and references therein.
For simple molecules, there is a fully theoretical approach without any adjustable parameters utilizing the first principle quantum mechanics calculations, see for example Slaviček et al. (2003) and references therein.

6.3 The three-body potential

The three-body intermolecular interactions are caused by polarizablilities of molecules. The simplest and the most often used is the Axilrod-Teller-Muto term

$$u(r,s,t) = \frac{\nu}{rst} \left(3\cos\theta_1\cos\theta_2\cos\theta_3 + 1\right) , \tag{101}$$

where ν is a strength parameter. It is a first term (DDD, dipole-dipole-dipole) in the multipole expansion. Analytical formulae and corresponding strength parameters are known for higher order terms (DDQ, dipole-dipole-quadrupole, DQQ, dipole-quadrupole-quadrupole,...) as well.
More accurate three-body potentials can be obtained using quantum chemical *ab initio* calculations Malijevský et al. (2007).

7. The virial equation of state

The virial equation of state in the statistical thermodynamics is an expansion of the compressibility factor $z = \frac{pV}{RT}$ in powers of density $\rho = \frac{N}{V}$

$$z = 1 + B_2\rho + B_3\rho^2 + \cdots , \tag{102}$$

where B_2 is the second virial coefficient, B_3 the third, *e.t.c.* The virial coefficients of pure gases are functions of temperature only. For mixtures they are functions of temperature and composition.

The first term in equation (102) gives the equation of state of ideal gas, the first two terms or three give corrections to non-ideality. Higher virial coefficients are not available experimentally. However, they can be determined from knowledge of intermolecular forces. The relations among the intermolecular forces and the virial coefficients are exact, the pair and the three-body of potentials are subjects of uncertainties, however.

7.1 Second virial coefficient

For the second virial coefficient of spherically symmetric molecules we find

$$B = -2\pi \int_0^\infty f(r)\, r^2 \mathrm{d}r = -2\pi \int_0^\infty \left(e^{-\beta u(r)} - 1\right) r^2 \mathrm{d}r , \tag{103}$$

where

$$f(r) = \exp[-\beta u(r)] - 1$$

is the *Mayer function*. For linear molecules we have

$$B = -\frac{1}{4}\int_0^\infty \int_0^\pi \int_0^\pi \int_0^{2\pi} \left[e^{-\beta u(r,\theta_1,\theta_2,\phi)} - 1\right] r^2 \sin\theta_1 \sin\theta_2 \mathrm{d}r\, \mathrm{d}\theta_1 \mathrm{d}\theta_2 \mathrm{d}\phi . \tag{104}$$

For general non-spherical molecules we obtain

$$B = -\frac{2\pi}{\int_{\vec{\omega}_1}\int_{\vec{\omega}_2} \mathrm{d}\vec{\omega}_1 \mathrm{d}\vec{\omega}_2} \int_0^\infty \int_{\omega_1}\int_{\omega_2} \left[e^{-\beta u(r,\vec{\omega}_1,\vec{\omega}_2)} - 1\right] r^2 \mathrm{d}r \mathrm{d}\vec{\omega}_1 \mathrm{d}\vec{\omega}_2 . \tag{105}$$

7.2 Third virial coefficient

The third virial coefficient may be written for spherically symmetric molecules as

$$C = C_{\mathrm{add}} + C_{\mathrm{nadd}} , \tag{106}$$

where

$$C_{\mathrm{add}} = -\frac{8}{3}\pi^2 \int_0^\infty \int_0^r \int_{|r-s|}^{r+s} \left(e^{-\beta u(r)} - 1\right)\left(e^{-\beta u(s)} - 1\right)\left(e^{-\beta u(t)} - 1\right) r\, s\, t\, \mathrm{d}r\, \mathrm{d}s\; \mathrm{d}t , \tag{107}$$

and

$$C_{\mathrm{nadd}} = \frac{8}{3}\pi^2 \int_0^\infty \int_0^r \int_{|r-s|}^{r+s} e^{-\beta u(r)} e^{-\beta u(s)} e^{-\beta u(t)} \left\{\exp[-\beta u_3(r,s,t)] - 1\right\} r\, s\, t\, \mathrm{d}r\, \mathrm{d}s\; \mathrm{d}t , \tag{108}$$

where $u_3(r,s,t)$ is the three-body potential. Analogous equations hold for non-spherical molecules.

7.3 Higher virial coefficients

Expressions for higher virial coefficients become more and more complicated due to an increasing dimensionality of the corresponding integrals and their number. For example, the ninth virial coefficient consists of 194 066 integrals with the Mayer integrands, and their dimensionalities are up to 21 Malijevský & Kolafa (2008) in a simplest case of spherically symmetric molecules. For hard spheres the virial coefficients are known up to ten, which is at the edge of a present computer technology Labík et al. (2005).

7.4 Virial coefficients of mixtures

For binary mixture of components 1 and 2 the second virial coefficient reads

$$B_2 = x_1^2 B_2(11) + 2x_1 x_2 B_2(12) + x_2^2 B_2(22)\,, \tag{109}$$

where x_i are the mole fractions, $B_2(ii)$ the second virial coefficients of pure components and $B_2(12)$ the crossed virial coefficient representing an influence of the interaction between molecule 1 and molecule 2.
The third virial coefficient reads

$$B_3 = x_1^3 B_3(111) + 3x_1^2 x_2 B_3(112) + 3x_1 x_2^2 B_3(122) + x_2^3 B_3(222)\,. \tag{110}$$

Extensions of these equations on multicomponent mixtures and higher virial coefficients is straightforward.

8. Dense gas and liquid

Determination of thermodynamic properties from intermolecular interactions is much more difficult for dense fluids (for gases at high densities and for liquids) than for rare gases and solids. This fact can be explained using a definition of the Helmholtz free energy

$$A = U - TS\,. \tag{111}$$

Free energy has a minimum in equilibrium at constant temperature and volume. At high temperatures and low densities the term TS dominates because not only temperature but also entropy is high. A minimum in A corresponds to a maximum in S and system, thus, is in the gas phase. Ideal gas properties may be calculated from a behavior of individual molecules only. At somewhat higher densities thermodynamic quantities can be expanded from their ideal-gas values using the virial expansion.
At low temperatures the energy term in equation (111) dominates because not only temperature but also entropy is small. For solids we may start from a concept of the ideal crystal.
No such simple molecular model as the ideal gas or the ideal crystal is known for liquid and dense gas. Theoretical studies of liquid properties are difficult and uncompleted up to now.

8.1 Internal structure of fluid

There is no internal structure of molecules in the ideal gas. There is a long-range order in the crystal. The fluid is between of the two extremal cases: it has a local order at short intermolecular distances (as crystal) and a long-range disorder (as gas).
The fundamental quantity describing the internal structure of fluid is the pair distribution function $g(r)$

$$g(r) = \frac{\rho(r)}{\rho}, \tag{112}$$

where $\rho(r)$ is local density at distance r from the center of a given molecule, and ρ is the average or macroscopic density of system. Here and in the next pages of this section we assume spherically symmetric interactions and the rule of the pair additivity of the intermolecular potential energy.

The pair distribution function may be written in terms of the intermolecular interaction energy u_N

$$g(r) = V^2 \frac{\int_{(V)} \cdots \int_{(V)} e^{-\beta u_N(\vec{r}_1, \vec{r}_2, \ldots, \vec{r}_N)} \mathrm{d}\vec{r}_3 \ldots \mathrm{d}\vec{r}_N}{\int_{(V)} \cdots \int_{(V)} e^{-\beta u_N(\vec{r}_1, \vec{r}_2, \ldots, \vec{r}_N)} \mathrm{d}\vec{r}_1 \ldots \mathrm{d}\vec{r}_N}. \tag{113}$$

It is related to the thermodynamic quantities using the pressure equation

$$z \equiv \frac{pV}{RT} = 1 - \frac{2}{3}\pi\rho\beta \int_0^\infty \frac{du(r)}{\mathrm{d}r} g(r) r^3 \mathrm{d}r, \tag{114}$$

the energy equation

$$\frac{U}{RT} = \frac{U^0}{RT} + 2\pi\rho\beta \int_0^\infty u(r) g(r) r^2 \mathrm{d}r, \tag{115}$$

where U^0 internal energy if the ideal gas, and the compressibility equation

$$\beta \left(\frac{\partial p}{\partial \rho} \right)_\beta = \left\{ 1 + 4\pi\rho \int_0^\infty \left[g(r) - 1 \right] r^2 \mathrm{d}r \right\}^{-1}. \tag{116}$$

Present mainstream theories of liquids can be divided into two large groups: perturbation theories and integral equation theories Hansen & McDonald (2006), Martynov (1992).

8.2 Perturbation theories

A starting point of the perturbation theories is a separation of the intermolecular potential into two parts: a harsh, short-range repulsion and a smoothly varying long-range attraction

$$u(r) = u^0(r) + u^p(r). \tag{117}$$

The term $u^0(r)$ is called *the reference potential* and the term $u^p(r)$ *the perturbation potential*. In the simplest case of the first order expansion of the Helmholtz free energy in the perturbation potential it holds

$$\frac{A}{RT} = \frac{A^0}{RT} + 2\pi\rho\beta \int_0^\infty u^p(r) g^0(r) r^2 \, \mathrm{d}r, \tag{118}$$

where A^0 is the Helmholtz free energy of a reference system.

In the perturbation theories knowledge of the pair distribution function and the Helmholtz free energy of the reference system is supposed. On one hand the reference system should be simple (the ideal gas is too simple and brings nothing new; a typical reference system is a fluid of hard spheres), and the perturbation potential should be small on the other hand. As a result of a battle between a simplicity of the reference potential (one must know its structural and thermodynamic properties) and an accuracy of a truncated expansion, a number of methods have been developed.

8.3 Integral equation theories

Among the integral equation theories the most popular are those based on the Ornstein-Zernike equation

$$h(r_{12}) = c(r_{12}) + \rho \int_{(V)} h(r_{13})c(r_{32})\,\mathrm{d}\vec{r}_3\,. \tag{119}$$

where $h(r) = g(r) - 1$ is the total correlation function and $c(r)$ the direct correlation function. This equation must be closed using a relation between the total and the direct correlation functions called *the closure* to the Ornstein-Zernike equation. From the diagrammatic analysis it follows

$$h(r) = \exp[-\beta u(r) + \gamma(r) + B(r)] - 1\,, \tag{120}$$

where

$$\gamma(r) = h(r) - c(r)$$

is the indirect (chain) correlation function and $B(r)$ is the bridge function, a sum of elementary diagrams. Equation (120) does not yet provide a closure. It must be completed by an approximation for the bridge function. The mostly used closures are in listed in Malijevský & Kolafa (2008). The simplest of them are the hypernetted chain approximation

$$B(r) = 0 \tag{121}$$

and the Percus-Yevick approximation

$$B(r) = \gamma(r) - \ln[\gamma(r) + 1]\,. \tag{122}$$

Let us compare the perturbation and the integral equation theories. The first ones are simpler but they need an extra input - the structural and thermodynamic properties of a reference system. The accuracy of the second ones depends on a chosen closure. Their examples shown here, the hypernetted chain and the Percus-Yevick, are too simple to be accurate.

8.4 Computer simulations

Besides the above theoretical approaches there is another route to the thermodynamic quantities called the computer experiments or pseudoexperiments or simply simulations. For a given pair intermolecular potential they provide values of thermodynamic functions in the dependence on the state variables. In this sense they have characteristics of real experiments. Similarly to them they do not give an explanation of the bulk behavior of matter but they serve as tests of approximative theories. The thermodynamic values are free of approximations, or more precisely, their approximations such as a finite number of molecules in the basic box or a finite number of generated configurations can be systematically improved Kolafa et al. (2002). The computer simulations are divided into two groups: the Monte Carlo simulations and the molecular dynamic simulations. The Monte Carlo simulations generate the ensemble averages of structural and thermodynamic functions while the molecular dynamics simulations generate their time averages. The methods are described in detail in the monograph of Allen and Tildesley Allen & Tildesley (1987).

9. Interpretation of thermodynamic laws

In Section 2 the axioms of the classical or phenomenological thermodynamics have been listed. The statistical thermodynamics not only determines the thermodynamic quantities from knowledge of the intermolecular forces but also allows an interpretation of the phenomenological axioms.

9.1 Axiom on existence of the thermodynamic equilibrium

This axiom can be explained as follows. There is a very, very large number of microscopic states that correspond to a given macroscopic state. At unchained macroscopic parameters such as volume and temperature of a closed system there is much more equilibrium states then the states out of equilibrium. Consequently, a spontaneous transfer from non-equilibrium to equilibrium has a very, very high probability. However, a spontaneous transfer from an equilibrium state to a non-equilibrium state is not excluded.
Imagine a glass of whisky on rocks. This two-phase system at a room temperature transfers spontaneously to the one-phase system - a solution of water, ethanol and other components. It is not excluded but it is highly improbable that a glass with a dissolved ice will return to the initial state.

9.2 Axiom of additivity

This axiom postulates that the internal energy of the macroscopic system is a sum of its two macroscopic parts

$$U = U_1 + U_2 \,. \tag{123}$$

However, if we consider a macroscopic system as an assembly of molecules, equation (123) does not take into account intermolecular interactions among molecules of subsystem 1 and molecules of subsystem 2. Correctly the equation should be

$$U = U_1 + U_2 + U_{12} \,. \tag{124}$$

Due to the fact that intermolecular interactions vanish at distances of the order of a few molecule diameters, the term U_{12} is negligible in comparison with U.

9.3 The zeroth law of thermodynamics and the negative absolute temperatures

The statistical thermodynamics introduces temperature formally as parameter $\beta = \frac{1}{k_B T}$ in the expression (11) for the partition function

$$Q = \sum_i \exp(-\beta E_i) \,.$$

As energies of molecular systems are positive and unbounded, temperature must be positive otherwise the equation diverges. For systems with bounded energies

$$E_{\min} \leq E_i \leq E_{\max}$$

both negative and positive temperatures are allowed. Such systems are in lasers, for example.

9.4 The second law of thermodynamics

From equation (3) it follows that entropy of the adiabatically isolated system either grows for spontaneous processes or remains constant in equilibrium

$$\mathrm{d}S \geq 0 \,. \tag{125}$$

Entropy in the statistical thermodynamics is connected with probability via equation (20)

$$S = -k_B \sum_i P_i \ln P_i \,.$$

Thus, entropy may spontaneously decrease but with a low probability.

9.5 Statistical thermodynamics and the arrow of time

Direction of time from past to future is supported by three arguments

- **Cosmological time**
  The cosmological time goes according the standard model of Universe from the Big Bang to future.
- **Psychological time**
  We as human beings remember (as a rule) what was yesterday but we do not "*remember*" what will be tomorrow.
- **Thermodynamic time**
  Time goes in the direction of the growth of entropy in the direction given by equation (125).

The statistical thermodynamics allows due to its probabilistic nature a change of a direction of time "**from coffin to the cradle**" but again with a very, very low probability.

9.6 The third law of thermodynamics

Within the statistical thermodynamics the third law may be easily derived from equation (20) relating entropy and the probabilities. The state of the ideal crystal at $T = 0$ K is one. Its probability $P_0 = 1$. By substituting to the equation we get $S = 0$.

10. Acknowledgement

This work was supported by the Ministry of Education, Youth and Sports of the Czech Republic under the project No. 604 613 7316.

List of symbols

A	Helmholtz free energy
B	second virial coefficient
B_i	i-th virial coefficient
$B(r)$	bridge function
C_V	isochoric heat capacity
C_p	isobaric capacity
$c(r)$	direct correlation function
E	energy
ϵ	energy of molecule
G	Gibbs free energy
$g(r)$	pair distribution function
H	enthalpy
h	Planck constant
$h(r)$	total correlation function
k_B	Boltzmann constant
N	number of molecules, Avogadro number
P	probability
p	pressure
Q	heat
Q	partition function
q	partition function of molecule
R	(universal) gas constant (8.314 in SI units)
S	entropy
T	temperature

τ time
U internal energy
u_N potential energy of N particles
u pair potential
W work
W number of accessible states
X measurable thermodynamic quantity
x mole fraction

11. References

Feynman, R. P.; Leghton, R. B. & Sands, M. (2006). *The Feynman lectures on physics*,Vol. 1, Pearson, ISBN 0-8053-9046-4, San Francisco.

Lucas K. (1991). *Applied Statistical Thermodynamics*, Springer-Verlag, ISBN 0-387-52007-4, New York, Berlin, Heidelberg.

Aziz, R.A. (1984). Interatomic Potentials for Rare-Gases: Pure and Mixed Interactions, In: *Inert Gases* M. L. Klein (Ed.), 5 - 86, Springer-Verlag, ISBN 3-540-13128-0, Berlin, Heidelberg.

Slaviček, P.; Kalus, R.; Paška, P.; Odvárková, I.; Hobza, P. & Malijevský, A. (2003). State-of-the-art correlated ab initio potential energy curves for heavy rare gas dimers:Ar_2, Kr_2, and Xe_2. *Journal of Chemical Physics*, 119, 4, 2102-2119, ISSN:0021-9606.

Malijevský, Alexandr; Karlický, F.; Kalus, R. & Malijevský, A. (2007). Third Viral Coeffcients of Argon from First Principles. *Journal of Physical Chemistry C*, 111, 43, 15565 - 15568, ISSN:1932-7447.

Malijevský, A. & Kolafa, J. (2008). Introduction to the thermodynamics of Hard Spheres and Related Systems In: *Theory and Simulation of Hard-Sphere Fluids and Related Systems* A. Mulero (Ed.) 27 - 36, Springer-Verlag, ISBN 978-3-540-78766-2, Berlin, Heidelberg.

Labík, S.; Kolafa, J. & Malijevský, A. (2005). Virial coefficients of hard spheres and hard disks up to the ninth. *Physical Review E*, 71, 2-1, 021105/1-021105/8, ISSN:1539-3755.

Allen, M. P. & Tildesley, D. J. (1987). *Computer Simulation of Liquids*, Claredon Press, ISBN: 0-19-855645-4, Oxford.

Hansen, J.-P. & McDonald, I. R. (2006). *Theory of Simple Fluids*, Elsevier, ISBN: 978-0-12-370535-8, Amsterdam.

Martynov, G. A. (1992). *Fundamental Theory of Liquids*, Adam Hilger, ISBN: 0-7503-0069-8, Bristol, Philadelphia and New York.

Malijevský, A. & Kolafa, J. (2008). Structure of Hard Spheres and Related Systems In: *Theory and Simulation of Hard-Sphere Fluids and Related Systems* A. Mulero (Ed.) 1 - 26, Springer-Verlag, ISBN 978-3-540-78766-2, Berlin, Heidelberg.

Kolafa, J.; Labík & Malijevský, A. (2002). The bridge function of hard spheres by direct inversion of computer simulation data. *Molecular Physics*, 100, 16, 2629 - 2640, ISSN: 0026-8976.

Baus, M.& Tejero, C., F. (2008). *Equilibrium Statistical Physics: Phases of Matter and Phase Transitions*, Springer, ISBN: 978-3-540-74631-7, Heidelberg.

Ben-Naim, A. (2010). *Statistical Thermodynamics Based on Information: A Farewell to Entropy*, World-Scientific, ISBN: 978-981-270-707-9, Philadelphia, New York.

Plischke, M. & Bergsen, B. (2006). *Equilibrium Statistical Physics*, University of British Columbia, Canada, ISBN: 978-981-256-048-3.

6

On the Chlorination Thermodynamics

Brocchi E. A. and Navarro R. C. S.
Pontifical Catholic University of Rio de Janeiro
Brazil

1. Introduction

Chlorination roasting has proven to be a very important industrial route and can be applied for different purposes. Firstly, the chlorination of some important minerals is a possible industrial process for producing and refining metals of considerable technological importance, such as titanium and zirconium. Also, the same principle is mentioned as a possible way of recovering rare earth from concentrates (Zang et al., 2004) and metals, of considerable economic value, from different industrial wastes, such as, tailings (Cechi et al., 2009), spent catalysts (Gabalah & Djona, 1995), slags (Brocchi & Moura, 2008) and fly ash (Murase et al., 1998). The chlorination processes are also presented as environmentally acceptable (Neff, 1995, Mackay, 1992).

In general terms the chlorination can be described as reaction between a starting material (mineral concentrate or industrial waste) with chlorine in order to produce some volatile chlorides, which can then be separated by, for example, selective condensation. The most desired chloride is purified and then used as a precursor in the production of either the pure metal (by reacting the chloride with magnesium) or its oxide (by oxidation of the chloride).

The chlorination reaction has been studied on respect of many metal oxides (Micco et. al., 2011; Gaviria & Bohe, 2010; Esquivel et al., 2003; Oheda et al., 2002) as this type of compound is the most common in the mentioned starting materials. Although some basic thermodynamic data is enclosed in these works, most of them are related to kinetics aspects of the gas - solid reactions. However, it is clear that the understanding of the equilibrium conditions, as predicted by classical thermodynamics, of a particular oxide reaction with chlorine can give strong support for both the control and optimization of the process. In this context, the impact of industrial operational variables over the chlorination efficiency, such as the reaction temperature and the reactors atmosphere composition, can be theoretically appreciated and then quantitatively predicted. On that sense, some important works have been totally devoted to the thermodynamics of the chlorination and became classical references on the subject (Kellog, 1950; Patel & Jere, 1960; Pilgrim & Ingraham, 1967; Sano & Belton, 1980).

Originally, the approach applied for the study of chemical equilibrium studies was based exclusively on ΔG_r^o x T and predominance diagrams. Nowadays, however, advances in computational thermodynamics enabled the development of softwares that can perform more complex calculations. This approach, together with the one accomplished by simpler techniques, converge to a better understanding of the intimate nature of the equilibrium states for the reaction system of interest. Therefore, it is understood that the time has come for a review on chlorination thermodynamics which can combine its basic aspects with a now available new kind of approach.

The present chapter will first focus on the thermodynamic basis necessary for understanding the nature of the equilibrium states achievable through chlorination reactions of metallic oxides. Possible ways of graphically representing the equilibrium conditions are discussed and compared. Moreover, the chlorination of V_2O_5, both in the absence as with the presence of graphite will be considered. The need of such reducing agent is clearly explained and discussed. Finally, the equilibrium conditions are appreciated through the construction of graphics with different levels of complexity, beginning with the well known ΔG_r^o x T diagrams, and ending with gas phase speciation diagrams, rigorously calculated through the minimization of the total Gibbs energy of the system.

2. Chemical reaction equilibria

The equilibrium state achieved by a system where a group of chemical reactions take place simultaneously can be entirely modeled and predicted by applying the principles of classical thermodynamics.

Supposing that we want to react some solid transition metal oxide, say M_2O_5, with gaseous Cl_2. Lets consider for simplicity that the reaction can result in the formation of only one gaseous chlorinated specie, say MCl_5. The transformation is represented by the following equation:

$$M_2O_5(s)+5Cl_2(g)\rightarrow 2MCl_5(g)+\frac{5}{2}O_2(g) \tag{1}$$

In this system there are only two phases, the pure solid oxide M_2O_5 and a gas phase, whose composition is characterized by definite proportions of Cl_2, O_2, and MCl_5. If temperature, total pressure, and the total molar amounts of O, Cl, and M are fixed, the chemical equilibrium is calculated by finding the global minimum of the total Gibbs energy of the system (Robert, 1993).

$$G = n_{M_2O_5}^{s} g_{M_2O_5}^{s} + G^{g} \tag{2}$$

Where $g_{M_2O_5}^{s}$ represents the molar Gibbs energy of pure solid M_2O_5 at reaction's temperature and total pressure, $n_{M_2O_5}^{s}$ the number of moles of M_2O_5 and G^g the molar Gibbs energy of the gaseous phase, which can be computed through the knowledge of the *chemical potential* of all molecular species present ($\mu_{Cl_2}^{g}$, $\mu_{O_2}^{g}$, $\mu_{MCl_5}^{g}$):

$$G^{g} = n_{Cl_2}^{g}\mu_{Cl_2}^{g} + n_{O_2}^{g}\mu_{O_2}^{g} + n_{MCl_5}^{g}\mu_{MCl_5}^{g}$$
$$\mu_{Cl_2}^{g} = \left(\frac{\partial G^{g}}{\partial n_{Cl_2}^{g}}\right)_{T,P,n_{O_2},n_{MCl_5}} \tag{3}$$

The minimization of function (2) requires that for the restrictions imposed to the system, the first order differential of G must be equal to zero. By fixing the reaction temperature (T) and pressure (P) and total amount of each one of the elements, this condition can be written according to equation (4) (Robert, 1993).

$$
\begin{gathered}
dG_{T,P,n_{O},n_{Cl},n_{M}} = 0 \\
dG = g^{s}_{M_2O_5} dn_{M_2O_5} + \mu^{g}_{Cl_2} dn_{Cl_2} + \mu^{g}_{O_2} dn_{O_2} + \mu^{g}_{MCl_5} dn_{MCl_5} \\
g^{s}_{M_2O_5} dn_{M_2O_5} + \mu^{g}_{Cl_2} dn_{Cl_2} + \mu^{g}_{O_2} dn_{O_2} + \mu^{g}_{MCl_5} dn_{MCl_5} = 0
\end{gathered}
\tag{4}
$$

The development of the chlorination reaction can be followed through introduction of a reaction coordinate called *degree of reaction* (ε),whose first differential is computed by the ratio of its molar content variation of each specie participating in the reaction and the stoichiometric coefficient(Eq. 1).

$$
d\varepsilon = \frac{dn_{Cl_2}}{(-5)} = \frac{dn_{M_2O_5}}{(-1)} = \frac{dn_{O_2}}{(+5/2)} = \frac{dn_{MCl_5}}{(+2)}
\tag{5}
$$

The numbers inside the parenthesis in the denominators of the fractions contained in equation (5) are the stoichiometric coefficient of each specie multiplied by "-1" if it is represented as a reactant, or "+1" if it is a product. The equilibrium condition (Eq. 4) can now be rewritten in the following mathematical form:

$$
\begin{gathered}
-g^{s}_{M_2O_5} d\varepsilon - 5\mu^{g}_{Cl_2} d\varepsilon + \frac{5}{2}\mu^{g}_{O_2} d\varepsilon + 2\mu^{g}_{MCl_5} d\varepsilon = 0 \\
\left(-g^{s}_{M_2O_5} - 5\mu^{g}_{Cl_2} + \frac{5}{2}\mu^{g}_{O_2} + 2\mu^{g}_{MCl_5}\right) d\varepsilon = 0
\end{gathered}
\tag{6}
$$

At the desired equilibrium state the condition defined by Eq. (6) must be valid for all possible values of the differential $d\varepsilon$. This can only be accomplished if the term inside the parenthesis is equal to zero. This last condition is the simplest mathematical representation for the chemical equilibrium associated with reaction (1).

$$
-g^{s}_{M_2O_5} - 5\mu^{g}_{Cl_2} + \frac{5}{2}\mu^{g}_{O_2} + 2\mu^{g}_{MCl_5} = 0
\tag{7}
$$

The chemical potentials can be computed through knowledge of the molar Gibbs energy of each pure specie in the gas phase, and its chemical activity. For the chloride MCl_5, for example, the following function can be used (Robert, 1993):

$$
\mu^{g}_{MCl_5} = g^{g}_{MCl_5} + RT \ln a^{g}_{MCl_5}
\tag{8}
$$

Where $a^{g}_{MCl_5}$ represents the chemical activity of the component MCl_5 in the gas phase. By introducing equations analogous to Eq. (8) for all components of the gas phase, Eq. (7) can be rewritten according to Eq. (9). There, the activity of M_2O_5 is not present in the term located at the left hand side because, as this oxide is assumed to be pure, its activity must be equal to one (Robert, 1993).

$$
\ln\left(\frac{a^{2}_{MCl_5} a^{5/2}_{O_2}}{a^{5}_{Cl_2}}\right) = -\frac{\left(2g^{g}_{MCl_5} + \frac{5}{2} g^{g}_{O_2} - 5g^{g}_{Cl_2} - g^{s}_{M_2O_5}\right)}{RT} = -\frac{\Delta G_r}{RT}
\tag{9}
$$

The numerator of the right side of Eq. (9) represents the molar Gibbs energy of reaction (1). It involves only the molar Gibbs energies of the species participating in the reaction as pure substances, at T and P established in the reactor. The molar Gibbs energy of a pure component is only a function of T and P (Eq. 10), so the same must be valid for the reactions Gibbs energy (Robert, 1993).

$$dg(T,P) = -sdT + vdP \tag{10}$$

Where s and v denote respectively the molar entropy and molar volume of the material, which for a pure substance are themselves only a function of T and P.

It is a common practice in treating reactions involving gaseous species to calculate the Gibbs energy of reaction not at the total pressure prevailing inside the reactor, but to fix it at 1 atm. This is in fact a reference pressure, and can assume any suitable value we desire. The molar Gibbs energy of reaction is in this case referred to as the *standard molar Gibbs energy of reaction*. According to this definition, the standard Gibbs energy of reaction must depend only on the reactor's temperature.

By assuming that the total pressure inside the reactor (P) is low enough for neglecting the effect of the interactions among the species present in the gas phase, Eq. (9) can be rewritten in the following form:

$$\frac{{P_{MCl_5}}^2 {P_{O_2}}^{5/2}}{{P_{Cl_2}}^5} = \exp\left(-\frac{\left(2g^g_{MCl_5} + \frac{5}{2} g^g_{O_2} - 5g^g_{Cl_2} - g^s_{M_2O_5} \right)}{RT} \right) \tag{11}$$

The activities were calculated as the ratio of the partial pressure of each component and the reference pressure chosen (P = 1 atm). This proposal is based on the thermodynamic description of an ideal gas (Robert, 1993). For MCl_5, for example, the chemical activity is calculated as follows:

$$a_{MCl_5} = \frac{P_{MCl_5}}{1} = P_{MCl_5} = x^g_{MCl_5} P \tag{12}$$

Where $x^g_{MCl_5}$ stands for the mol fraction of MCl_5 in the gas phase. Similar relations hold for the other species present in the reactor atmosphere. The activity is then expressed as the product of the mol fraction of the specie and the total pressure exerted by the gaseous solution.

The right hand side of Eq. (11) defines the equilibrium constant (K) of the reaction in question. This quantity can be calculated as follows:

$$K = \exp\left(-\frac{\Delta G^o_r}{RT} \right) \tag{13}$$

The symbol "o" is used to denote that The molar reaction Gibbs energy (ΔG^o_r) is calculated at a reference pressure of 1 atm.

At this point, three possible situations arise. If the standard molar Gibbs energy of the reaction is negative, then $K > 1$. If it is positive, $K < 1$ and if it is equal to zero $K = 1$. The first situation defines a process where in the achieved equilibrium state, the atmosphere tends to be richer in the desired products. The second situation characterizes a reaction where the reactants are present in higher concentration in equilibrium. Finally, the third possibility defines the situation where products and reactants are present in amounts of the same order of magnitude.

2.1 Thermodynamic driving force and ΔG_r^o vs. T diagrams

Equation (6) can be used to formulate a mathematical definition of the thermodynamic driving force for a chlorination reaction. If the reaction proceeds in the desired direction, then $d\varepsilon$ must be positive. Based on the fact that by fixing T, P, n(O), n(Cl), and n(M) the total Gibbs energy of the system is minimum at the equilibrium, the reaction will develop in the direction of the final equilibrium state, if and only if, the value of G reduces, or in other words, the following inequality must then be valid:

$$-g_{M_2O_5}^{s} - 5\mu_{Cl_2}^{g} + \frac{5}{2}\mu_{O_2}^{g} + 2\mu_{MCl_5}^{g} < 0 \tag{14}$$

The left hand side of inequality (14) defines the thermodynamic driving force of the reaction ($\Delta\mu_r$).

$$\Delta\mu_r = -g_{M_2O_5}^{s} - 5\mu_{Cl_2}^{g} + \frac{5}{2}\mu_{O_2}^{g} + 2\mu_{MCl_5}^{g} \tag{15}$$

If $\Delta\mu_r$ is negative, classical thermodynamics says that the process will develop in the direction of obtaining the desired products. However, a positive value is indicative that the reaction will develop in the opposite direction. In this case, the formed products react to regenerate the reactants. By using the mathematical expression for the chemical potentials (Eq. 8), it is possible to rewrite the driving force in a more familiar way:

$$\Delta\mu_r = \Delta G_r^o + RT\ln\left(\frac{P_{MCl_5}^2 P_{O_2}^{5/2}}{P_{Cl_2}^5}\right) = \Delta G_r^o + RT\ln Q \tag{16}$$

According to Eq. (16), the ratio involving the partial pressure of the components defines the so called *reaction coefficient* (Q). This parameter can be specified in a given experiment by injecting a gas with the desired proportion of O_2 and Cl_2. The partial pressure of MCl_5, on the other hand, would then be near zero, as after the formation of each species, the fluxing gas removes it from the atmosphere in the neighborhood of the sample.

At a fixed temperature and depending on the value of Q and the standard molar Gibbs energy of the reaction considered, the driving force can be positive, negative or zero. In the last case the reaction ceases and the equilibrium condition is achieved. It is important to note, however, that by only evaluating the reactions Gibbs energy one is not in condition to predict the reaction path followed, then even for positive values of ΔG_r^o, it is possible to find a value Q that makes the driving force negative. This is a usual situation faced in industry, where the desired equilibrium is forced by continuously injecting reactants, or removing products. In all cases, however, for computing reaction driving forces it is vital to know the temperature dependence of the reaction Gibbs energy.

2.1.1 Thermodynamic basis for the construction of ΔG_r^o x T diagrams

To construct the ΔG_r^o x T diagram of a particular reaction we must be able to compute its standard Gibbs energy in the whole temperature range spanned by the diagram.

$$\Delta G_r^o = \Delta H_r^o - T\Delta S_r^o$$

$$\Delta H_r^o = \Delta H_{298} + \int_{298.15\text{ K}}^{T} \Delta C_P^o dT$$

$$\Delta S_r^o = \Delta S_{298} + \int_{298.15\text{ K}}^{T} \frac{\Delta C_P^o}{T} dT \tag{17}$$

$$\Delta C_P^o = \frac{d\Delta H_r^o}{dT} = 2C_{P,MCl_5}^{o,g} + \frac{5}{2}C_{P,O_2}^{g} - 5C_{P,Cl_2}^{g} - C_{P,M_2O_5}^{s}$$

$$\Delta H_{298} = 2H_{298,\,MCl_5}^{o} + \frac{5}{2}H_{298,O_2}^{g} - 5H_{298,Cl_2}^{g} - H_{298,M_2O_5}^{s}$$

$$\Delta S_{298} = 2S_{298,\,MCl_5}^{o} + \frac{5}{2}S_{298,O_2}^{g} - 5S_{298,Cl_2}^{g} - S_{298,M_2O_5}^{s}$$

For accomplishing this task one needs a mathematical model for the molar standard heat capacity at constant pressure, valid for each participating substance for T varying between 298.15 K and the final desired temperature, its molar enthalpy of formation (H_{298}^o) and its molar entropy of formation (S_{298}^o)at 298.15 K

For the most gas - solid reactions both the molar standard enthalpy (ΔH_r^o) and entropy of reaction (ΔS_r^o) do not depend strongly on temperature, as far no phase transformation among the reactants and or products are present in the considered temperature range. So, the observed behavior is usually described by a line (Fig. 1), whose angular coefficient gives us a measurement of ΔS_r^o and ΔH_r^o is defined by the linear coefficient.

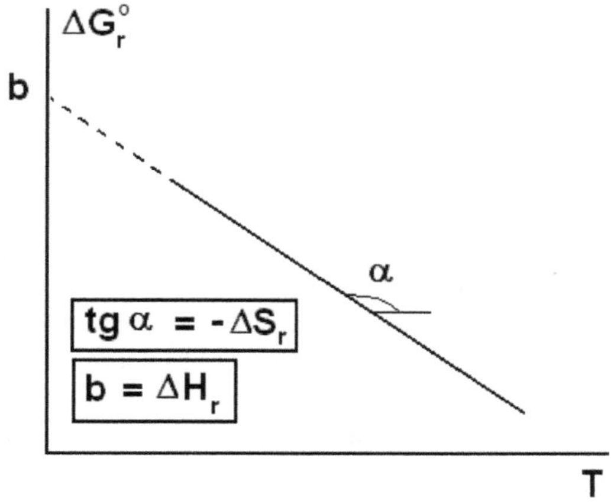

Fig. 1. Hypothetical ΔG_r^o x T diagram

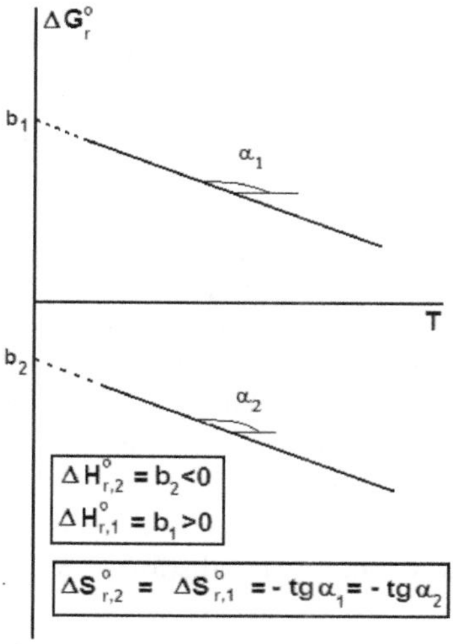

Fig. 2. Endothermic and exothermic reactions

Further, for a reaction defined by Eq. (1) the number of moles of gaseous products is higher than the number of moles of gaseous reactants, which, based on the ideal gas model, is indicative that the chlorination leads to a state of grater disorder, or greater entropy. In this particular case then, the straight line must have negative linear coefficient ($-\Delta S_r^o < 0$), as depicted in the graph of Figure (1).

The same can not be said about the molar reaction enthalpy. In principle the chlorination reaction can lead to an evolution of heat (exothermic process, then $\Delta H_r^o < 0$) or absorption of heat (endothermic process, then $\Delta H_r^o > 0$). In the first case the linear coefficient is positive, but in the later it is negative. Hypothetical cases are presented in Fig. (2) for the chlorination of two oxides, which react according to equations identical to Eq. (1). The same molar reaction entropy is observed, but for one oxide the molar enthalpy is positive, and for the other it is negative.

Finally, it is worthwhile to mention that for some reactions the angular coefficient of the straight line can change at a particular temperature value. This can happen due to a phase transformation associated with either a reactant or a product. In the case of the reaction (1), only the oxide M_2O_5 can experience some phase transformation (melting, sublimation, or ebullition), all of them associated with an increase in the molar enthalpy of the phase. According to classical thermodynamics, the molar entropy of the compound must also increase (Robert, 1993).

$$\Delta S_t = \frac{\Delta H_t}{T_t} \tag{18}$$

Where ΔS_t, ΔH_t and T_t represent respectively, the molar entropy, molar enthalpy and temperature of the phase transformation in question. So, to include the effect for melting of M_2O_5 at a temperature T_t, the molar reaction enthalpy and entropy must be modified as follows.

$$\Delta H_r^o = \int_{298.15}^{T_t} \Delta C_P^o dT - \Delta H_{t,M_2O_5} + \int_{T_t}^{T} \Delta C_P^o dT$$
$$\Delta S_r^o = \int_{298.15}^{T_t} \frac{\Delta C_P^o}{T} dT - \frac{\Delta H_{t,M_2O_5}}{T_t} + \int_{T_t}^{T} \frac{\Delta C_P^o}{T} dT \quad (19)$$

It should be observed that the molar entropy and enthalpy associated with the phase transition experienced by the oxide M_2O_5 were multiplied by its stoichiometric number "-1", which explains the minus sign present in both relations of Eq. (19).

An analogous procedure can be applied if other phase transition phenomena take place. One must only be aware that the mathematical description for the molar reaction heat capacity at constant pressure (ΔC_P^o) must be modified by substituting the heat capacity of solid M_2O_5 for a model associated with the most stable phase in each particular temperature range. If, for example, in the temperature range of interest M_2O_5 melts at T_t, for $T > T_t$, the molar heat capacity of solid M_2O_5 must be substituted for the model associated with the liquid state (Eq. 20).

$$\Delta C_P^o = 2C_{P,MCl_5}^{o,g} + \frac{5}{2}C_{P,O_2}^{g} - 5C_{P,Cl_2}^{g} - C_{P,M_2O_5}^{s} \quad (T < T_t)$$
$$\Delta C_P^o = 2C_{P,MCl_5}^{o,g} + \frac{5}{2}C_{P,O_2}^{g} - 5C_{P,Cl_2}^{g} - C_{P,M_2O_5}^{l} \quad (T > T_t) \quad (20)$$

The effect of a phase transition over the geometric nature of the ΔG_r^o x T curve can be directly seen. The melting of M_2O_5 makes it's molar enthalpy and entropy higher. According to Eq. (19), such effects would make the molar reaction enthalpy and entropy lower. So the curve should experience a decrease in its first order derivative at the melting temperature (Figure 3).

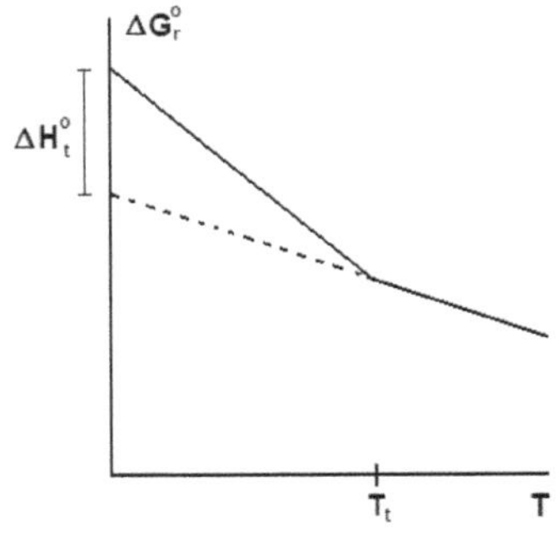

Fig. 3. Effect of M_2O_5 melting over the ΔG_r^o x T diagram

Based on the definition of the reaction Gibbs energy (Eq. 17), similar transitions involving a product would produce an opposite effect. The reaction Gibbs energy would in these cases dislocate to more negative values. In all cases, though, the magnitude of the deviation is proportional to the magnitude of the molar enthalpy associated with the particular transition observed. The effect increases in the following order: melting, ebullition and sublimation.

2.2 Multiple reactions

In many situations the reaction of a metallic oxide with Cl_2 leads to the formation of a family of chlorinated species. In these cases, multiple reactions take place. In the present section three methods will be described for treating this sort of situation, the first of them is of qualitative nature, the second semi-qualitative, and the third a rigorous one, that reproduces the equilibrium conditions quantitatively.

The first method consists in calculating ΔG_r^o x T diagrams for each reaction in the temperature range of interest. The reaction with the lower molar Gibbs energy must have a greater thermodynamic driving force. The second method involves the solution of the equilibrium equations independently for each reaction, and plotting on the same space the concentration of the desired chlorinated species. Finally, the third method involves the calculation of the thermodynamic equilibrium by minimizing the total Gibbs energy of the system. The concentrations of all species in the phase ensemble are then simultaneously computed.

2.2.1 Methods based on ΔG_r^o x T diagrams

It will be supposed that the oxide M_2O_5 can generate two gaseous chlorinated species, MCl_4 and MCl_5:

$$\begin{aligned} &M_2O_5(s)+5Cl_2(g)\rightarrow 2MCl_5(g)+\frac{5}{2}O_2(g) \\ &M_2O_5(s)+4Cl_2(g)\rightarrow 2MCl_4(g)+\frac{5}{2}O_2(g) \end{aligned} \tag{21}$$

The first reaction is associated with a reduction of the number of moles of gaseous species (Δn_g = -0.5), but in the second the same quantity is positive (Δn_g = 0.5). If the gas phase is described as an ideal solution, the first reaction should be associated with a lower molar entropy than the second. The greater the number of mole of gaseous products, the greater the gas phase volume produced, and so the greater the entropy generated. By plotting the molar Gibbs energy of each reaction as a function of temperature, the curves should cross each other at a specific temperature (T_C). For temperatures greater than T_C the formation of MCl_4 becomes thermodynamically more favorable (see Figure 4).

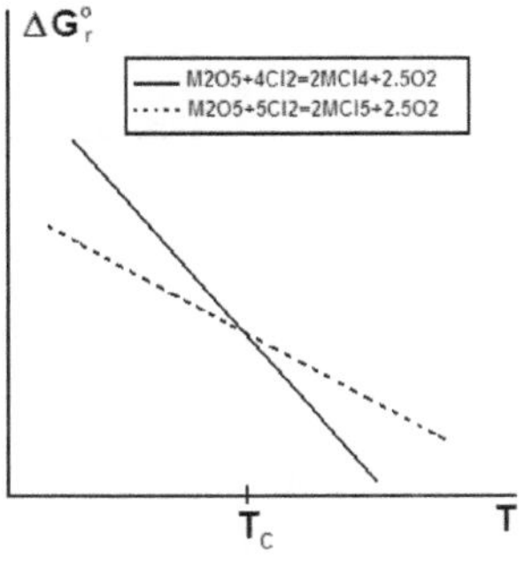

Fig. 4. Hypothetical ΔG_r^o x T curves with intercept.

An interesting situation occurs, if one of the chlorides can be produced in the condensed state (liquid or solid). Let's suppose that the chloride MCl_5 is liquid at lower temperatures.

The ebullition of MCl_5, which occur at a definite temperature (T_t), dislocates the curve to lower values for temperatures higher than T_t. Such an effect would make the production of MCl_5 in the gaseous state thermodynamically more favorable even for temperatures greater than T_c (Figure 5). Such fact the importance of considering phase transitions when comparing ΔG_r^o x T curves for different reactions.

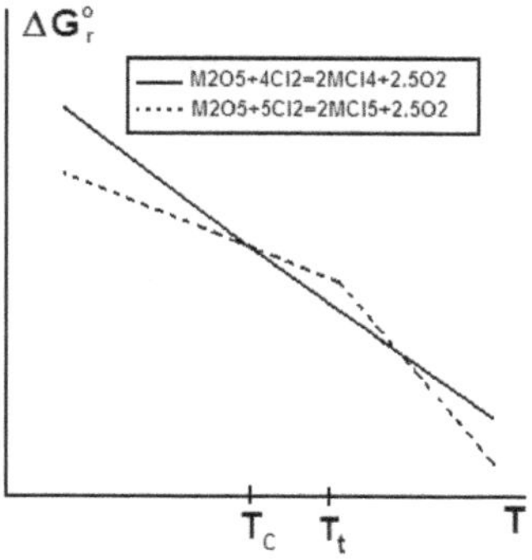

Fig. 5. Effect of MCl_5 boiling temperature

Although simple, the method based on the comparison of ΔG_r^o x T diagrams is of limited application. The problem is that for discussing the thermodynamic viability of a reaction one must actually compute the thermodynamic driving force (Eq. 15 and 16), and by doing so, one must fix values for the concentration of Cl_2 and O_2 in the reactor's atmosphere, which, in the end, define the value of the reaction coefficient.

If the ΔG_r^o x T curves of two reactions lie close to one another (difference lower than 10 KJ/mol), it is impossible to tell, without a rigorous calculation, which chlorinated specie should have the highest concentration in the gaseous state, as the computed driving forces will lie very close from each other. In these situations, other methods that can address the direct effect of the reactor's atmosphere composition should be applied.

Apart from its simplicity, the ΔG_r^o x T diagrams have another interesting application in relation to the proposal of reactions mechanisms. From the point of view of the kinetics, the process of forming higher chlorinated species by the "collision" of one molecule of the oxide M_2O_5 and a group of molecules of Cl_2, and vise versa, shall have a lower probability than the one defined by the first formation of a lower chlorinated specie, say MCl_2, and the further reaction of it with one or two Cl_2 molecules (Eq. 22).

Let's consider that M can form the following chlorides: MCl, MCl_2, MCl_3, MCl_4, and MCl_5. The synthesis of MCl_5 can now be thought as the result of the coupled reactions represented by Eq. (22).

$$\begin{aligned} &M_2O_5 + Cl_2 \rightarrow 2MCl + 2.5O_2 \\ &MCl + 0.5Cl_2 \rightarrow MCl_2 \quad MCl_2 + 0.5Cl_2 \rightarrow MCl_3 \\ &MCl_3 + 0.5Cl_2 \rightarrow MCl_4 \quad MCl_4 + 0.5Cl_2 \rightarrow MCl_5 \end{aligned} \tag{22}$$

By plotting the ΔG_r^o x T diagrams of all reactions presented in Eq. (22) it is possible to evaluate if the thermodynamic stability of the chlorides follows the trend indicated by the proposed reaction path. If so, the curves should lay one above the other. The standard reaction Gibbs energy would then grow in the following order: MCl, MCl_2, MCl_3, MCl_4 and MCl_5 (Figure 6).

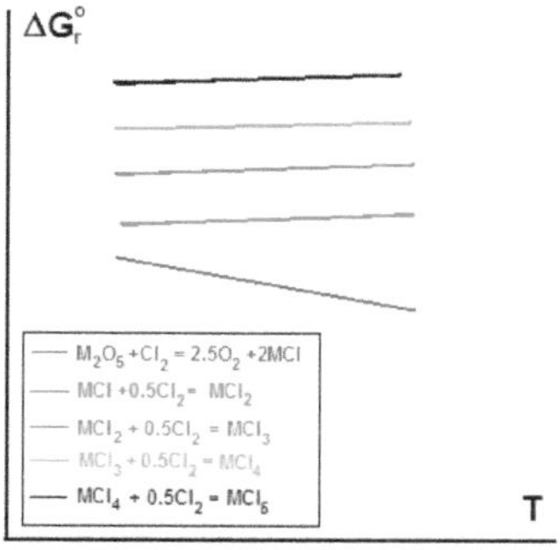

Fig. 6. Hypothetic ΔG_r^o x T curves for successive chlorination reactions

Another possibility is that the curve for the formation of one of the higher chlorinated species is associated with lower Gibbs energy values in comparison with the curve of a lower chlorinated compound. A possible example thereof is depicted on Figure (7), where the ΔG_r^o x T curve for the production of MCl_3 lies bellow the curve associated with the formation of MCl_2.

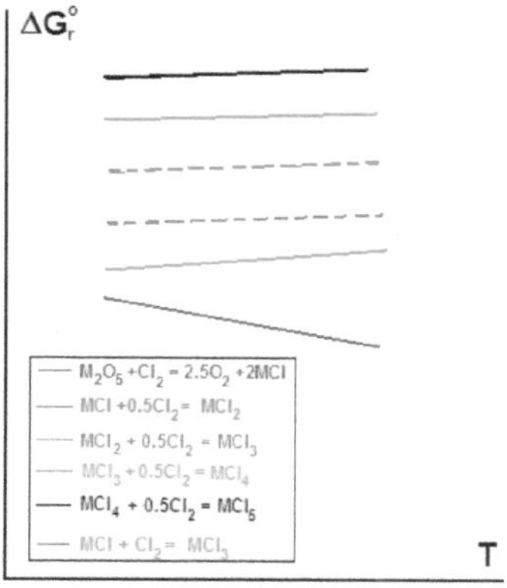

Fig. 7. Successive chlorination reactions – direct formation of MCl_3 from MCl

The formation of the species MCl_2 would be thermodynamically less favorable, and MCl_3 is preferentially produced directly from MCl ($MCl + Cl_2 = MCl_3$). In this case, however, for the diagram to remain thermodynamically consistent, the curves associated with the formation of MCl_2 from MCl and MCl_3 from MCl (broken lines) should be substituted for the curve associated with the direct formation of MCl_3 from MCl for the entire temperature range.
The same effect could originate due to the occurrence of a phase transition. Let's suppose that in the temperature range considered MCl_3 sublimates at T_s. Because of this

phenomenon the curve for the formation of MCl_2 crosses the curve for the formation of the last chloride at T_c, so that for $T > T_c$ its formation is associated with a higher thermodynamic driving force (Figure 8). So, for $T > T_c$, MCl_3 is formed directly from MCl, resulting in the same modification in the reaction mechanism as mentioned above.

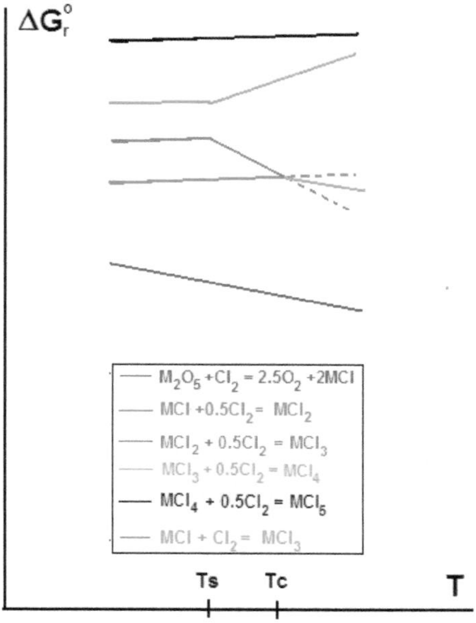

Fig. 8. Direct formation of MCl_3 from MCl stimulated by MCl_3 sublimation

For temperatures higher than T_c, the diagram of Figure (8) looses its thermodynamic consistency, as, according to what was mentioned in the last paragraph, the formation of MCl_2 from MCl is impossible in this temperature range. The error can be corrected if, for $T >$ Tc, the curves associated with the formation of MCl_2 and MCl_3 (broken lines) are substituted for the curve associated with the formation of MCl_3 directly from MCl.
A direct consequence of that peculiar thermodynamic fact, as described in Figures (7) and (8), is that under these conditions, a predominance diagram would contain a straight line showing the equilibrium between MCl and MCl_3, and the field corresponding to MCl_2 would not appear.

2.2.2 Method of Kang and Zuo

Kang & Zuo (1989) introduced a simple method for comparing the thermodynamic tendencies of formation of compounds obtained by gas – solid reactions, in that each equilibrium equation is solved independently, and the concentration of the desired species plotted as a function of the gas phase concentration and or temperature. The method will be illustrated for the reactions defined by Eq. (21). The concentrations of MCl_4 and MCl_5 in the gaseous phase can be computed as a function of temperature, partial pressure of Cl_2, and partial pressure of O_2.

$$P_{MCl_5} = \sqrt{\frac{P_{Cl_2}{}^5}{P_{O_2}{}^{5/2}} \exp\left(-\frac{\left(2g^g_{MCl_5} + \frac{5}{2} g^g_{O_2} - 5g^g_{Cl_2} - g^s_{M_2O_5}\right)}{RT} \right)}$$
$$P_{MCl_4} = \sqrt{\frac{P_{Cl_2}{}^4}{P_{O_2}{}^{5/2}} \exp\left(-\frac{\left(2g^g_{MCl4} + \frac{5}{2} g^g_{O2} - 4g^g_{Cl2} - g^s_{M2O5}\right)}{RT} \right)} \tag{23}$$

Next, two intensive properties must be chosen, whose values are fixed, for example, the partial pressure of Cl_2 and the temperature. The partial pressure of each chlorinated species becomes in this case a function of only the partial pressure of O_2.

$$P_{MCl_5} = f_{MC_5}\left(T, P_{Cl_2}\right) P_{O_2}{}^{5/2}$$
$$P_{MCl_4} = f_{MCl_4}\left(T, P_{Cl_2}\right) P_{O_2}^{5/2}$$
$$f_{MCl_5}\left(T, P_{Cl_2}\right) = P_{Cl_2}{}^{-5/2} \exp\left(-\frac{\left(2g^g_{MCl_5} + \frac{5}{2} g^g_{O_2} - 5g^g_{Cl_2} - g^s_{M_2O_5}\right)}{2RT} \right)$$
$$f_{MCl_4}\left(T, P_{Cl_2}\right) = P_{Cl_2}{}^{-2} \exp\left(-\frac{\left(2g^g_{MCl_4} + \frac{5}{2} g^g_{O_2} - 4g^g_{Cl_2} - g^s_{M_2O_5}\right)}{2RT} \right) \tag{24}$$

By fixing T and $P(Cl_2)$ the application of the natural logarithm to both sides of Eq. (24) results in a linear behavior.

$$\ln P_{MCl_5} = \ln f_{MCl_4} + 2.5 \ln P_{O_2}$$
$$\ln P_{MCl_4} = \ln f_{MCl_5} + 2.5 \ln P_{O_2} \tag{25}$$

The lines associated with the formation of MCl_4 and MCl_5 would have the same angular coefficient, but different linear coefficients. If the partial pressure of Cl_2 is equal to one (pure Cl_2 is injected into the reactor), the differences in the standard reaction Gibbs energy controls the values of the linear coefficients observed. If the lowest Gibbs energy values are associated with the formation of MCl_5, its line would have the greatest linear coefficient (Figure 9).

An interesting situation occurs if the curves obtained for the chlorinated species of interest cross each other (Figure 10). This fact would indicate that for some critical value of $P(O_2)$ there would be a different preference for the system to generate each one of the chlorides. One of them prevails for higher partial pressure values and the other for values of $P(O_2)$ lower than the critical one. Such a behavior could be exemplified if the chlorination of M also generates the gaseous oxychloride $MOCl_3$ ($M_2O_5 + 2Cl_2 = 2MOCl_3 + 1.5O_2$).

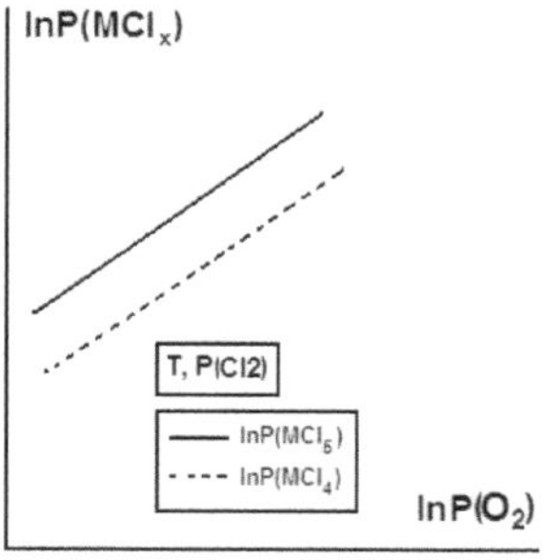

Fig. 9. Concentrations of MCl_4 and MCl_5, as a function of $P(O_2)$

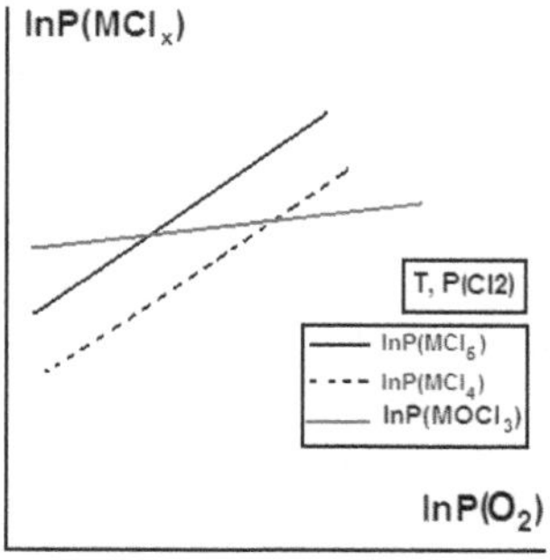

Fig. 10. Concentrations of $MOCl_3$, MCl_4 and MCl_5 as a function of $P(O_2)$

$$\ln P_{\mathrm{MOCl_3}} = \ln f_{\mathrm{MOCl_3}} + 1.5 \ln P_{\mathrm{O_2}} \tag{26}$$

The linear coefficient of the line associated with the $MOCl_3$ formation is higher for the initial value of $P(O_2)$ than the same factor computed for MCl_4 and MCl_5. As the angular coefficient is lower for $MOCl_3$, The graphic of Figure (10) depicts a possible result.

According to Figure (10), three distinct situations can be identified. For the initial values of $P(O_2)$, the partial pressure of $MOCl_3$ is higher than the partial pressure of the other chlorinated compounds.

By varying $P(O_2)$, a critical value is approached after which $P(MCl_5)$ assumes the highest value, being followed by $P(MOCl_3)$ and then $P(MCl_4)$. A second critical value of $P(O_2)$ can be identified in the graphic above. For $P(O_2)$ values higher than this, the atmosphere should be more concentrated in MCl_5 and less concentrated in $MOCl_3$, MCl_4 assuming a concentration value in between.

2.2.3 Minimization of the total gibbs energy

The most general way of describing equilibrium is to fix a number of thermodynamic variables (physical parameters that can be controlled in laboratory), and to chose an appropriate thermodynamic potential, whose maxima or minima describe the possible equilibrium states available to the system.

By fixing *T*, *P*, and total amounts of the components M, O, and Cl ($n(O)$, $n(M)$, and $n(Cl)$), the global minimum of the total Gibbs energy describes the equilibrium state of interest, which is characterized by a proper phase ensemble, their amounts and compositions. This method is equivalent to solve all chemical equilibrium equations at the same time, so, that the compositions of the chlorinated species in each one of the phases present are calculated simultaneously.

For treating the equilibrium associated with the chlorination processes, two type of diagrams are important: *predominance diagrams*, and *phase speciation diagrams*. The first sort of diagram describes the equilibrium phase ensemble as a function of temperature, and or partial pressure of Cl_2 or O_2. The second type describes how the composition of individual phases varies with temperature and or concentration of Cl_2 or O_2.

The first step is to change the initial constraint vector (T, P, n(O), n(M), n(Cl)), by modifying the definition of the components. Instead considering as components the elements O, M, and Cl, we can describe the global composition of the system by specifying amounts of M, Cl_2 and O_2 (T, P, $n(O_2)$, n(M), $n(Cl_2)$).

According to the phase-rule (Eq. 27) applied to a system with three components (M, Cl_2 and O_2), by specifying five degrees of freedom (intensive variables or restriction equations) the equilibrium calculation problem has a unique solution:

$$\begin{gathered} L = C + 2 - F \\ F = 0 \quad C = 3 \\ L = 5 \end{gathered} \tag{27}$$

Where F denotes the number of phases present (as we do not know the nature of the phase ensemble, $F = 0$ at the beginning), C is the number of components, and L defines the number of degrees of freedom (equations and or intensive variables) to be specified. So, with $L = 5$, the constraint vector must have five coordinates (T, P, $n(O_2)$, n(M), $n(Cl_2)$).

In reality, the chlorination system is described as an open system, where a gas flux of definite composition is established. The constraint vector defined so far is consistent with the definition of a closed system, which by definition does not allow matter to cross its boundaries. The calculation can become closer to the physical reality of the process if we specify the chemical activities of Cl_2 and O_2 in the gas phase, instead of fixing their global molar amounts. Such a restriction would be analogous as fixing the inlet gas composition. Further, if the gas is considered to behave ideally, the chemical activities can be replaced by the respective values of the partial pressure of the gaseous components. So, the final constraint vector should be defined as follows: T, P, n(M), $P(Cl_2)$, $P(O_2)$.

The two types of computation mentioned in the first paragraph can now be discussed. For generating a speciation diagram, only one of the parameters T, $P(Cl_2)$, or $P(O_2)$ is varied in a definite range. The composition of some phase of interest, for example the gas, can then be plotted as a function of the thermodynamic coordinate chosen. On the other hand, by systematically varying two of the parameter defined in the group T, $P(Cl_2)$, or $P(O_2)$, a predominance diagram can be constructed (Figure 11). The diagram is usually drawn in space $P(Cl_2)$ x $P(O_2)$ and is composed by cells, which describe the stability limits of individual phases. A line describes the equilibrium condition involving two phases, and a point the equilibrium involving three phases.

Let's take a closer look in the nature of a predominance diagram applied to the case studied so far. In this situation, one must consider the gas phase, the solid metal M, and possible oxides, MO, MO_2, and M_2O_5, obtained through oxidation of the element M at different oxygen potentials. The equilibrium involving two oxides defines a unique value of the partial pressure of O_2, which is independent of the Cl_2 concentration.

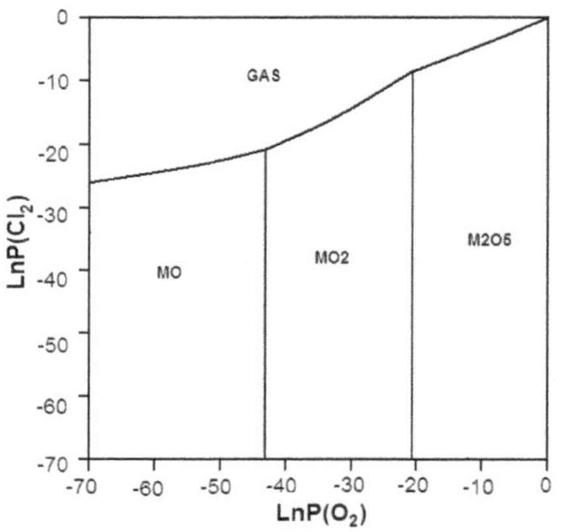

Fig. 11. Hypothetical predominance diagram chlorides mixed in the gas phase

For the equilibrium between MO and MO_2, for example, Eq. (28) enables the determination of the $P(O_2)$ value, which is fixed by choosing T and is independent of the Cl_2 partial pressure. As a consequence, such equilibrium states are defined by a vertical line.

$$P_{O_2} = \exp\left(-\frac{\left(g^{s}_{MO_2} - g^{s}_{MO} - 0.5 g^{g}_{O_2} \right)}{RT} \right) \tag{28}$$

The equilibrium when the phase ensemble is defined by the gas and one of the metal oxides, say MO_2, is also defined by a line, whose inclination is determined by fixing T, P, n(M) and $P(O_2)$. This time the concentration of Cl_2, MCl_4 and MCl_5 are computed by solving the group of non-linear equations presented bellow (Eq. 29). The first equation defines the restriction that the molar quantity of M is constant (mass conservative restriction). The second equation represents the conservation of the total mass of the gas phase (the summation of all mol fractions must be equal to one).

$$\begin{aligned} n_M &= n_{MO_2} + n^{g}\left(x^{g}_{MCl_4} + x^{g}_{MCl_5} \right) \\ 1 &= x^{g}_{MCl_4} + x^{g}_{MCl_5} + x^{g}_{O_2} + x^{g}_{Cl_2} \\ & \mu^{g}_{MCl_5} - \mu^{g}_{MCl_4} - \frac{\mu^{g}_{Cl_2}}{2} = 0 \\ & 2\mu^{g}_{MCl5} + 2.5\mu^{g}_{O_2} - g^{s}_{M_2O_5} - 5\mu^{g}_{Cl_2} = 0 \\ & 2\mu^{g}_{MCl_4} + 2.5\mu^{g}_{O_2} - g^{s}_{M_2O_5} - 4\mu^{g}_{Cl_2} = 0 \end{aligned} \tag{29}$$

The other three relations define, respectively, the equilibrium conditions for the following group of reactions:

$$\begin{aligned} & MCl_4(g) + 0.5Cl_2(g) \rightarrow MCl_5(g) \\ & MO_2(s) + 2.5Cl_2(g) \rightarrow MCl_5(g) + O_2(g) \\ & MO_2(s) + 2Cl_2(g) \rightarrow MCl_4(g) + O_2(g) \end{aligned} \tag{30}$$

So, we have five equations and five unknowns (n^g, n_{MO_2}, $x^g_{MCl_4}$, $x^g_{MCl_5}$, $x^g_{Cl_2}$), indicating that the equilibrium calculation admits a unique solution.
Finally by walking along a vertical line associated with the coexistence of two metallic oxides, for example MO and MO_2, a condition is achieved where the gaseous chlorides are formed. The equilibrium between the two oxides and the gas phase is defined by a point. In other words by fixing *T* and *P*, all equilibrium properties are uniquely defined. The equation associated with the coexistence of MO and MO_2 (Eq. 28) is added and the partial pressure of O_2 is allowed to vary, resulting in six variables and six equations (Eq. 31).
Equations (30) and (31) were presented here only with a didactic purpose. In praxis, the majority of the thermodynamic software (*Thermocalc*, for example) are designed to minimize the total Gibbs energy of the system. The algorithm varies systematically the composition of the equilibrium phase ensemble until the global minimum is achieved. By doing so the same algorithm can be implemented for dealing with all possible equilibrium conditions, eliminating at the end the difficulty of proposing a group of linear independent chemical equations, which for a system with a great number of components can become a complicated task.

$$\begin{aligned} & n_M = n_{MO_2} + n^g\left(x^g_{MCl_4} + x^g_{MCl_5}\right) \\ & 1 = x^g_{MCl_4} + x^g_{MCl_5} + x^g_{O_2} + x^g_{Cl_2} \\ & \mu^g_{MCl_5} - \mu^g_{MCl_4} - 0.5\mu^g_{Cl_2} = 0 \\ & 2\mu^g_{MCl_5} + 2.5\mu^g_{O_2} - g^s_{M_2O_5} - 5\mu^g_{Cl_2} = 0 \\ & 2\mu^g_{MCl_4} + 2.5\mu^g_{O_2} - g^s_{M_2O_5} - 4\mu^g_{Cl_2} = 0 \\ & g^s_{MO_2} - g^s_{MO} - 0.5\mu^g_{O_2} = 0 \end{aligned} \tag{31}$$

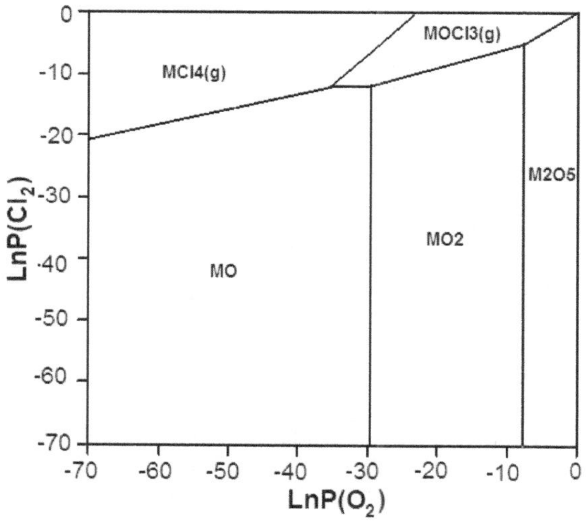

Fig. 12. Hypothetical predominance diagram: pure gaseous chlorides

A simplified version of the predominance diagram of Figure (11) can be achieved through considering each possible gaseous chloride as a pure substance. In this case, the field representing the gas phase will be divided into sub-regions, each one representative of the stability of each gaseous chlorinated compound. By considering, that, besides MCl_5 and MCl_4, gaseous $MOCl_3$ can also be formed, a diagram similar to the one presented on Figure (12) would represent possible stability limits found in equilibrium.
The diagram of Figure (12) is associated with a temperature value where gaseous MCl_5 can not be present in equilibrium for any suitable value of $P(Cl_2)$ and $P(O_2)$ chosen. It is interesting to note, that in this sort of diagram, there is a direct relation between the inclination of a line representative of the equilibrium between a gaseous chloride or oxychloride and an oxide, with the stoichiometric coefficients of the chemical reaction behind the transformation.
According to Eq. (32), the inclination of the line associated with the equilibrium between $MOCl_3$ and MO_2 should be lower than the one associated with the equilibrium between $MOCl_3$ and M_2O_5. On the other hand, in the case of the equilibrium between MO and $MOCl_3$, the line is horizontal (does not depend on $P(O_2)$), as the same number of oxygen atoms is present in the reactant and products, so O_2 does not participate in the reaction.

$$\begin{aligned} \ln P_{Cl_2} &= \frac{1}{3}\ln P_{O_2} - \frac{2}{3}\ln K_{MO_2}(T) \\ \ln P_{Cl_2} &= \frac{1}{2}\ln P_{O_2} - \frac{1}{3}\ln K_{M_2O_5}(T) \\ \ln P_{Cl_2} &= -\frac{2}{3}\ln K_{MO}(T) \end{aligned} \tag{32}$$

Where, K_{MO_2}, $K_{M_2O_5}$ and K_{MO} represent respectively the equilibrium constants for the formation of $MOCl_3$ from MO_2, M_2O_5 and MO (Eq. 33).

$$\begin{aligned} &MO_2 + 1.5Cl_2 \rightarrow MOCl_3 + 0.5O_2 \\ &M_2O_5 + 3Cl_2 \rightarrow 2MOCl_3 + 1.5O_2 \\ &MO + 1.5Cl_2 \rightarrow MOCl_3 \end{aligned} \tag{33}$$

The diagrams of Figures (11) and (12) depict a behavior, where no condensed chlorinated phases are present. For many oxides, however, there is a tendency of formation of solid or liquid chlorides and or oxychlorides, which must appear in the predominance diagram as fields between the pure oxides and the gas phase regions. Such a behavior can be observed in the equilibrium states accessible to the system V - O - Cl.

3. The system V – O – Cl

Vanadium is a transition metal that can form a variety of oxides. At ambient temperature and oxygen potential, the form V_2O_5 is the most stable. It is a solid stoichiometric oxide, where vanadium occupies the +5 oxidation state. By lowering the partial pressure of O_2, the valence of vanadium varies considerably, making it is possible to produce a family of stoichiometric oxides: V_2O_4, V_3O_5, V_4O_7, VO, VO_2 and V_2O_3. Recently, it has been discovered that vanadium can also form a variety of non-stoichiometric oxygenated compounds (Brewer & Ebinghaus, 1988), however, to simplify the treatment of the present chapter, these phases will not be included in the data-base used for the following computations. Additionally, it was considered that the concentration of the oxides in gas phase is low enough to be neglected. Further, on what touches the computations that follows, the software *Thermocalc* was used in all cases, and it will always be assumed that equilibrium is achieved, or in other words, kinetic effects can be neglected.

The relative stability of the possible vanadium oxides can be assessed through construction of a predominance diagram in the space T - $P(O_2)$ (see Figure 13). As thermodynamic constraints we have $n(V)$ (number of moles of vanadium metal - it will be supposed that $n(V)$ =1), T, P and $P(O_2)$. The reaction temperature will be varied in the range between 1073 K and 1500 K and the partial pressure of O_2 in the range between 8.2.10^{-40}atm and 1atm.

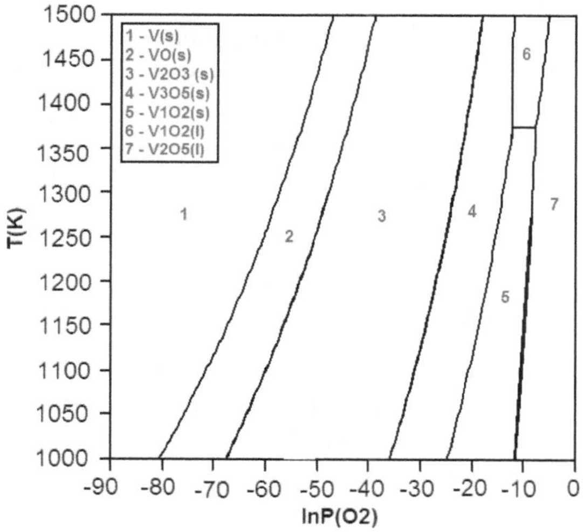

Fig. 13. Predominance diagram for the system V - O

The total pressure was fixed at 1atm. It can be seen that for the temperature range considered and a partial pressure of O_2 in the neighborhood of 1atm, V_2O_5 is formed in the liquid state. Through lowering the oxygen potential, crystalline vanadium oxides precipitate, VO_2 being formed first, followed by V_2O_3, VO, and finally V. The horizontal line between fields "5" and "6" indicates the melting of V_1O_2, which according to classical thermodynamics must occur at a fixed temperature. Next it will be considered the species formed by vanadium, chlorine and oxygen.

3.1 Vanadium oxides and chlorides

The already identified species formed between vanadium, chlorine and oxygen are: VCl, VCl_2, VCl_3, VCl_4, VOCl, $VOCl_2$, $VOCl_3$, VO_2Cl.

On Table (1) it was included information regarding the physical states at ambient conditions and some references related to phase equilibrium studies conducted on samples of specific vanadium chlorinated compounds.

Only a few studies were published in literature in relation to the thermodynamics of vanadium chlorinated phases. On Table (1) some references are given for earlier investigations associated with measurements of the vapor pressure for the sublimation of VCl_2 and VCl_3, and the boiling of $VOCl_3$ and VCl_4. There are also evidences for the occurrence of specific thermal decomposition reactions (Eq. 34), such as those of VCl_3, $VOCl_2$ and VO_2Cl (Oppermann, 1967).

Chloride	Physical state	Equilibrium data	Reference
VCl	-	-	-
VCl_2	Solid	Sublimation(	McCarley & Roddy (1964)
VCl_3	Solid	Sublimation/ Thermal decomposition	McCarley & Roddy (1964)
VCl_4	Liquid	Ebulition	Oppermann (1962a)
$VOCl_3$	Liquid	Ebulition(	Oppermann (1967)
VO_2Cl	Solid	Thermal decomposition(	Oppermann (1967)
$VOCl_2$	Solid	Thermal decomposition(	Oppermann (1967)
VOCl	Solid	Synthesis and characterization(	Schäffer at al. (1961)

Table 1. Physical nature and phase equilibrium data for vanadium chlorinated compounds

$$\begin{aligned} &2VCl_3(s) \rightarrow VCl_2(s) + VCl_4(g) \\ &3VO_2Cl(s) \rightarrow VOCl_3(g) + V_2O_5(s) \\ &2VOCl_2(s) \rightarrow VOCl_3(g) + VOCl(s) \end{aligned} \tag{34}$$

Chromatographic measurements conducted recently confirmed the possible formation of VCl, VCl_2, VCl_3, and VCl_4 in the gas phase (Hildenbrand et al., 1988). In this study the molar Gibbs energy models for the mentioned chlorides were revised, and new functions proposed. In the case vanadium oxychlorides, models for the molar Gibbs energies of gaseous VOCl, $VOCl_3$, and $VOCl_2$ have already been published (Hackert et al., 1996).
For gaseous VO_2Cl, on the other hand, no thermodynamic model exists, indicating the low tendency of this oxychloride to be stabilized in the gaseous state.

3.1.1 The $V - O_2 - Cl_2$ stability diagram

The relative stability of the possible chlorinated compounds of vanadium can be assessed through construction of predominance diagrams by fixing the temperature and systematic varying the values of $P(Cl_2)$ and $P(O_2)$.
For the temperature range usually found in chlorination praxis, three temperatures were considered, 1073 K, 1273 K and 1573 K. The partial pressure of Cl_2 and O_2 were varied in the range between $3.98.10^{-31}$atm and 1atm. All chlorinated species are considered to be formed at the standard state (pure at 1atm). The predominance diagrams can be observed on Figures (14), (15) and (16).
The stability field of VCl_2(l) grows in relation to those associated to VCl_4 and $VOCl_3$. At 1573 K the VCl_2(l) area is the greatest among the chlorides and the VCl_3(g) field appears. So, as temperature achieves higher values the concentration of VCl_3 in the gas phase should increase in comparison with the other chlorinated species, including VCl_2. This behavior agrees with the one observed during the computation of the gas phase speciation and will be better discussed on topic (3.1.3.2).

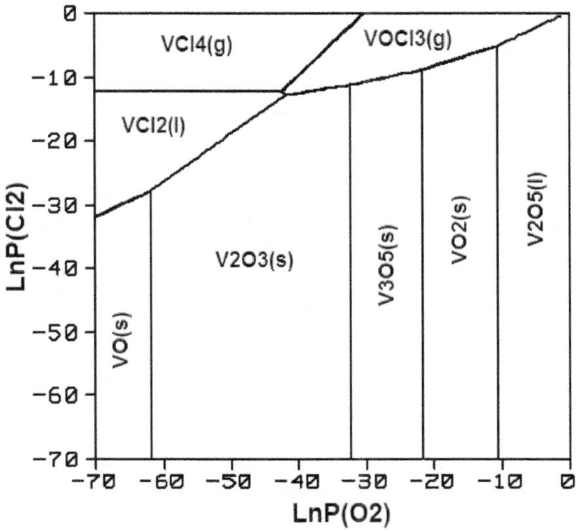

Fig. 14. Predominance diagram for the system V – O – Cl at 1073 K

Finally, by starting in a state inside a field representing the formation of VCl_4 or VCl_3 and by making $P(O_2)$ progressively higher, a value is reached, after which $VOCl_3$(g) appears. So, the mol fraction of VCl_4 and VCl_3 in gas should reduce when $P(O_2)$ achieves higher values. This is again consistent with the speciation computations developed on topic (3.1.3.2).

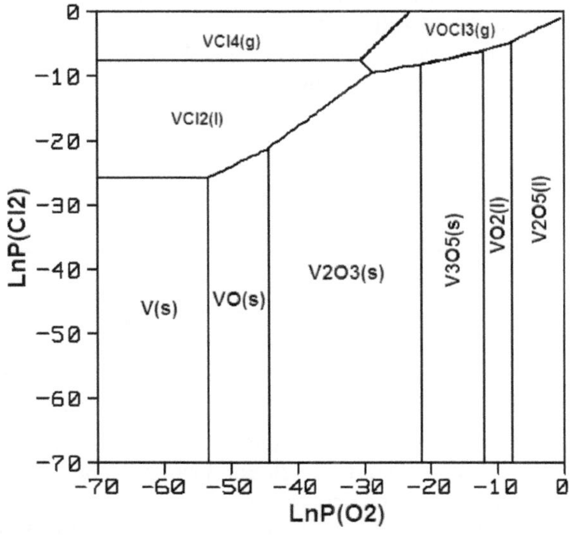

Fig. 15. Predominance diagram for the system V – O – Cl at 1273 K

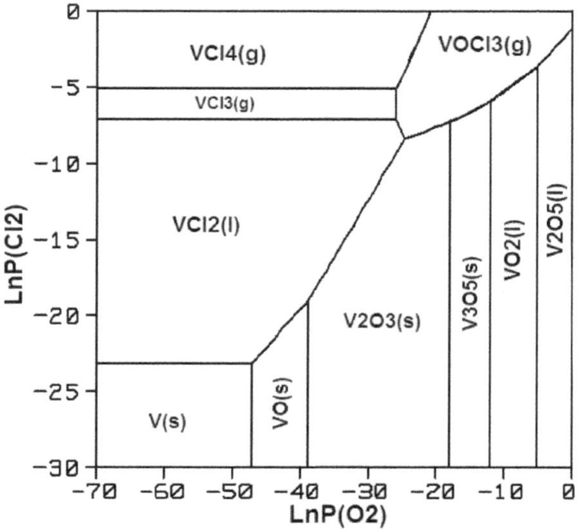

Fig. 16. Predominance diagram for the system V – O – Cl at at 1573 K

3.1.2 V_2O_5 direct chlorination and the effect of the reducing agent

The direct chlorination of V_2O_5 is a process, which consists in the reaction of a V_2O_5 sample with gaseous Cl_2.

$$V_2O_5 + Cl_2 = \text{Chloride/Oxychloride} + O_2 \tag{35}$$

In praxis, temperature lies usually between 1173 K and 1473 K. The chlorination equilibrium could then be dislocated in the direction of the formation of chlorides and oxychlorides if one removes O_2 and or adds Cl_2 to the reactors atmosphere. So, for low $P(O_2)$ ($< 10^{-20}$ atm) and high $P(Cl_2)$ (between 0.1 and 1 atm) values, according to the predominance diagrams of Figures (14) and (15), VCl_4 should be the most stable vanadium chloride, which is produced according to Eq. (36).

$$V_2O_5 + 2Cl_2 = 2VCl_4 + 2.5O_2 \tag{36}$$

K	T (K)
$1.76257.10^{-13}$	1173
$5.82991.10^{-11}$	1273
$1.0397.10^{-08}$	1473

Table 2. Equilibrium constant for the reaction represented by Eq. (37)

The equilibrium constant for reaction represented by Eq. (36) is associated with very low values between 1173 K and 1473 K (see Table 2). So, it can be concluded that the formation of VCl_4 has a very low thermodynamic driving force in the temperature range considered. One possibility to overcome this problem is to add to the reaction system some carbon bearing compound (Allain et al., 1997, Gonzallez et al., 2002a; González et al., 2002b; Jena et al., 2005). The compound decomposes producing graphite, which reacts with oxygen dislocating the chlorination equilibrium in the desired direction. A simpler route, however, would be to admit carbon as graphite together with the oxide sample into the reactor. If graphite is present in excess, the O_2 concentration in the reactor's atmosphere is maintained at very low values, which are achievable through the formation of carbon oxides (Eq. 37)

$$\begin{aligned} 2C + O_2 &= 2CO \\ C + O_2 &= CO_2 \end{aligned} \tag{37}$$

So, for the production of VCl_4 in the presence of graphite, the reaction of C with O_2 can lead to the evolution of gaseous CO or CO_2 (Eq. 38).

$$\begin{aligned} V_2O_5(s,l) + 4Cl_2(g) + 2.5C(s) &\rightarrow 2VCl_4(l,g) + 2.5CO_2(g) \\ V_2O_5(s,l) + 4Cl_2(g) + 5C(s) &\rightarrow 2VCl_4(l,g) + 5CO(g) \end{aligned} \tag{38}$$

The effect of the presence of graphite over the ΔG_r^o x T curves for the formation of VCl_4 can be seen in the diagram of Figure (17). As a matter of comparison, the plot for the formation of the same species in the absence of graphite is also shown, together with the curves for the reactions associated with the formation of CO and CO_2 for one mole of O_2 (Eq. 37).

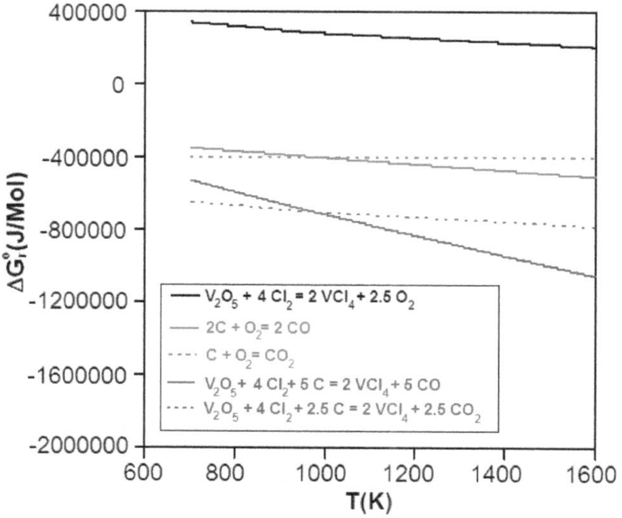

Fig. 17. ΔG_r^o vs. T for for the formation of VCl_4

It can be readily seen that graphite strongly reduces the standard molar Gibbs energy of reaction, promoting in this way considerably the thermodynamic driving force associated with the chlorination process. The presence of graphite has also an impact over the standard molar reaction enthalpy. The direct action of Cl_2 is associated with an endothermic reaction (positive linear coefficient), but by adding graphite the processes become considerably exothermic (negative linear coefficient).

The curves associated with the VCl_4 formation in the presence of the reducing agent cross each other at 973 K, the same temperature where the curves corresponding to the formation of CO and CO_2 have the same Gibbs energy value. This point is defined by the temperature, where the Gibbs energy of the Boudouard reaction ($C + CO_2 = 2CO$) is equal to zero.
The equivalence of this point and the intersection associated with the curves for the formation of VCl_4 can be perfectly understood, as the Boudouard reaction can be obtained through a simple linear combination, according to Eq. (39). So, the molar Gibbs energy associated with the Boudouard reaction is equal to the difference between the molar Gibbs energy of the VCl_4 formation with the evolution of CO and the same quantity for the reaction associated with the CO_2 production. When the curves for the formation of VCl_4 crosses each other, the difference between their molar Gibbs energies is zero, and according to Eq. (39) the same must happen with the molar Gibbs energy of the Boudouard reaction.

$$\begin{aligned}
&1)\ V_2O_5(s,l) + 4Cl_2(g) + 5C(s) \xrightarrow{\Delta G_1} 2VCl_4(l,g) + 5CO(g) \\
&- \\
&2)\ V_2O_5(s,l) + 4Cl_2(g) + 2.5C(s) \xrightarrow{\Delta G_2} 2VCl_4(l,g) + 2.5CO_2(g) \\
&= \\
&3)\ C(s) + CO_2(g) \xrightarrow{\Delta G_3} 2CO(g)
\end{aligned} \tag{39}$$

$$\Delta G_3 = \Delta G_1 - \Delta G_2$$
$$\lim_{T \to 973K}(\Delta G_3) = \lim_{T \to 973K}(\Delta G_1 - \Delta G_2) = \Delta G - \Delta G = 0$$

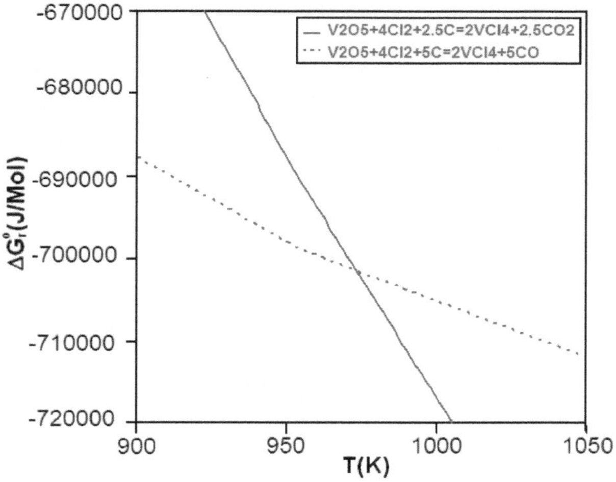

Fig. 18. ΔG_r^o vs. T the formation of VCl_4 – melting of V_2O_5

The inflexion point present on the curves of Figure (17) is associated with the melting of V_2O_5. This inflexion is better evidenced on the graphic of Figure (18). As V_2O_5 is a reactant, according to the concepts developed on topic (2.2.1), the curve should experience a

reduction of its inclination at the melting temperature of the oxide. However, the presence of the inflexion point is much more evident for the reactions with the lowest variation of number of moles of gaseous reactants, as is the case for the direct action of Cl_2, which leads to the evolution of CO_2 ($\Delta n_g = 0.5$).
The quantity Δn_g controls the molar entropy of the reaction. By lowering the magnitude Δn_g the value of the reaction entropy reduces, and the effect of melting of V_2O_5 over the standard molar reaction Gibbs energy becomes more evident.
Based on the predominance diagrams of topic (3.1.1), $VOCl_3$ should be formed for $P(Cl_2)$ close to 1atm as $P(O_2)$ gets higher. The presence of graphite has the same effect over the molar Gibbs energy of formation of $VOCl_3$, promoting in this way the thermodynamic driving force for the reaction. Its curve is compared with the one for the formation of VCl_4 on Figure (19). The inflexion around 954 K is again associated with the melting of V_2O_5. As the reaction associated with the formation of VCl_4, the formation of $VOCl_3$ has a negative molar reaction enthalpy. So, if the gas phase is considered ideal, for the production of both chlorinated compounds the system should transfer heat to its neighborhood (exothermic reaction).

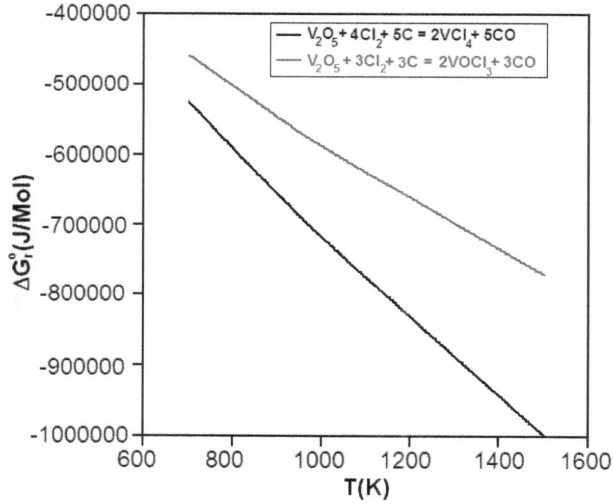

Fig. 19. ΔG_r^o vs. T for the formation of $VOCl_3$ and VCl_4

On what touches the molar reaction entropy, the graphic of Figure (19) indicates, that the reaction associated with the formation of VCl_4 should generate more entropy (more negative angular coefficient for the entire temperature range). This can be explained by the fact, that in the case of VCl_4 the variation of the number of mole of gaseous reactants and products ($\Delta n_g = 3$) is higher than the value for the formation of $VOCl_3$ ($\Delta n_g = 2$). This illustrates how important the magnitude of Δn_g is for the molar entropy of a gas - solid reaction.
Finally, it should be pointed out that the standard molar Gibbs energy has the same order of magnitude for both chlorinated species considered. So, only by appreciating the ΔG_r^o x T curves of these chlorides it is impossible to tell case which species should be found in the gas with the highest concentration. This problem will be covered on topic (3.2).

3.1.2.1 Successive chlorination steps

As discussed on topic (2.2.1), the standard free energy vs. temperature diagram is a valuable tool for suggesting possible reactions paths. Let's consider first the formation of VCl_4. Such a process could be thought as the result of three stages. In the first one, a lower chlorinated compound (VCl) is formed. The precursor then reacts with Cl_2 resulting in higher chlorinated species (Eq. 40).

$$\begin{aligned} &V_2O_5 + Cl_2 + C \rightarrow VCl + CO_2 / CO \\ &VCl + 0.5Cl_2 \rightarrow VCl_2 \\ &VCl_2 + 0.5Cl_2 \rightarrow VCl_3 \\ &VCl_3 + 0.5Cl_2 \rightarrow VCl_4 \end{aligned} \tag{40}$$

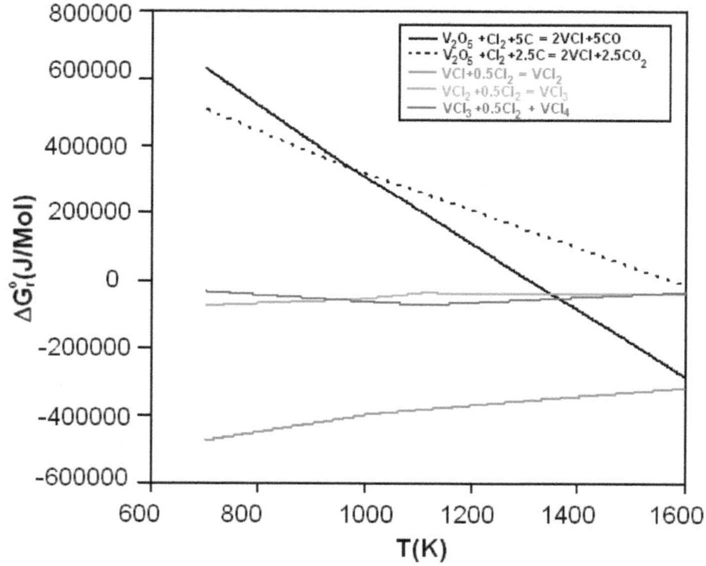

Fig. 20. ΔG_r^o x T for reaction paths of Eq. (38)

The ΔG_r^o x T plots associated with reactions paths represented by mechanisms of Eq. (40) were included on Figure (20). Two inflexion points are evidenced in the diagram of Figure (20). The first one around 1000 K is associated with VCl_2 melting. The second one, around 1100 K, is associated with the sublimation of VCl_3. It can be deduced that only for temperatures greater than 1600 K the path described by Eq. (40) would be possible. For lower temperatures, the molar Gibbs energy of the first step is higher than the one associated with the second.

Another mechanism can be thought for the production of VCl_4. This time, VCl_2 is formed first, which then reacts to give VCl_3 and finally VCl_4 (Eq. 41). The characteristic ΔG_r^o x T curves for the reactions defined in Eq. (41) are presented on Figures (21) and (22).

$$\begin{gathered} V_2O_5 + Cl_2 + C \rightarrow VCl_2 + CO_2 / CO \\ VCl_2 + 0.5Cl_2 \rightarrow VCl_3 \\ VCl_3 + 0.5Cl_2 \rightarrow VCl_4 \end{gathered} \quad (41)$$

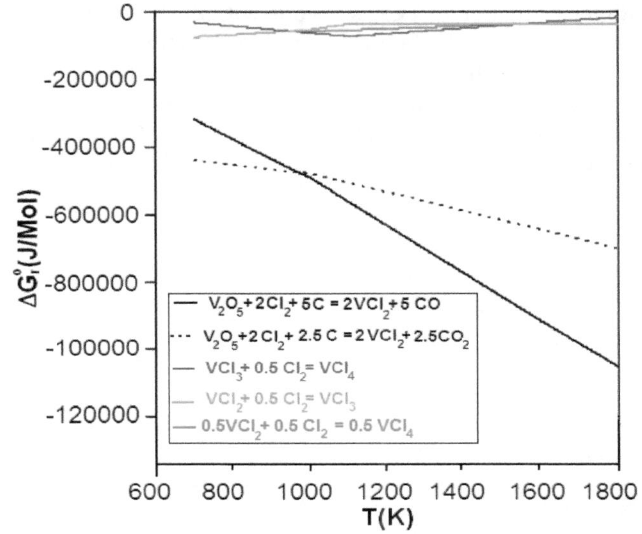

Fig. 21. ΔG_r^o x T for reaction paths of Eq. (41)

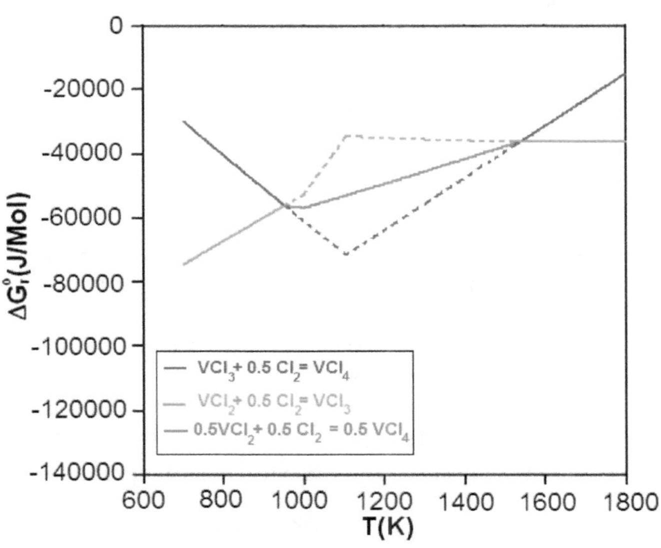

Fig. 22. ΔG_r^o x T for reaction paths of Eq. (41)

The inflexion points have the same meaning as described for diagram of Figure (20). It can be seen that the first step has a much higher thermodynamic tendency as the other. Also, for temperatures lower than 953 K the second step leads to the formation of VCl_3, which then reacts to give VCl_4. However, for temperatures higher than 953 K and lower than 1539 K, the step associated with the formation of VCl_4 is the one with the lowest standard Gibbs energy. So, in this temperature range, VCl_4 should be formed directly from VCl_2, as suggested by Eq. (42). In order to achieve thermodynamic consistency in the mentioned temperature interval, the curves associated with the formation of VCl_3 and VCl_4 according to Eq. (41) should be substituted for the curve associated with reaction defined by Eq. (42), which was represented with red color in the plots presented on Figures (21) and (22).

$$VCl_2 + Cl_2 = VCl_4 \tag{42}$$

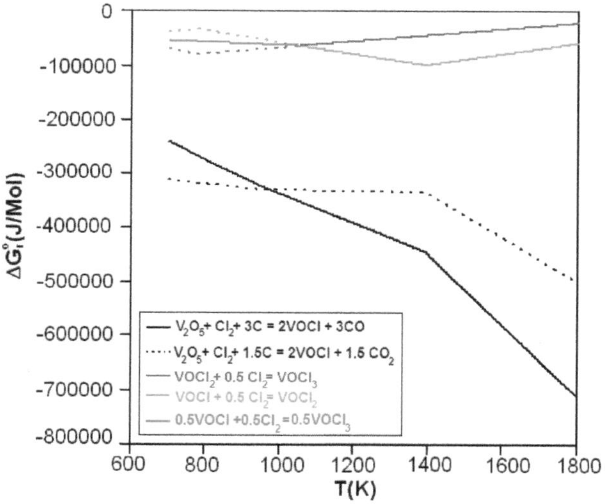

Fig. 23. ΔG_r^o x T for reaction paths of Eq. (43)

For temperatures higher than 1539 K, however, the mechanism is again described by Eq. (41), VCl_3 being formed first, which then reacts leading to VCl_4. It is also interesting to recognize that the sublimation of VCl_3 is responsible for the inversion of the behavior for temperatures higher than approximately 1400 K, where the second reaction step is again the one with the second lowest Gibbs energy of reaction.

On what touches the synthesis of $VOCl_3$, a reaction path can be proposed (Eq. 43), in that VOCl is formed first, which then reacts to give $VOCl_2$, which by itself then reacts to form $VOCl_3$. The ΔG_r^o x T diagrams associated with these reactions are presented on Figure (23).

$$\begin{gathered} V_2O_5 + Cl_2 + C \rightarrow VOCl + CO/CO_2 \\ VOCl + 0.5Cl_2 \rightarrow VOCl_2 \\ VOCl_2 + 0.5Cl_2 \rightarrow VOCl_3 \end{gathered} \tag{43}$$

The inflexion point around 800 K is associated with the sublimation of $VOCl_2$, and around 1400 K with the sublimation of VOCl. According to the ΔG_r^o x T curves presented on Figure (23), it can be deduced that the reaction steps will follow the proposed order only for temperatures higher than 1053 K. At lower temperatures $VOCl_2$ should be formed directly from VOCl (Eq. 44). It is interesting to note that the sublimation of $VOCl_2$ is the phenomenon responsible for the described inversion of behavior. Again, to attain thermodynamic consistency for temperatures higher than 1053 K, the curves associated with the formation of $VOCl_2$ and $VOCl_3$ according to Eq. (43) must be substituted for the curve associated with reaction represented by Eq. (44), which was drawn with red color in the diagram plotted on Figure (23). It should be mentioned indeed, that the reaction equations compared must be written with the same stoichiometric coefficient for Cl_2, or equivalently, the Gibbs energy of reaction (44) must be multiplied by 1/2.

$$VOCl + Cl_2 = VOCl_3 \tag{44}$$

Finally, some remarks may be constructed about the possible reaction order values in relation to Cl_2. According to the discussion developed so far, for the temperature range between 1100 K and 1400 K, Eq. (45) describes the most probable reactions paths for the formation of VCl_4 and $VOCl_3$. As a result, depending on the nature of the slowest step, the reaction order in respect with Cl_2 can be equal to one, two or ½.

$$\begin{aligned} &V_2O_5 + 2Cl_2 + 5C \rightarrow 2VCl_2 + 5CO \\ &VCl_2 + Cl_2 \rightarrow VCl_4 \\ &V_2O_5 + Cl_2 + 5C \rightarrow 2VOCl + 5CO \\ &VOCl + 0.5Cl_2 \rightarrow VOCl_2 \\ &VOCl_2 + 0.5Cl_2 \rightarrow VOCl_3 \end{aligned} \tag{45}$$

3.1.3 Relative stability of VCl_4 and $VOCl_3$

As is evident from the discussion developed on topic (3.1.2), the chlorinated compounds VCl_4 and $VOCl_3$ are the most stable species in the gas phase as the atmosphere becomes concentrated in Cl_2. The relative stability of these two chlorinated compounds will be first accessed on topic (3.1.3.1) by applying the method introduced by Kang & Zuo (1989) and secondly on topic (3.1.3.2) through computing some speciation diagrams for the gas phase.

3.1.3.1 Method of Kang and Zuo

As shown in thon topic (2.2.2) the concentrations of VCl_4 and $VOCl_3$ can be directly computed by considering that each chlorinated compound is generated independently. It will be assumed that the inlet gas is composed of pure Cl_2 ($P(Cl_2)$ = 1 atm). Further, two temperature values were investigated, 1073 K and 1373 K. At these temperatures, the presence of graphite makes the atmosphere richer in CO, so that for the computations the following reactions will be considered:

$$\begin{aligned} &V_2O_5 + 4Cl_2 + 5C \rightarrow 2VCl_4 + 5CO \\ &V_2O_5 + 3Cl_2 + 5C \rightarrow 2VOCl_3 + 3CO \end{aligned} \tag{46}$$

The concentrations of $VOCl_3$ and VCl_4 can then be expressed as a function of P(CO) and temperature according to Eq. (47).

$$P_{VCl_4} = \sqrt{P_{CO}^5 K_1} \rightarrow \ln P_{VCl_4} = \frac{\ln K_1}{2} + \frac{5}{2}\ln P_{CO}$$
$$P_{VOCl_3} = \sqrt{P_{CO}^3 K_2} \rightarrow \ln P_{VOCl_3} = \frac{\ln K_2}{2} + \frac{3}{2}\ln P_{CO} \quad (47)$$

Where K_1 and K_2 represent, respectively, the equilibrium constants for the reactions associated with the formation of VCl_4 and $VOCl_3$ (Eq. 46). By applying Eq. (47) the partial pressure of VCl_4 and $VOCl_3$ were computed as a function of $P(CO)$. The results were plotted on graphic contained in Figure (24). The significant magnitude of the partial pressure values computed for $VOCl_3$ and VCl_4 is a consequence of the huge negative standard Gibbs energy of reaction associated with the formation of these species in the temperature range considered (see Figure 19).

According to Figure (24), VCl_4 is the chloride with the highest partial pressure for both specified temperatures. Also, for both temperatures, $P(VOCl_3)$ becomes higher than $P(VCl_4)$ only for significant values of $P(CO)$. At 1073 K, for example, the partial pressures of the species have equal values only for $P(CO)$ equal to $2.98.10^3$atm, and at 1373 K the same happens for $P(CO)$ equal to $8.91.10^3$atm.

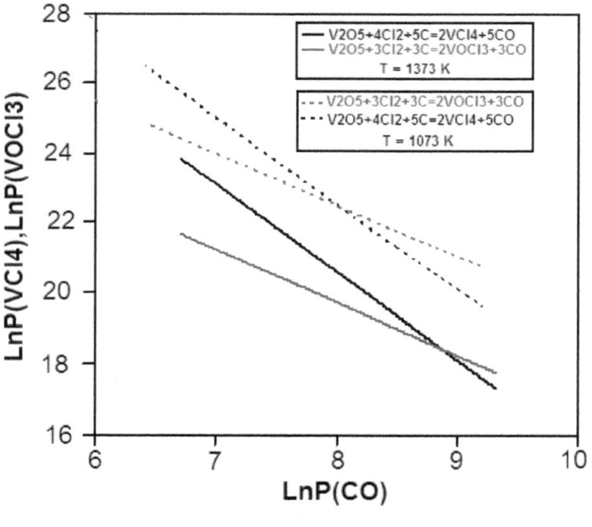

Fig. 24. Ln(P(VCl4)) and Ln(P(VOCl3)) as a function of Ln(P(CO))

As a result, it is expected that the formation of gaseous VCl_4 should have a much greater tendency of occurrence in the temperature range studied. These results will be confirmed through construction of speciation diagrams for the gas phase, a task that will be accomplished on topic (3.1.3.2).

3.1.3.2 Gas phase speciation

The construction of speciation diagrams for the gas phase enables the elaboration of a complementary picture of the chlorination process in question. The word "speciation" means the concentration of all species in gas. This brings another level of complexity to the quantitative description of equilibrium, as the species build a solution, and as so, their concentrations must be determined at the same time. This sort of information can only arises if one solves the system o equilibrium equations associated to all possible chemical reactions involving the species that form the gas. For the present system (V - O - Cl - C) this task becomes very tricky, as the number of possible species present is pretty significant (ex. CO, CO_2, O_2, VCl_2, VCl_3, VCl, VCl_4, $VOCl_3$, VOCl), and so the number of possible chemical reactions connecting them. So, we must think in another route for simultaneously computing the concentration of the gaseous species produced by our chlorination process. The only possible way consists in minimizing the total Gibbs energy of the system (see topic 2.2.3).

The equilibrium state is defined by fixing *T*, *P*, $n(V_2O_5)$, $P(O_2)$ and $P(Cl_2)$. The number of moles of V_2O_5 is fixed at one. If graphite is present in excess, the partial pressure of O_2 is controlled by according to the Boudouard equilibrium (Eq. 39), so that, its presence forces $P(O_2)$ to attain very low values (typically lower than 10^{-20}atm). The total pressure is fixed at 1atm and *T* varies in the range between 1073 K and 1473 K.

An excess of graphite is desirable, so that the chlorination reactions can achieve a considerable driving force at the desired conditions. Computationally speaking, this can be done in two ways. One possibility is to define an amount of carbon much greater than the number of moles of V_2O_5. Other possibility, which has been made accessible through modern computational thermodynamic software, consists in defining the phase "solid graphite" as fixed with a definite amount. The equilibrium compositions (intensive variables) are not a function of the amount of phases present (size of the system), depending only of temperature and total pressure. So we are free to choose any suitable value we desire, such for example zero ($n_{graphite}$ = 0). This last alternative was implemented in the computations conducted in the present topic.

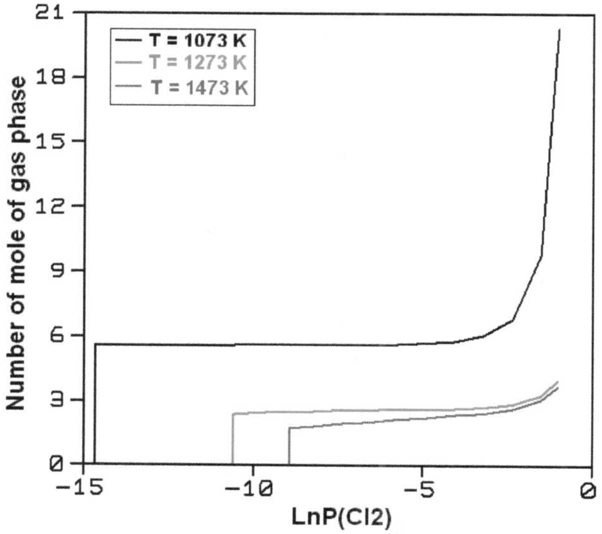

Fig. 25. Number of moles of gas as a function of $P(Cl_2)$

On Figure (25), the number of moles of gas produced was plotted as a function of $P(Cl_2)$ for T equal to 1073 K, 1273 K, and 1473 K. The partial pressure of O_2 was fixed at $1.93.10^{-22}$ atm, and the partial pressure of Cl_2 is varied between $3.6.10^{-7}$atm and 0.61 atm.

Each curve is defined by three stages. First, for very low values of $P(Cl_2)$, no gas is formed. At this conditions VCl_2(l) is present in equilibrium with graphite. The equilibrium ensemble does not experience any modification until a critical $P(Cl_2)$ value is reached, at which a discontinuity can be evidenced. The gas phase appears in equilibrium and for any $P(Cl_2)$ higher than the critical one, the number of moles VCl_2(l) becomes equal to zero. This condition defines the second stage, where for higher $P(Cl_2)$ values the gas composition changes accordingly, through forming of chlorides and oxychlorides. Finally, a $P(Cl_2)$ value is reached, where all capacity of the system for forming chlorinated compounds is exhausted, and the effect of adding more Cl_2 is only the dilution of the chlorinated species formed. As a consequence, the number of mole of gas phase experiences a significant elevation. At 1073 K, for example, Figure (26) describes the effect of $P(Cl_2)$ over the gas phase composition during the second and third stages. We see that the mol fraction of VCl_4 raises (second stage) and after achieving a maximum value starts to decrease (third stage). The concentration variations during the second sage can be ascribed to the occurrence of reactions represented by Eq. (48), which have at 1073 K equilibrium constants much higher than unity (Table 3). The reduction of the mol fraction of $VOCl_3$ can be understood as a dilution effect, which is motivated by the elevation of the mol fractions of VCl_4 and Cl_2.

$$\begin{aligned} VCl_2 + Cl_2 &\rightarrow VCl_4 \\ VCl_3 + 0.5Cl_2 &\rightarrow VCl_4 \end{aligned} \qquad (48)$$

K	Chemical reaction
$1.95.10^5$	$VCl_2 + Cl_2 \rightarrow VCl_4$
$2.12.10^3$	$VCl_3 + 0.5Cl_2 \rightarrow VCl_4$

Table 3. Equilibrium constants at 1073 K for the reactions represented by Eq. (48)

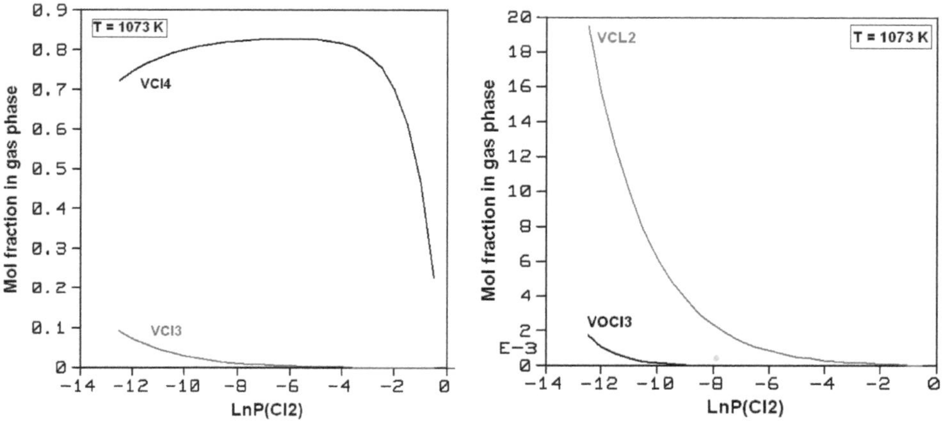

Fig. 26. Concentration of vanadium chlorides and oxychlorides as a function of $P(Cl_2)$ at 1073 K

Besides $P(Cl_2)$, temperature should also have an effect over the composition of the gas phase. This was studied as follows. Six temperature values were chosen in the range between 1073 K and 1473 K. Next, for each temperature the critical $P(Cl_2)$ value (the one associated with the formation of the first gas molecules) is identified. The composition of the most stable gaseous species is then computed and is presented in Table (4). During the calculations the partial pressure of O_2 was fixed at $1.93.10^{-22}$atm.

T(K)	X(CO)	X(CO_2)	X($VOCl_3$)	X(VCl_2)	X(VCl_3)	X(VCl_4)
1073	0.16	$3.64.10^{-3}$	$1.95.10^{-2}$	$1.74.10^{-3}$	$9.27.10^{-2}$	0.72
1100	0.12	$1.23.10^{-3}$	$1.02.10^{-2}$	$2.61.10^{-3}$	0.12	0.75
1200	$4.21.10^{-2}$	$3.36.10^{-5}$	$1.08.10^{-3}$	$9.87.10^{-3}$	0.26	0.69
1300	$1.76.10^{-2}$	$1.60.10^{-6}$	$1.45.10^{-4}$	$2.98.10^{-2}$	0.43	0.52
1373	$1.00.10^{-2}$	$2.29.10^{-7}$	$3.69.10^{-5}$	$5.97.10^{-2}$	0.56	0.37
1400	$8.26.10^{-3}$	$1.17.10^{-7}$	$2.27.10^{-5}$	$7.6.10^{-2}$	0.59	0.32
1473	$5.07.10^{-3}$	$2.18.10^{-8}$	$6.22.10^{-6}$	0.14	0.66	0.19

Table 4. Composition of the "first" gas formed as a function of temperature

As expected, the mol fraction of CO is greater than the mol fraction of CO_2 for the entire temperature range studied. Also, the chloride VCl_4 has the highest concentration at 1073 K, a phase which occupies a large area of the predominance diagram at this temperature (Figure 14). As temperature attains higher values, the mol fraction of VCl_4 and $VOCl_3$ become progressive lower and the atmosphere more concentrated in VCl_2 and VCl_3. So, at 1473 K the situation is significant different from the equilibrium state observed at 1073 K. Such behavior is again consistent with the information contained on the predominance diagrams (Figures 14, 15 and 16) where can be seen that the stability fields of VCl_4(g) and $VOCl_3$(g) shrink while the area representing the phase VCl_3(g) grows. At 1573 K it occupies a visible amount of the diagrams space (Figure 16).

It is worthwhile to mention that a more detailed look on the results seems to incorporate apparent inconsistencies. i) The minimum partial pressure of Cl_2 for the formation of pure VCl_4(g) at 1073 K (Figure 14) is higher than the critical pressure for the formation of the first gaseous species at this temperature (Figure 25). ii) Measurable amounts of VCl_3 (greater or equal to 0.1) were detected for temperatures higher than 1100 K (Table 3) but no VCl_3(g) field was observed in the predominance diagram computed at 1273 K (Figure 15). iii) Also, no field associated with the formation of VCl_2(g) could be detected even at 1573 K (Figure 16) but the speciation computation predicts its presence in measurable amounts at the last temperature ($x(VCl_2)$ = 0.14) (Table 3). All these thermodynamic values differences are a consequence of the fact that the pure molar Gibbs energy of each component is higher than its chemical potential in the ideal gas solution, the former model being used for the predominance diagrams construction while the later is applied to the speciation calculations. Therefore, the driving force for the formation of the gaseous compounds is reduced accordingly to Eq. (49).

$$\mu^{g}_{VCl_3} - g^{g}_{VCl_3} = RT\ln x^{g}_{VCl_2} < 0 \tag{49}$$

Another possible type of computation is to study the effect of $P(O_2)$ over the composition of the gas phase. This variable is restricted by the fact that the amount of graphite phase is fixed. So there is a maximum value of $P(O_2)$ at each temperature for which the thermodynamic modeling remains consistent and the computation can be performed. By fixing the temperature at 1373K, the upper limit for $P(O_2)$ was equal to to $1.56.10^{-18}$atm and the value of $P(Cl_2)$ associated with the appearance of the first gaseous molecules is identified as $2.05.10^{-4}$atm. The composition of the gas is then computed by fixing $P(Cl_2)$ at $2.05.10^{-4}$atm. Three different $P(O_2)$ levels were studied, $1.3.10^{-24}$, $5.4.10^{-22}$ and $1.56.10^{-18}$atm. The results are presented on Table (5).

$P(O_2)$(atm)	$X(CO)$	$X(CO_2)$	$X(VOCl_3)$	$X(VCl_2)$	$X(VCl_3)$	$X(VCl_4)$
$1.30.10^{-24}$	$8.22.10^{-4}$	$1.54.10^{-9}$	$3.05.10^{-6}$	0.0597	0.56	0.38
$5.24.10^{-22}$	$1.65.10^{-2}$	$6.22.10^{-7}$	$6.03.10^{-5}$	0.0588	0.55	0.37
$1.56.10^{-18}$	0.902	$1.85.10^{-3}$	$3.21.10^{-4}$	$5.72.10^{-3}$	0.054	0.036

Table 5. Gas phase speciation as function of $P(O_2)$

The mol fractions of CO and CO_2 gets higher for higher values of $P(O_2)$. This is consistent with the dislocation of the equilibrium represented by Eq. (50) in the direction of the formation of the two carbon oxides.

$$\begin{aligned} C + O_2 &\rightarrow CO_2 \\ 2C + O_2 &\rightarrow 2CO \end{aligned} \tag{50}$$

Also, the Boudouard equilibrium demands that at the chosen temperature (1373 K) the atmosphere is more concentrated in CO. This was indeed observed for each equilibrium state investigated. It is interesting to observe that for $P(O_2)$ varying between $5.24.10^{-22}$atm and $1.56.10^{-18}$atm the mol fraction of CO and CO_2 experience a much higher variation in comparison with the one observed for lower $P(O_2)$ values.

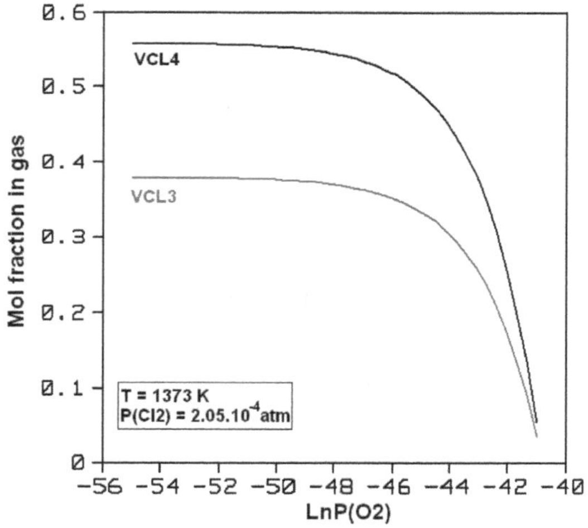

Fig. 27. Mol fractions of VCl_3 and VCl_4 as a function of $P(O_2)$

In the case of the vanadium chlorides and oxychlorides an interesting trend is evidenced. The concentration of $VOCl_3$ grows and of VCl_2, VCl_3 and VCl_4 reduce appreciably for the same O_2

partial pressure range. The concentration variations associated with the vanadium chlorinated compounds is analogous to the variations observed in the concentrations of CO and CO_2.

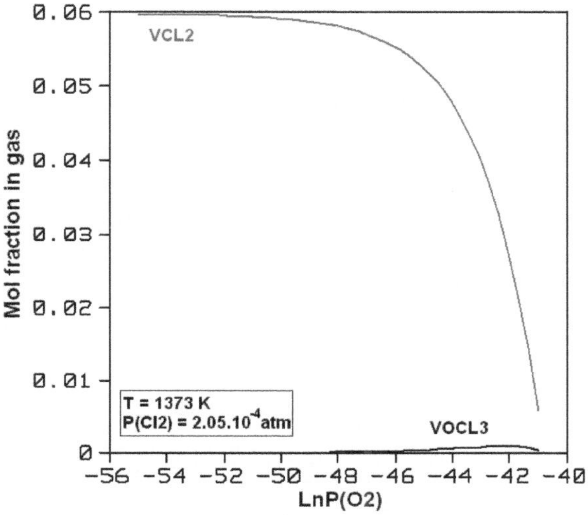

Fig. 28. Mol fractions of $VOCl_3$ and VCl_2 as a function $P(O_2)$

For $P(O_2)$ lower than $5.24.10^{-22}$atm the variations are much less significant. To get a better picture of the trend observed for the chlorides and oxychlorides, their concentrations were plotted as a function of $P(O_2)$, which was varied in the range spanned by the data of Table (5) (Figures 27, 28 and 29)

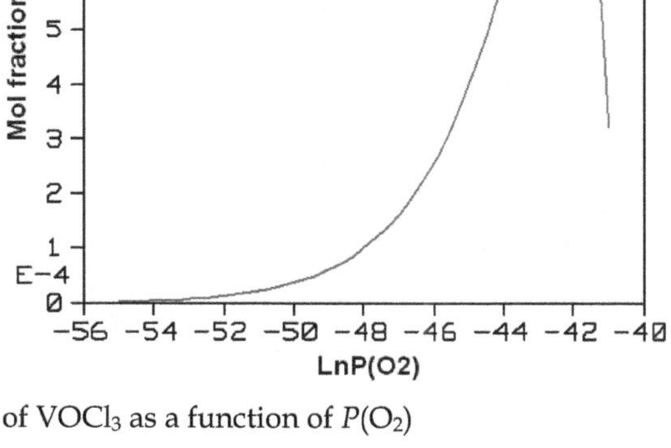

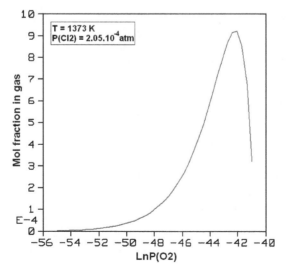

Fig. 29. Mol fraction of $VOCl_3$ as a function of $P(O_2)$

The variations depicted on Figures (27), (28) and (29) are consistent with the occurrence of reactions represented by Eq. (51). As $P(O_2)$ achieves higher values, it reacts with VCl_3, VCl_2 and or VCl_4 resulting in $VOCl_3$. Such phenomena could explain the significant reduction of VCl_3, VCl_4 and VCl_2 concentrations, and the concomitant elevation of the $VOCl_3$ mol fraction.

$$\begin{aligned} VCl_3 + 0.5O_2 &= VOCl_3 \\ VCl_2 + O_2 + VCl_4 &= 2VOCl_3 \\ VCl_4 + 0.5O_2 &= VOCl_3 + 0.5Cl_2 \end{aligned} \tag{51}$$

The participation of VCl_4 in the second reaction is supported by the fact that its equilibrium concentration lowering is more sensible to $LnP(O_2)$ than observed for VCl_3 (Figure 27). The consumption of VCl_4 by the second reaction is also consistent with the maximum observed in the curve obtained for $VOCl_3$ concentration (Figure 29). As less VCl_4 is available, less $VOCl_3$ can be produced.

K	Chemical reaction
$8.01.10^6$	$VCl_3 + 0.5O_2 = VOCl_3$
$1.89.10^{13}$	$VCl_2 + O_2 + VCl_4 = 2VOCl_3$
$1.01.10^5$	$VCl_4 + 0.5O_2 = VOCl_3 + 0.5Cl_2$

Table 6. Equilibrium constants at 1373 K for reactions represented by Eq. (51)

The occurrence of reactions represented by Eq. (51) is supported by classical thermodynamics, as the equilibrium constant (*K*) computed at 1373 K for all chemical reactions above assume values appreciably greater than unity (see Table 6).

3.1.3.3 V2O5 chlorination enthalpy

For the implementation of an industry process based on chemical reaction is fundamental to know the amount of heat generated or absorbed from that. Exothermic processes (heat is released) reach higher temperatures, and frequently demand engineering solutions for protecting the oven structure against the tremendous heat generated by the chemical phenomena. In this context, endothermic processes (heat is absorbed) are easier controlled, but the energy necessary to stimulate the reaction must be continuously supplied, making the energy investment larger.

The variation of the total enthalpy of the system for the chlorination process in question was calculated as a function of $P(Cl_2)$ (Figure 30). The partial pressure of O_2 was forced to be equal to $1.93.10^{-22}$atm and four temperature levels were studied, 1000 K, 1100 K, 1300 K and 1700 K. It can be seen that the total enthalpy for the process conducted at 1000 K reduces with the advent of the chlorination reactions, indicating that the chlorination process is exothermic. However, the molar enthalpy magnitude is progressively lower up to a certain temperature where it is zero. Above that, the molar reaction enthalpy becomes positive, and Figure (30) illustrates its value for 1700 K.

This is perfectly consistent with the results presented on Table (3). As temperature gets higher, the mol fractions of VCl_4 and $VOCl_3$ reduce and that for VCl_3 and VCl_2 experience a significant elevation. For some temperature between 1300 K and 1373 K the mol fractions of VCl_4 and VCl_3 assume equal values. This point is related to the condition where the chlorination enthalpy is zero. For higher temperatures, where $x(VCl_4) < x(VCl_3)$ the process becomes progressively more endothermic. It is interesting to see that he explained behavior is consistent with the fact that the global formation reactions of VCl_3 and VCl_2 are associated with positive molar reaction enthalpies and that of $VOCl_3$ and VCl_4 with negative molar reactions enthalpies (Figure 31).

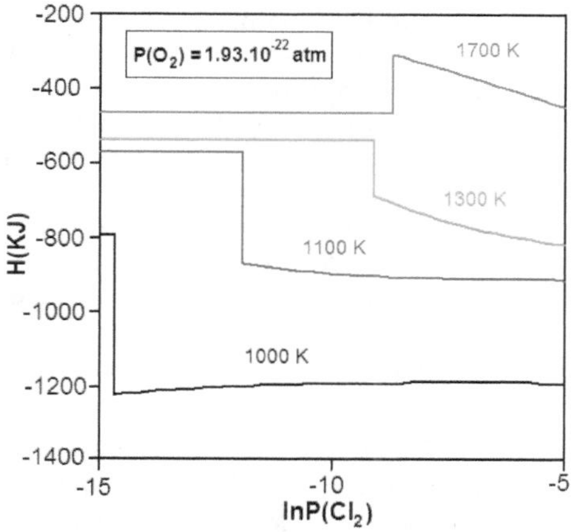

Fig. 30. Total as a function of $P(Cl_2)$

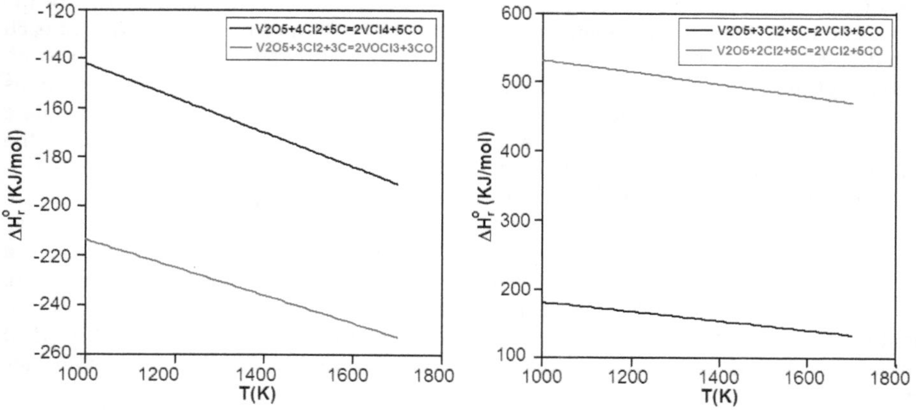

Fig. 31. Molar reaction enthalpy for the formation of gaseous VCl_3, VCl_2, VCl_4 and $VOCl_3$ as a function of temperature

4. Final remarks

In this chapter three different approaches to the chlorination equilibrium study of an oxide were presented. The first two are based on the construction of ΔG_r^o x T diagrams (topic 2.2.1) and on the calculations, first introduced by Kang & Zuo (1989) (topic 2.2.2), respectively. Both of them take into consideration that each chlorinated compound is produced independently. The third one has its fundamental based on the total Gibbs energy minimization of the reaction system and the gas phase equilibrium composition is calculated considering that the formed species are produced simultaneously (topic 2.2.3).

The method based on the construction of ΔG_r^o x T was applied on topic (3.1.2) for studying the thermodynamic viability of the reaction between gaseous Cl_2 and V_2O_5. The discussion evidenced that the chlorination is thermodynamically feasible only in the presence of a reducing agent (graphite in the case of the present work) and was initially focused on the production of VCl_4 and $VOCl_3$. The same approach was employed for studying the possible mechanisms associated with the formation of VCl_4 and $VOCl_3$. According to the results (topic 3.1.2.1), the synthesis of these two compounds is subdivided in different stages, which can vary in nature, depending on the temperature range considered. In global terms though, both $VOCl_3$ and VCl_4 have molar reaction Gibbs energies of the same magnitude order. So, it was not possible to clearly identify, which one of them should be produced in greater quantities. The problem of the relative stability between $VOCl_3$ and VCl_4 was then addressed by the implementation of the method of Kang & Zuo (1989) (topic 3.1.3.1). The results indicated that VCl_4 should have a higher concentration in comparison with $VOCl_3$ in the temperature range between 1073 K and 1373 K.

It can de said that both, the method based on the ΔG_r^o x T diagrams construction as well as the Kang & Zuo method (1989), incorporate some simplifications and are very easy to implement. However, they lead to only a superficial knowledge of the true nature of the equilibrium state achievable. Thanks to the development of computational thermodynamic software, of which *Thermocalc* is a good example, more complex computations can be realized. For example, by allowing the chlorides and oxychlorides to build a gaseous solution, the minimization of the total Gibbs energy of the system results in the direct computation of the mol fraction of each chlorinated species present in the gas phase (topic 3.1.3.2). This method can be seen as an improvement of the idea put forward by Kang & Zuo (1989), in that all equilibrium equations are solved simultaneously, with the further advantage that one does not need to formulate a group of independent reactions that cover all possible chemical interactions among the components, a task that can become very complex for metals, as in case of vanadium, which can produce a family of chlorides and oxychlorides.

The conclusion that graphite strongly promotes the thermodynamic driving force necessary to chlorination and that VCl_4 should be formed preferentially in relation to $VOCl_3$ are perfectly consistent with the results based on the total Gibbs energy minimization. However, by the application of this last method it was possible to go a little further, through investigation of the effect of $P(Cl_2)$ over the chlorination enthalpy and by studying the effect of temperature, Cl_2 and O_2 partial pressures over the concentrations of vanadium chlorides and oxychlorides in the gas phase.

The predictions associated with the effect of temperature over the gas phase speciation (Table 4) indicate that the mol fractions of VCl_2 and VCl_3 grow significantly in the range between 1073 K and 1473 K and, as a result, the concentrations of $VOCl_3$, VCl_4, CO and CO_2 exhibit a significant reduction (topic 3.1.3.2). This finding agrees with the tendency depicted by the predominance diagrams constructed for the system V - O - Cl, where the VCl_4(g) and $VOCl_3$(g) fields shrink and that of VCl_3(g) grows (Figures 14, 15 and 16). Also, the calculated mol fraction of CO is at all temperatures much higher than the mol fraction of CO_2, a fact that is consistent with the establishment of the Bourdouard equilibrium for temperatures higher than 973 K, where the concentrations of the two mentioned carbon oxides have the same magnitude.

The fact that the speciation computation indicates appreciable amounts of VCl_3 for temperatures higher than 1100 K and of VCl_2 for temperatures higher than 1473 K, apparently contradicting the information contained in the predominance diagrams of Figures (15) and (16), is a mere consequence of the fact that on topic (3.1.3.2) the gaseous chlorides build an ideal gas solution. The chemical potentials of VCl_3 and also of VCl_2 are lower than their pure molar Gibbs energies. The species become more stable in the gaseous solution in relation to the pure state, and their mol fractions assume higher values for the same temperature imposed. The same idea explains why VCl_4 is formed in significant amounts at 1073 K for a $P(Cl_2)$ value lower than the one observed in the predominance diagram of Figure (14) for the equilibrium between VCl_2(l) and VCl_4(g).

The effect of adding more Cl_2 after all vanadium has been converted to gaseous chlorinated compounds is also consistent with the expectations. At 1073 K the results indicate that the mol fraction of VCl_4 grows while all other relevant chlorinated species reduces (topic 3.1.3.2). This can be explained by the reaction of VCl_2 and VCl_3 with Cl_2 resulting in VCl_4, which have a significant negative driving force at 1073 K (Table 3).

The study of the impact of varying $P(O_2)$ over the gas phase composition at 1373 K indicated that the mol fractions of CO and CO_2 experience significant elevation as $P(O_2)$ becomes higher, a fact that is also observed in the case of $VOCl_3$ (Table 4). The concentration of all other chlorinated compounds reduces for the same studied range of $P(O_2)$. The influence of the oxygen chemical activity over the gas phase speciation can be explained by a group of proposed reactions between VCl_4, VCl_3 and VCl_2 with O_2 resulting in $VOCl_3$ (topic 3.1.3.2). All these reactions have equilibrium constants much higher than one, indicating an expressive thermodynamic driving force at 1373 K (Table 6).

The conclusions about the exothermic nature of the chlorination process in the temperature range between 1000 K and 1300 K and the observation that it becomes progressively more endothermic as 1700 K is approached (topic 3.1.3.3), are perfectly consistent with the fact that the atmosphere becomes progressively diluted in VCl_4 and $VOCl_3$, whose formations are associated with negative molar enthalpies and becomes richer in VCl_2 and VCl_3, whose molar enthalpy of formation are considerably positive (Figure 31).

Finally, we can conclude that the study of the equilibrium states achievable through the reaction between a transition metal oxide and gaseous Cl_2, can be now approached through the implementation of methods of different complexity levels. The most general one, in which the total Gibbs energy of the reaction system is minimized, enables the construction of a more detailed picture of the equilibrium state. However, as it is evident from the comparisons explained above, the most general method must be consistent with the tendencies predicted by simpler calculations.

5. References

Allain E., Djona M., Gaballah I. Kinetics of Chlorination and Carbochlorination of Pure Tantalum and Niobium Pentoxides. Metallurgical and materials transactions B, v. 28, p. 223 - 232, 1997.

Brewer L., Ebinghaus, B. B. The thermodynamics of the solid oxides of vanadium. Thermochimica Acta, v. 129, p. 49 – 55, 1988.

Brocchi, E. A.; Moura, F. J. Chlorination methods applied to recover refractory metals from tin slags. Minerals Engineering, v. 21, n. 2, p. 150-156, 2008.

Cecchi, E. et al. A feasibility study of carbochlorination of chrysotile tailings. International Journal of Mineral Processing, v. 93, n. 3 - 4, p. 278-283, 2009.

Esquivel, M. R., Bohé, A. E., Pasquevich, D. M. Carbochlorination of samarium sesquioxide. Thermoquimica Acta, v. 403, p. 207 – 218, 2003.

Gaballah, I., Djona, M. Recovery of Co, Ni, Mo, and V from unroasted spent hydrorefining catalysts by selective chlorination. Metallurgical and Materials Transactions B, v. 26, n. 1, p. 41-50, 1995.

Gaviria, J. P., Bohe, A. E. Carbochlorination of yttrium oxide. Thermochimica Acta, v. 509, n. 1-2, p. 100-110, 2010.

Gonzallez J. et al. ß-Ta_2O_5 Carbochlorination with different types of carbon. Canadian Metallurgical Quarterly, v. 41, n.1, p. 29 - 40, 2002.

Hackert A., Plies V. and Gruehn R. Nachweis und thermochemische Charakterisierung des Gasphasenmolekuls VOCl. Zeitschrift für anorganische und allgemeine Chemie, v. 622, p. 1651-1657, 1996.

Hildenbrand, D. L. et al. Thermochemistry of the Gaseous Vanadium Chlorides VCl, VCl_2, VCl_3, and VCl_4. Journal of Physical Chemistry A, v. 112, p. 9978 - 9982, 2008.

Jena, P. K.; Brocchi, E. A.; Gonzalez, J. Kinetics of low-temperature chlorination of vanadium pentoxide by carbon tetrachloride vapor. Metallurgical and Materials Transactions B, v. 36, n. 2, p. 195-199, 2005.

Kang, S. X.; Zuo, Y. Z. Chloridizing roasting of complex material containing low tin and high iron at high temperature. Kunming Metall. Res. Inst. Report, v. 89, n. 3, 1989.

Kellog, H. H. Thermodynamic relationships in chlorine metallurgy. Journal of metals, v. 188, p. 862 - 872, 1950.

Mackay, D. Is chlorine the evil element? Enviromental Science & Engineering, p. 49 - 52, 1992.

McCarley R. E., Roddy J. W. The Vapor Pressures of Vanadium(II) Chloride, Vanadium(III) Chloride, Vanadium(II) Bromide, and Vanadium(III) Bromide by Knudsen Effusion. Inorganic Chemistry, v. 3. n. 1, p. 60 - 63, 1964.

Micco, G.; Bohe, A. E.; Sohn, H. Y. Intrinsic Kinetics of Chlorination of WO_3 Particles With Cl_2 Gas Between 973 K and 1223 K (700°C and 950 °C. Metallurgical and Materials Transactions B, v. 42, n. 2, p. 316-323, 2011.

Murase, K. et al. Recovery of vanadium, nickel and magnesium from a fly ash of bitumen-in-water emulsion by chlorination and chemical transport. Journal of Alloys and Compounds, v. 264, n. 1 - 2, p. 151-156, 1998.

Neff, D. V. Environmentally acceptable chlorination processes. Aluminium Cast House Technology: Theory & Practice, Australasian, Asian, Pacific Conference, p. 211-225, 1995.

Oheda, M. W., Rivarola, J. B., Quiroga, O. D. Study of the chlorination of molybdenum trioxide mixed with carbon black. Minerals Engineering, v. 15, p. 585 - 591, 2002.

Oppermann, H. Preparation and properties of VCl_4, VCl_3, and $VOCl_3$. Zeitschrift für Chemie, v. 2, p. 376 - 377, 1962.

Oppermann H. Gleichgewichte mit $VOCl_3$, VO_2Cl, und $VOCl_2$. Zeitschrift für anorganishce und algemeine Chemie, v. 331, n. 3 - 4, p. 113 - 224, 1967.

Patel, C. C., Jere, G. V. Some Thermodynamical considerations in the chlorination of Ilmetite. Transactions of the metallurgical society of AIME, v. 218, p. 219 - 225, 1960.

Pilgrim, R. F., Ingraham, T. R. Thermodynamics of chlorination of iron, cobalt, nickel and copper sulphides. Canadian Metallurgical Quarterly, v. 6, n. 4, p. 333 - 346, 1967.

Robert, T. D. Thermodynamics in materials science, 1993.

Sano N., Belton, G. The thermodynamics of chlorination of Vanadium Pentoxide. Transactions of the Japan Institute of Metals, v.21, n. 9, p. 597 - 600, 1980.

Schäffer H., Wartenpfuhl H. Über die Herstellung von VOCl. Journal of the less common metals, v. 3, p. 29 - 33, 1961.

Zhang, L. et al. Rare earth extraction from bastnaesite concentrate by stepwise carbochlorination - chemical vapor transport-oxidation. Metallurgical and Materials Transactions B, v. 35, n. 2, p. 217-221, 2004.

7

Thermodynamics of Reactions Among Al_2O_3, CaO, SiO_2 and Fe_2O_3 During Roasting Processes

Zhongping Zhu, Tao Jiang, Guanghui Li, Yufeng Guo and Yongbin Yang
School of Minerals Processing & Bioengineering, Central South University, Changsha, Hunan 410083, China

1. Introduction

The thermodynamic of the chemical reactions among Al_2O_3, CaO, SiO_2 and Fe_2O_3 in the roasting processes was investigated in this chapter. The chemical reactions are classified into SiO_2-Al_2O_3 system, Fe_2O_3-Al_2O_3 system, SiO_2-Fe_2O_3 system, CaO-Al_2O_3 system, SiO_2-CaO system, SiO_2-calcium aluminates system, CaO-Fe_2O_3 system, Al_2O_3-calcium ferrites system and Al_2O_3-CaO-SiO_2-Fe_2O_3 system. When the roasting temperature is over 1100K, $3Al_2O_3 \cdot 2SiO_2$ is preferentially formed in SiO_2-Al_2O_3 system; $FeO \cdot Al_2O_3$ can be formed in Fe_2O_3-Al_2O_3 system; ferric oxide and SiO_2 could not generate iron silicate; $12CaO \cdot 7Al_2O_3$ is preferentially formed in CaO-Al_2O_3 system when one mole Al_2O_3 reacts with CaO; $2CaO \cdot SiO_2$ is preferentially formed in SiO_2-CaO system; except for $CaO \cdot 2Al_2O_3$ and $CaO \cdot Al_2O_3$, the other calcium aluminates can transform into calcium silicate by reacting with SiO_2 in SiO_2-calcium aluminates system; $2CaO \cdot Fe_2O_3$ is preferentially formed in CaO-Fe_2O_3 system; alumina is unable to form $3CaO \cdot Al_2O_3$ with calcium ferrites($2CaO \cdot Fe_2O_3$ and $CaO \cdot Fe_2O_3$), but able to form $12CaO \cdot 7Al_2O_3$ with $2CaO \cdot Fe_2O_3$; when CaO, Fe_2O_3, Al_2O_3, SiO_2 coexist, they are more likely to form ternary compound $2CaO \cdot Al_2O_3 \cdot SiO_2$ and $4CaO \cdot Al_2O_3 \cdot Fe_2O_3$.

2. Binary compounds

2.1 Fe_2O_3-Al_2O_3-$CaCO_3$ system

Fe_2O_3 and Al_2O_3 can all react with limestone during roasting to generate corresponding aluminates and ferrites. In Fe_2O_3-Al_2O_3-CaO system, the reaction Fe_2O_3 and Al_2O_3 with $CaCO_3$ coexist, and the reactions equations are as followed:

Reactions	A, J/mol	B, J/K.mol	Temperature, K
$CaCO_3+Al_2O_3=CaO \cdot Al_2O_3+CO_2$	161088.3	-244.1	298~1200
$CaCO_3+Fe_2O_3=CaO \cdot Fe_2O_3+CO_2$	151677.8	-220.9	298~1200

Table 1. The ΔG_T^θ of Fe_2O_3-Al_2O_3-$CaCO_3$ system （$\Delta G_T^\theta = A + BT$, J/mol; P_{CO_2} =30Pa, i.e., the partial pressure of CO_2 in the air)

The relationships between Gibbs free energy (ΔG_T^θ) and temperature (T) are as shown in figure 1.

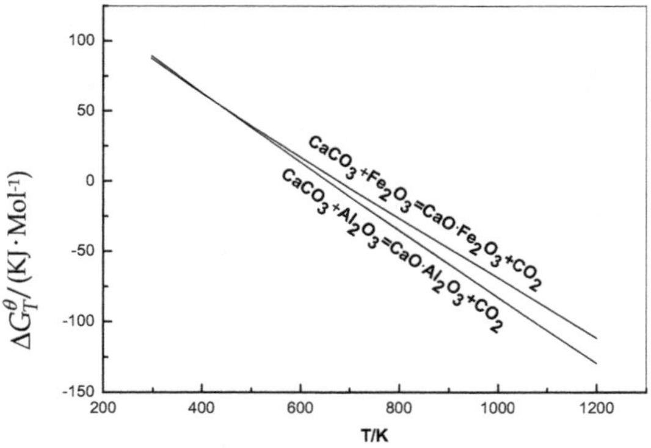

Fig. 1. Relationships between ΔG_T^θ and temperature in Fe_2O_3-Al_2O_3-$CaCO_3$ system

Figure 1 shows that, the Gibbs free energy of reactions on Fe_2O_3 and Al_2O_3 with $CaCO_3$ decreased with the rise of temperature in normal roasting process (due to decomposition of $CaCO_3$ over 1200K, so the curve has no drawing above 1200K), and the reactions all can automatically react to generate the corresponding calcium aluminate and calcium ferrite. The ΔG_T^θ of reaction with Al_2O_3 is more negative than the ΔG_T^θ of reaction with Fe_2O_3 at the same temperature. $CaCO_3$ has actually decomposed at 1473~1673K industrial roasting temperature, therefore, only CaO is taken into account on the following analysis.

2.2 SiO_2-Al_2O_3 system

SiO_2 mainly comes from the ore and coke ash in the roasting process. SiO_2 reacts with Al_2O_3 to form aluminium silicates. The aluminium silicates mainly include $Al_2O_3{\cdot}2SiO_2(AS_2)$, $Al_2O_3{\cdot}SiO_2$(AS,andalusite), AS(kyanite), AS(fibrolite), $3Al_2O_3{\cdot}2SiO_2(A_3S_2)$. Thermodynamic calculation indicates that, AS_2 can not be formed from the reaction of Al_2O_3 and SiO_2 under the roasting condition. The others equations are shown in table 2.

Reactions	A, J/mol	B, J/K.mol	Temperature, K
$Al_2O_3+SiO_2=Al_2O_3{\cdot}SiO_2$ (kyanite)	-8469.3	9.0	298~1696
$Al_2O_3+SiO_2=Al_2O_3{\cdot}SiO_2$ (fibrolite)	-4463.8	-0.9	298~1696
$Al_2O_3+SiO_2=Al_2O_3{\cdot}SiO_2$ (andalusite)	-6786.1	0.6	298~1696
$\frac{3}{2}Al_2O_3+SiO_2=(\frac{1}{2})3Al_2O_3\cdot 2SiO_2$	12764.7	-16.7	298~1696

Table 2. The ΔG_T^θ of Al_2O_3-SiO_2 system ($\Delta G_T^\theta = A + BT$, J/mol)

The relationships of ΔG_T^θ and temperature in Al_2O_3-SiO_2 system is shown in figure 2.

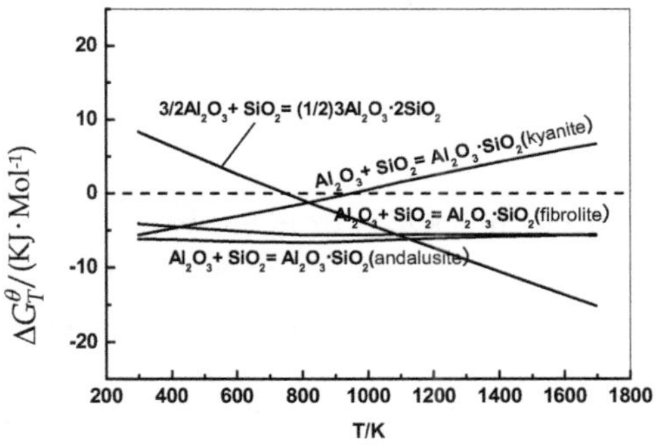

Fig. 2. Relationships of ΔG_T^θ and temperature in Al_2O_3-SiO_2 system

Figure 2 shows that, the ΔG_T^θ of kyanite is greater than zero at 1000~1700K, so the reaction cannot happen; the ΔG_T^θ of andalusite and fibrolite alter little with temperature changes; the ΔG_T^θ of A_3S_2 decreases with the rise of temperature. The thermodynamic order of forming aluminium silicates is A_3S_2, AS(andalusite), AS(fibrolite) at 1100~1700K.

2.3 Fe_2O_3-Al_2O_3 system

Al_2O_3 does not directly react with Fe_2O_3, but Al_2O_3 may react with wustite (FeO) produced during roasting process to form $FeO \cdot Al_2O_3$. No pure ferrous oxide (FeO) exists in the actual process. The ratio of oxygen atoms to iron atoms is more than one in wustite, which is generally expressed as Fe_xO(x=0.83~0.95), whose crystal structure is absence type crystallology. For convenience, FeO is expressed as wustite in this thesis. Al_2O_3 may react with wustite(FeO) to form $FeO \cdot Al_2O_3$ in the roasting process. The relationship of ΔG_T^θ and temperature is shown in figure 2, and the chemical reaction of the equation is as followed:

$$Al_2O_3+FeO=FeO \cdot Al_2O_3 \quad \Delta G_T^\theta = -30172.2+9.3T, \text{J.mol}^{-1} \quad 843\sim1650K \qquad (1)$$

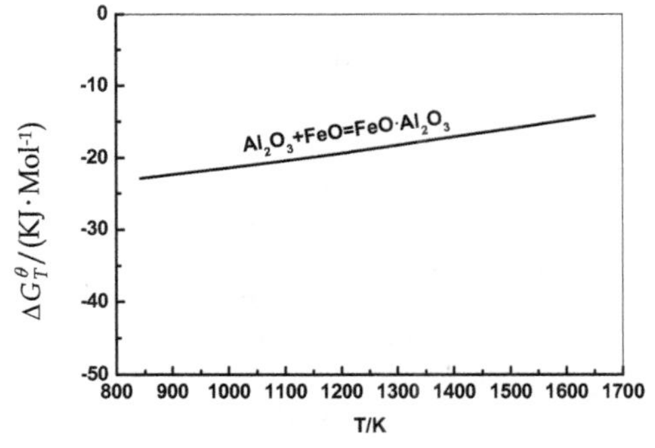

Fig. 3. Relationship of ΔG_T^θ and temperature in Fe_2O_3-Al_2O_3 system

Figure 3 shows that, the ΔG_T^θ is negative at 843~1650K, reaction can happen and generate $FeO \cdot Al_2O_3$; the ΔG_T^θ rises with the temperature, the higher temperature is, the lower thermodynamic reaction trends.

2.4 SiO_2-Fe_2O_3 system

SiO_2 also does not directly react with Fe_2O_3, but Al_2O_3 may react with wustite (FeO) to form $FeO \cdot SiO_2$ (FS) and $2FeO \cdot SiO_2(F_2S)$. The relationships of ΔG_T^θ and temperature is shown in figure 4, and the chemical reactions of the equations are shown in table 3.

Reactions	A, J/mol	B, J/K.mol	Temperature, K
$FeO+SiO_2 = FeO \cdot SiO_2$	26524.6	18.8	847~1413
$2FeO+SiO_2 = 2FeO \cdot SiO_2$	-13457.3	30.3	847~1493

Table 3. The ΔG_T^θ of SiO_2- Al_2O_3 system ($\Delta G_T^\theta = A + BT$, J/mol)

Figure 4 shows that, the ΔG_T^θ of SiO_2- Al_2O_3 system are above zero at 847~1500K, so all of the reactions can not happen to form ferrous silicates (FS and F_2S).

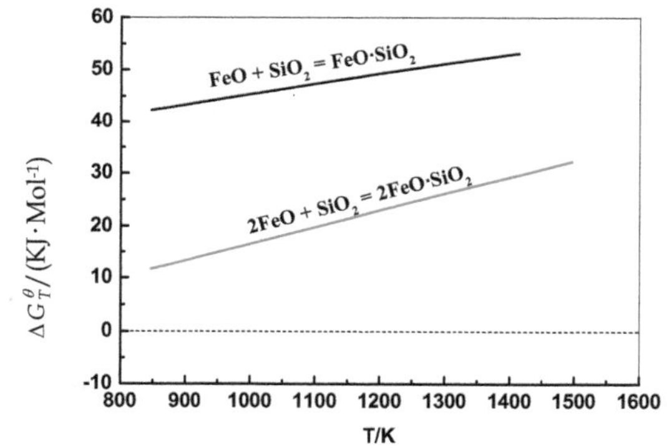

Fig. 4. Relationships of ΔG_T^θ and temperature in SiO_2-Fe_2O_3 system

2.5 CaO-Al_2O_3 system

Al_2O_3 can react with CaO to form calcium aluminates such as $3CaO \cdot Al_2O_3(C_3A)$, $12CaO \cdot 7Al_2O_3(C_{12}A_7)$, $CaO \cdot Al_2O_3$(CA) and $CaO \cdot 2Al_2O_3$ (CA_2). As regard as the calcium aluminates only $C_{12}A_7$ can be totally soluble in soda solution, C_3A and CA dissolve with a slow speed, and the other calcium aluminates such as CA_2 are completely insoluble. Equations that Al_2O_3 reacted with CaO to form C_3A, $C_{12}A_7$, CA and CA_2 are presented in table 4.

Figure 5 shows that, the ΔG_T^θ of reactions of Al_2O_3 with CaO decreases with the rise of temperature; all reactions automatically proceed to generate the corresponding calcium aluminates at normal roasting temperature (1473~1673K, same as follows); At the same roasting temperature, the thermodynamic order that one mole Al_2O_3 reacts with CaO to generate calcium aluminates such as $C_{12}A_7$, C_3A, CA, CA_2.

Reactions	A, J/mol	B, J/K.mol	Temperature, K
$3CaO + Al_2O_3 = 3CaO \cdot Al_2O_3$	-9.9	-28.4	298~1808
$\frac{12}{7}CaO + Al_2O_3 = (\frac{1}{7})12CaO \cdot 7Al_2O_3$	318.3	-44.5	298~1800
$CaO + Al_2O_3 = CaO \cdot Al_2O_3$	-15871.5	-18.1	298~1878
$\frac{1}{2}CaO + Al_2O_3 = (\frac{1}{2})CaO \cdot 2Al_2O_3$	-6667.2	-13.8	298~2023

Table 4. The ΔG_T^θ of Al_2O_3-CaO system （$\Delta G_T^\theta = A + BT$, J/mol)

The relationships between ΔG_T^θ and temperature (T) are shown in figure 5.

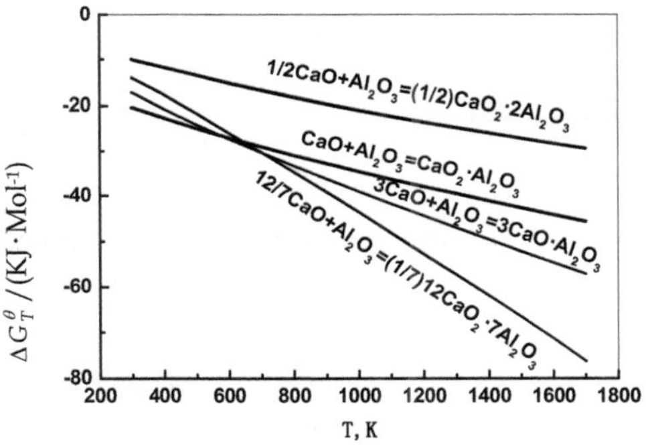

Fig. 5. Relationships between ΔG_T^θ and temperature in Al_2O_3-CaO system

Reactions	A, J/mol	B, J/K.mol	Temperature, K
$(\frac{4}{3})3CaO \cdot Al_2O_3 + Al_2O_3 = (\frac{1}{3})12CaO \cdot 7Al_2O_3$	13939.7	-65.8	298~1800
$(\frac{1}{2})3CaO \cdot Al_2O_3 + Al_2O_3 = (\frac{3}{2})CaO \cdot Al_2O_3$	-18843.8	-13.0	298~1878
$(\frac{1}{5})3CaO \cdot Al_2O_3 + Al_2O_3 = (\frac{3}{5})CaO \cdot 2Al_2O_3$	-6011.2	-10.9	298~2023
$(\frac{1}{5})12CaO \cdot 7Al_2O_3 + Al_2O_3 = (\frac{12}{5})CaO \cdot Al_2O_3$	-38544.8	18.8	298~1878
$(\frac{1}{17})12CaO \cdot 7Al_2O_3 + Al_2O_3 = (\frac{12}{17})CaO \cdot 2Al_2O_3$	-9541.1	-1.2	298~2023
$CaO \cdot Al_2O_3 + Al_2O_3 = CaO \cdot 2Al_2O_3$	2543.8	-9.5	298~2023

Table 5. The ΔG_T^θ of Al_2O_3-calcium aluminates system （$\Delta G_T^\theta = A + BT$, J/mol)

When CaO is insufficient, redundant Al_2O_3 may promote the newly generated high calcium-to-aluminum ratio (CaO to Al_2O_3 mole ratio) calcium aluminates to transform into lower calcium-to-aluminum ratio calcium aluminates. The reactions of the equations are presented in table 5:

The relationships between ΔG_T^θ of reactions of Al_2O_3-calcium aluminates system and temperature (T) are shown in figure 6.

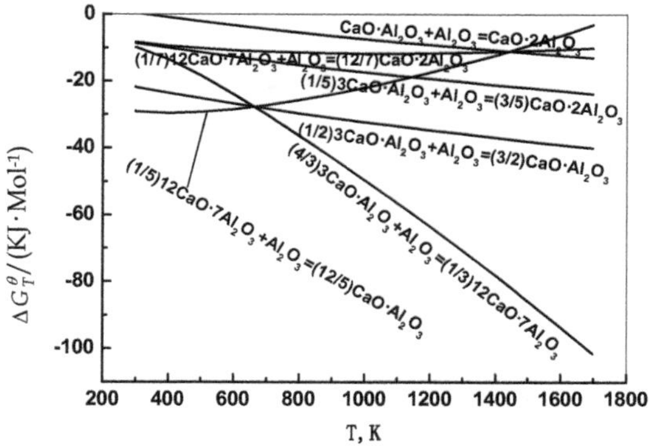

Fig. 6. Relationships between ΔG_T^θ of reactions Al_2O_3-calcium aluminates system and temperature

Figure 6 shows that, Gibbs free energy of the reaction of Al_2O_3-calcium aluminates system are negative at 400~1700K, and all the reactions automatically proceed to generate the corresponding low calcium-to-aluminum ratio calcium aluminates; Except for the reaction of Al_2O_3-$C_{12}A_7$, the ΔG_T^θ of the rest reactions decreases with the rise of temperature and becomes more negative. Comparing figure 4 with figure 5, it can be found that Al_2O_3 reacts with CaO easily to generate $C_{12}A_7$.

2.6 SiO_2- CaO system

SiO_2 can react with CaO to form $CaO\cdot SiO_2$ (CS), $3CaO\cdot 2SiO_2$ (C_3S_2), $2CaO\cdot SiO_2$ (C_2S) and $3CaO\cdot SiO_2$(C_3S) in roasting process. The reactions are shown in table 6, and the relationships between $\triangle G^0$ of the reactions of SiO_2 with CaO and temperature are shown in figure 7.

Reactions	A, J/mol	B, J/K.mol	Temperature, K
$CaO+SiO_2 = CaO\cdot SiO_2$(pseud-wollastonite)	-83453.0	-3.4	298~1817
$CaO+SiO_2 = CaO\cdot SiO_2$(wollastonite)	-89822.9	-0.3	298~1817
$\frac{3}{2}CaO+SiO_2=(\frac{1}{2})\ 3CaO\cdot 2SiO_2$	-108146.6	-3.1	298~1700
$3CaO+SiO_2 = 3CaO\cdot SiO_2$	-111011.9	-11.3	298~1800
$2CaO+SiO_2 = 2CaO\cdot SiO_2(\beta)$	-125875.1	-6.7	298~2403
$2CaO+SiO_2 = 2CaO\cdot SiO_2(\gamma)$	-137890.1	3.7	298~1100

Table 6. The ΔG_T^θ of SiO_2-CaO system($\Delta G_T^\theta = A + BT$, J/mol)

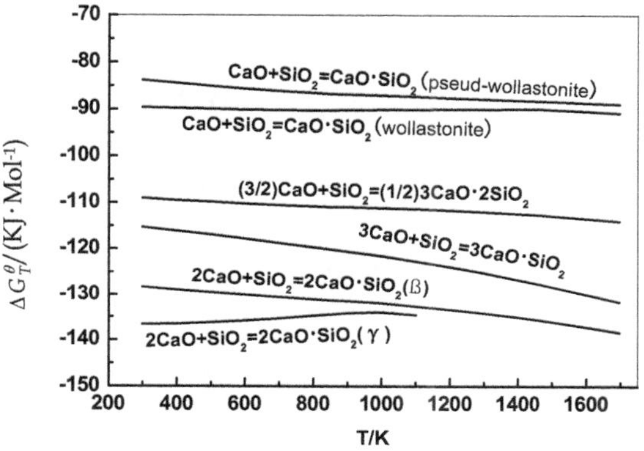

Fig. 7. Relationships between ΔG_T^θ and temperature

Figure7 shows that, SiO_2 reacts with CaO to form γ-C_2S when temperature below 1100K, but β-C_2S comes into being when the temperature above 1100K. At normal roasting temperature, the thermodynamic order of forming calcium silicate is C_2S, C_3S, C_3S_2, CS.
Figure 5 ~ figure 7 show that, CaO reacts with SiO_2 and Al_2O_3 firstly to form C_2S, and then $C_{12}A_7$. Therefore, it is less likely to form aluminium silicates in roasting process.

2.7 SiO_2- calcium aluminates system

In the CaO-Al_2O_3 system, if there exists some SiO_2, the newly formed calcium aluminates are likely to react with SiO_2 to transform to calcium silicates and Al_2O_3 because SiO_2 is more acidity than that of Al_2O_3. The reaction equations are presented in table 7, the relationships between ΔG_T^θ and temperature are shown in figure 8.

Figure 8 shows that, the ΔG_T^θ of all the reactions increases with the temperature increases; the reaction ($3CA_2+SiO_2=C_3S+6Al_2O_3$) can not happen when the roasting temperature is above 900K , i.e., the lowest calcium-to-aluminum ratio calcium aluminates cannot transform to the highest calcium-to-silicon ratio (CaO to SiO_2 molecular ratio) calcium silicate; when the temperature is above 1500K, the ΔG_T^θ of reaction($3CA+ SiO_2=C_3S+3Al_2O_3$) is also more than zero; but the other calcium aluminates all can react with SiO_2 to generate calcium silicates at 800~1700K. The thermodynamic sequence of calcium aluminates reaction with SiO_2 is firstly C_3A, and then $C_{12}A_7$, CA, CA_2.

Reactions	A, J/mol	B, J/K.mol	Temperature, K
$(3)CaO \cdot 2Al_2O_3 + SiO_2 = 3CaO \cdot SiO_2 + 6Al_2O_3$	-69807.8	70.8	298~1800
$(3)CaO \cdot Al_2O_3 + SiO_2 = 3CaO \cdot SiO_2 + 3Al_2O_3$	-62678.8	42.6	298~1800
$(\frac{1}{4})12CaO \cdot 7Al_2O_3 + SiO_2 = 3CaO \cdot SiO_2 + \frac{7}{4}Al_2O_3$	-111820.6	66.7	298~1800
$(2)CaO \cdot 2Al_2O_3 + SiO_2 = 2CaO \cdot SiO_2 + 4Al_2O_3$	-98418.8	48.1	298~1710
$(\frac{3}{2})CaO \cdot 2Al_2O_3 + SiO_2 = (\frac{1}{2})3CaO \cdot 2SiO_2 + 3Al_2O_3$	-87585.9	38.0	298~1700
$CaO \cdot 2Al_2O_3 + SiO_2 = CaO \cdot SiO_2 + 2Al_2O_3$	-76146.6	27.1	298~1817
$CaO \cdot Al_2O_3 + SiO_2 = CaO \cdot SiO_2 + Al_2O_3$	-73770.2	17.7	298~1817
$(\frac{3}{2})CaO \cdot Al_2O_3 + SiO_2 = (\frac{1}{2})3CaO \cdot 2SiO_2 + \frac{3}{2}Al_2O_3$	-84021.4	23.8	298~1700
$(2)CaO \cdot Al_2O_3 + SiO_2 = 2CaO \cdot SiO_2 + 2Al_2O_3$	-93666.1	29.2	298~1710
$(\frac{1}{12})12CaO \cdot 7Al_2O_3 + SiO_2 = CaO \cdot SiO_2 + \frac{7}{12}Al_2O_3$	-90150.8	25.7	298~1800
$(\frac{1}{8})12CaO \cdot 7Al_2O_3 + SiO_2 = (\frac{1}{2})3CaO \cdot 2SiO_2 + \frac{7}{8}Al_2O_3$	-108592.3	35.9	298~1700
$(\frac{1}{6})12CaO \cdot 7Al_2O_3 + SiO_2 = 2CaO \cdot SiO_2 + \frac{7}{6}Al_2O_3$	-126427.4	45.3	298~1710
$(\frac{1}{3})3CaO \cdot Al_2O_3 + SiO_2 = CaO \cdot SiO_2 + \frac{1}{3}Al_2O_3$	-86654.2	9.4	298~1808
$3CaO \cdot Al_2O_3 + SiO_2 = 3CaO \cdot SiO_2 + Al_2O_3$	-100774.6	16.9	298~1808
$(\frac{1}{2})3CaO \cdot Al_2O_3 + SiO_2 = (\frac{1}{2})3CaO \cdot 2SiO_2 + \frac{1}{2}Al_2O_3$	-103069.3	11.0	298~1700
$(\frac{2}{3})3CaO \cdot Al_2O_3 + SiO_2 = 2CaO \cdot SiO_2 + \frac{2}{3}Al_2O_3$	-119063.3	12.1	298~1710

Table 7. The ΔG_T^θ of the reactions SiO_2 with calcium aluminates($\Delta G_T^\theta = A + BT$, J/mol)

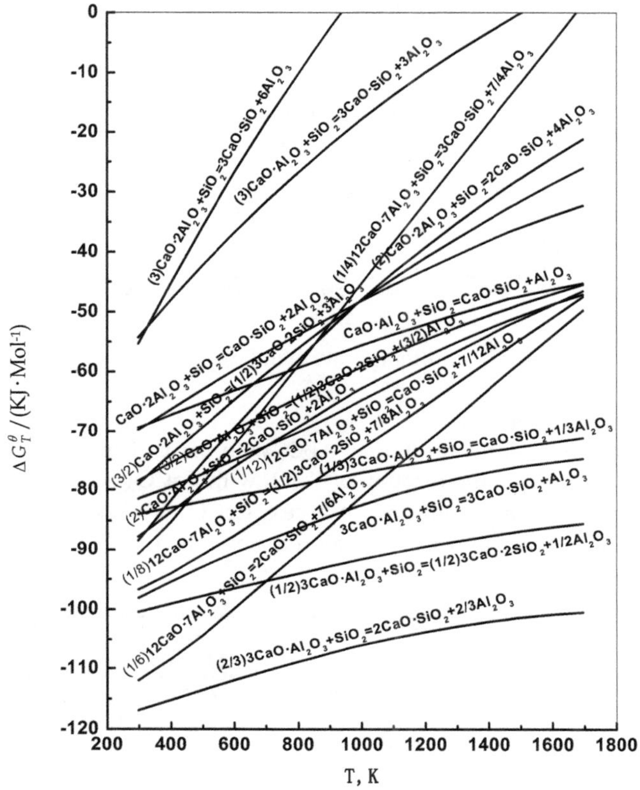

Fig. 8. Relationships between ΔG_T^θ and temperature in SiO_2-calcium aluminates system

2.8 CaO- Fe_2O_3 system

Fe_2O_3 can react with CaO to form CaO·Fe_2O_3(CF) and 2CaO·Fe_2O_3(C_2F). When Fe_2O_3 is used up, the newly formed C_2F can react with Fe_2O_3 to form CF. The reaction equations are shown in table 8, and the relationships between $\triangle G^0$ and temperature are shown in figure 9.

Figure 9 shows that, Fe_2O_3 reacts with CaO much easily to form C_2F; CF is not from the reaction of C_2F and Fe_2O_3, but from the directly reaction of Fe_2O_3 with CaO. When Fe_2O_3 is excess, C_2F can react with Fe_2O_3 to form CF.

Reactions	A, J/mol	B, J/K.mol	Temperature, K
$CaO+Fe_2O_3=CaO\cdot Fe_2O_3$	-19179.9	-11.1	298~1489
$2CaO+Fe_2O_3=2CaO\cdot Fe_2O_3$	-40866.7	-9.3	298~1723
$2CaO\cdot Fe_2O_3+Fe_2O_3=(2)CaO\cdot Fe_2O_3$	2340.8	-12.6	298~1489

Table 8. The ΔG_T^θ of Fe_2O_3-CaO system($\Delta G_T^\theta = A + BT$, J/mol)

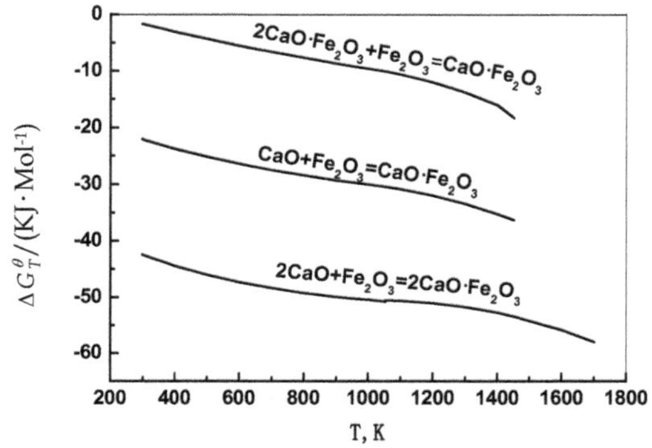

Fig. 9. Relationships between ΔG_T^{θ} and temperature in Fe_2O_3-CaO system

2.9 Al_2O_3- calcium ferrites system

Figure 1 shows that, the ΔG_T^{θ} of the reaction of Al_2O_3 with $CaCO_3$ is more negative than that of Fe_2O_3 with $CaCO_3$, therefore, the reaction of Fe_2O_3 with $CaCO_3$ occurs after the reaction of Al_2O_3 with $CaCO_3$ under the conditions of excess $CaCO_3$. The new generated calcium ferrites are likely to transform into calcium aluminates when $CaCO_3$ is insufficient, the reactions are as followed:

Reactions	A, J/mol	B, J/K.mol	Temperature, K
$(3)CaO\bullet Fe_2O_3 + Al_2O_3 = 3CaO\bullet Al_2O_3 + 3Fe_2O_3$	47922.7	4.5	298~1489
$(\frac{3}{2})2CaO\cdot Fe_2O_3 + Al_2O_3 = 3CaO\cdot Al_2O_3 + \frac{3}{2}Fe_2O_3$	49.6	-1.2×10^{-2}	298~1723
$(\frac{12}{7})CaO\cdot Fe_2O_3 + Al_2O_3 = (\frac{1}{7})12CaO\cdot 7Al_2O_3 + \frac{12}{7}Fe_2O_3$	32685.1	-24.5	298~1489
$(\frac{6}{7})2CaO\cdot Fe_2O_3 + Al_2O_3 = (\frac{1}{7})12CaO\cdot 7Al_2O_3 + \frac{6}{7}Fe_2O_3$	34514.4	-35.0	298~1723
$CaO\bullet Fe_2O_3 + Al_2O_3 = CaO\bullet Al_2O_3 + Fe_2O_3$	3626.6	-7.5	298~1489
$(\frac{1}{2})CaO\cdot Fe_2O_3 + Al_2O_3 = (\frac{1}{2})\ CaO\cdot 2Al_2O_3 + \frac{1}{2}Fe_2O_3$	3215.1	-8.8	298~1489
$(\frac{1}{4})2CaO\cdot Fe_2O_3 + Al_2O_3 = (\frac{1}{2})\ CaO\cdot 2Al_2O_3 + \frac{1}{4}Fe_2O_3$	3168.6	-11.0	298~1723
$(\frac{1}{2})2CaO\cdot Fe_2O_3 + Al_2O_3 = CaO\cdot Al_2O_3 + \frac{1}{2}Fe_2O_3$	4009.5	-12.8	298~1723

Table 9. The ΔG_T^{θ} of the reaction Al_2O_3 with calcium ferrites($\Delta G_T^{\theta} = A + BT$, J/mol)

The relationships between ΔG_T^{θ} and temperature (T) are shown in figure 10. Figure 10 shows that, Al_2O_3 cannot replace the Fe_2O_3 in calcium ferrites to generate C_3A, and also cannot replace the Fe_2O_3 in $CaO\bullet Fe_2O_3$(CF) to generate $C_{12}A_7$, but it can replace the Fe_2O_3 in $2CaO\bullet Fe_2O_3$(C_2F) to generate $C_{12}A_7$ when the temperature is above 1000K, the higher temperature is, the more negative Gibbs free energy is; Al_2O_3 can react with CF and C_2F to form CA or CA_2, the higher temperature, more negative ΔG_T^θ. Because Fe_2O_3 reacts with CaO more easily to generate C_2F (Fig.9), therefore, $C_{12}A_7$ is the reaction product at normal roasting temperature(1073~1673K) under the conditions that CaO is sufficent in batching and the ternary compounds are not considered.

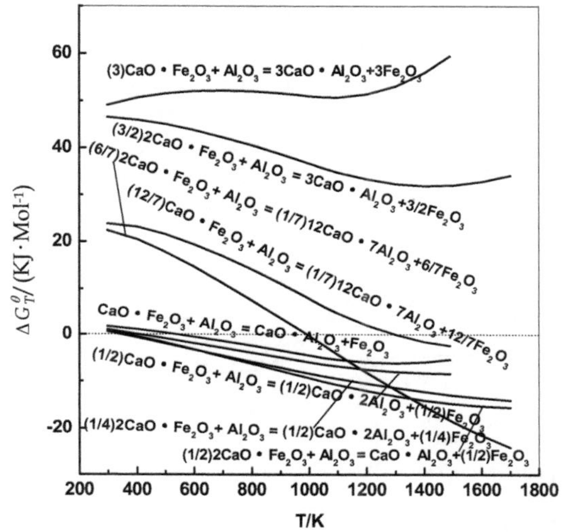

Fig. 10. Relationship between ΔG_T^θ and temperature in Al_2O_3- calcium ferrites system

3. Ternary compounds in Al_2O_3-CaO-SiO_2-Fe_2O_3 system

The ternary compounds formed by CaO, Al_2O_3 and SiO_2 in roasting process are mainly $2CaO\cdot Al_2O_3\cdot SiO_2(C_2AS)$, $CaO\cdot Al_2O_3\cdot 2SiO_2(CAS_2)$, $CaO\cdot Al_2O_3\cdot SiO_2(CAS)$ and $3CaO\cdot Al_2O_3\cdot 3SiO_2(C_3AS_3)$. In addition, ternary compound $4CaO\cdot Al_2O_3\cdot Fe_2O_3(C_4AF)$ is formed form CaO, Al_2O_3 and Fe_2O_3. The equations are shown in table 10:

Reactions	A, J/mol	B, J/K.mol	Temperature, K
$CaO\cdot SiO_2 + CaO\cdot Al_2O_3 = 2CaO\cdot Al_2O_3\cdot SiO_2$	-30809.41	0.60	298~1600
$\frac{1}{2}Al_2O_3 + \frac{1}{2}CaO + SiO_2 = (\frac{1}{2})CaO\cdot Al_2O_3\cdot 2SiO_2$	-47997.55	-7.34	298~1826
$Al_2O_3 + 2CaO + SiO_2 = 2CaO\cdot Al_2O_3\cdot SiO_2$	-50305.83	-9.33	298~1600
$Al_2O_3 + CaO + SiO_2 = CaO\cdot Al_2O_3\cdot SiO_2$	-72975.54	-9.49	298~1700
$\frac{1}{3}Al_2O_3 + CaO + SiO_2 = (\frac{1}{3})3CaO\cdot Al_2O_3\cdot 3SiO_2$	-112354.51	20.86	298~1700
$4CaO + Al_2O_3 + Fe_2O_3 = 4CaO\cdot Al_2O_3\cdot Fe_2O_3$	-66826.92	-62.5	298~2000
$Al_2O_3 + 2CaO + SiO_2 = 2CaO\cdot Al_2O_3\cdot SiO_2$ (cacoclasite)	-136733.59	-17.59	298~1863

Table 10. The ΔG_T^θ of forming ternary compounds ($\Delta G_T^\theta = A + BT$, J/mol)

The relationships between ΔG_T^θ and temperature (T) are shown in figure 11. Figure 11 shows that, except for C_3AS_3(Hessonite), all the ΔG_T^θ of the reactions get more negative with the temperature increasing; the thermodynamic order of generating ternary compounds at sintering temperature of 1473K is: C_2AS(cacoclasite) , C_4AF, CAS, C_3AS_3, C_2AS, CAS_2.
C_2AS may also be formed by the reaction of CA and CS, the curve is presented in figure 11. Figure 11 shows that, the ΔG_T^θ of reaction (Al_2O_3+CaO+SiO_2) is lower than that of reaction of CA and CS to generate C_2AS. So C_2AS does not form from the binary compounds CA and CS, but from the direct combination among Al_2O_3, CaO, SiO_2. Qiusheng Zhou thinks that, C_4AF is not formed by mutual reaction of calcium ferrites and sodium aluminates, but from the direct reaction of CaO, Al_2O_3 and Fe_2O_3. Thermodynamic analysis of figure 1~figure11 shows that, reactions of Al_2O_3, Fe_2O_3, SiO_2 and CaO are much easier to form C_2AS and C_4AF, as shown in figure 12.

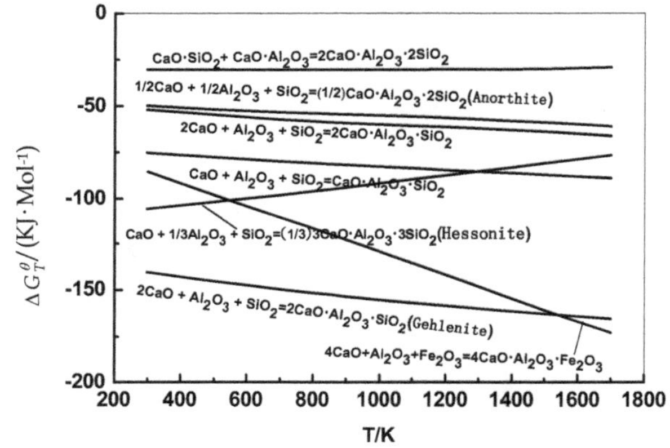

Fig. 11. Relationships between ΔG_T^θ of ternary compounds and temperature

Figure 12 shows that, in thermodynamics, C_2AS and C_4AF are firstly formed when Al_2O_3, Fe_2O_3, SiO_2 and CaO coexist, and then calcium silicates, calcium aluminates and calcium ferrites are generated.

4. Summary

1) When Al_2O_3 and Fe_2O_3 simultaneously react with CaO, calcium silicates are firstly formed, and then calcium ferrites. In thermodynamics, when one mole Al_2O_3 reacts with CaO, the sequence of generating calcium aluminates are 12CaO·7Al_2O_3, 3CaO·Al_2O_3, CaO·Al_2O_3, CaO·2Al_2O_3. When CaO is insufficient, redundant Al_2O_3 may promote the newly generated high calcium-to-aluminum ratio calcium aluminates to transform to lower calcium-to-aluminum ratio calcium aluminates. Fe_2O_3 reacts with CaO easily to form2CaO·Fe_2O_3, and CaO·Fe_2O_3 is not from the reaction of 2CaO·Fe_2O_3 and Fe_2O_3 but form the directly combination of Fe_2O_3 with CaO. Al_2O_3 cannot replace the Fe_2O_3 in calcium ferrites to generate 3CaO·Al_2O_3, and also cannot replace the Fe_2O_3 in CaO•Fe_2O_3 to generate 12CaO·7Al_2O_3, but can replace the Fe_2O_3 in 2CaO•Fe_2O_3 to generate 12CaO·7Al_2O_3 when the temperature is above 1000K; Al_2O_3 can react with calcium ferrites to form CaO·Al_2O_3 or CaO·2Al_2O_3.

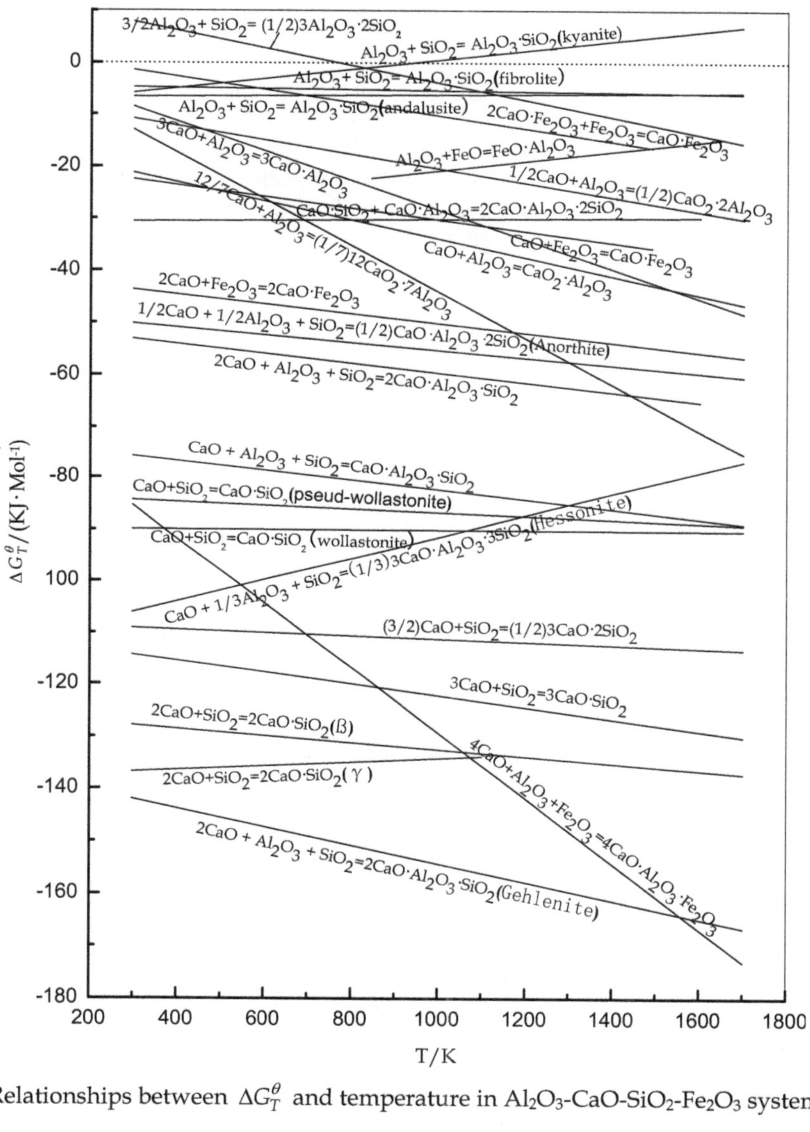

Fig. 12. Relationships between ΔG_T^θ and temperature in Al_2O_3-CaO-SiO_2-Fe_2O_3 system

2) One mole SiO_2 reacts with Al_2O_3 much easily to generate $3Al_2O_3 \cdot 2SiO_2$, Fe_2O_3 can not react with SiO_2 in the roasting process in the air. Al_2O_3 can not directly react with Fe_2O_3, but can react with wustite (FeO) to form $FeO \cdot Al_2O_3$.

3) In thermodynamics, the sequence of one mole SiO_2 reacts with CaO to form calcium silicates is $2CaO \cdot SiO_2$, $3CaO \cdot SiO_2$, $3CaO \cdot 2SiO_2$ and $CaO \cdot SiO_2$. Calcium aluminates can react with SiO_2 to transform to calcium silicates and Al_2O_3. $CaO \cdot 2Al_2O_3$ can not transform to $3CaO \cdot SiO_2$ when the roasting temperature is above 900K; when the temperature is above

1500K, $3CaO \cdot Al_2O_3$ can not transform to $3CaO \cdot SiO_2$; but the other calcium aluminates all can all react with SiO_2 to generate calcium silicates at 800~1700K.

4) Reactions among Al_2O_3, Fe_2O_3, SiO_2 and CaO easily form $2CaO \cdot Al_2O_3 \cdot SiO_2$ and $4CaO \cdot Al_2O_3 \cdot Fe_2O_3$. $2CaO \cdot Al_2O_3 \cdot SiO_2$ does not form from the reaction of $CaO \cdot Al_2O_3$ and $CaO \cdot SiO_2$, but from the direct reaction among Al_2O_3, CaO, SiO_2. And $4CaO \cdot Al_2O_3 \cdot Fe_2O_3$ is also not formed via mutual reaction of calcium ferrites and sodium aluminates, but from the direct reaction of CaO, Al_2O_3 and Fe_2O_3. In thermodynamics, when Al_2O_3, Fe_2O_3, SiO_2 and CaO coexist, $2CaO \cdot Al_2O_3 \cdot SiO_2$ and $4CaO \cdot Al_2O_3 \cdot Fe_2O_3$ are firstly formed, and then calcium silicates, calcium aluminates and calcium ferrites.

5. Symbols used

Thermodynamic temperature: T, K
Thermal unit: J
Amount of substance: mole
Standard Gibbs free energy: ΔG_T^θ ,J

6. References

Li, B.; Xu, Y. & Choi, J. (1996). Applying Machine Learning Techniques, *Proceedings of ASME 2010 4th International Conference on Energy Sustainability*, pp.14-17, ISBN 842-6508-23-3, Phoenix, Arizona, USA, May 17-22, 2010

Rayi H. S. ; Kundu N.(1986). Thermal analysis studies on the initial stages of iron oxide reduction, *Thermochimi, Acta.* 101:107~118,1986

Coats A.W. ; Redferm J.P.(1964). Kinetic parameters from thermogravimetric data, *Nature,* 201:68,1964

LIU Gui-hua, LI Xiao-bin, PENG Zhi-hong, ZHOU Qiu-sheng(2003). Behavior of calcium silicate in leaching process. *Trans Nonferrous Met Soc China,* January 213−216,2003

Paul S. ; Mukherjee S.(1992). Nonisothermal and isothermal reduction kinetics of iron ore agglomerates, *Ironmaking and steelmaking,* March 190~193, 1992

ZHU Zhongping, JIANG Tao, LI Guanghui, HUANG Zhucheng(2009). Thermodynamics of reaction of alumina during sintering process of high-iron gibbsite-type bauxite, *The Chinese Journal of Nonferrous Metals,* Dec 2243~2250, 2009

ZHOU Qiusheng, QI Tiangui, PENG Zhihong, LIU Guihua, LI Xiaobin(2007). Thermodynamics of reaction behavior of ferric oxide during sinter-preparing process, *The Chinese Journal of Nonferrous Metals,* Jun 974~978, 2007

Barin I., Knacke O.(1997). *Thermochemical properties of inorganic substances,* Berlin:Supplement, 1997

Barin I., Knacke O.(1973). *Thermochemical properties of inorganic substances,* Berlin: Springer, 1973

8

Thermodynamic Perturbation Theory of Simple Liquids

Jean-Louis Bretonnet
Laboratoire de Physique des Milieux Denses, Université Paul Verlaine de Metz
France

1. Introduction

This chapter is an introduction to the thermodynamics of systems, based on the correlation function formalism, which has been established to determine the thermodynamic properties of simple liquids. The article begins with a preamble describing few general aspects of the liquid state, among others the connection between the phase diagram and the pair potential $u(r)$, on one hand, and between the structure and the pair correlation function $g(r)$, on the other hand. The pair correlation function is of major importance in the theory of liquids at equilibrium, because it is required for performing the calculation of the thermodynamic properties of systems modeled by a given pair potential. Then, the article is devoted to the expressions useful for calculating the thermodynamic properties of liquids, in relation with the most relevant features of the potential, and provides a presentation of the perturbation theory developed in the four last decades. The thermodynamic perturbation theory is founded on a judicious separation of the pair potential into two parts. Specifically, one of the greatest successes of the microscopic theory has been the recognition of the quite distinct roles played by the repulsive and attractive parts of the pair potential in predicting many properties of liquids. Much attention has been paid to the hard-sphere potential, which has proved very efficient as natural reference system because it describes fairly well the local order in liquids. As an example, the Yukawa attractive potential is also mentioned.

2. An elementary survey

2.1 The liquid state

The ability of the liquids to form a free surface differs from that of the gases, which occupy the entire volume available and have diffusion coefficients ($\sim 0,5\ \mathrm{cm^2s^{-1}}$) of several orders of magnitude higher than those of liquids ($\sim 10^{-5}\ \mathrm{cm^2s^{-1}}$) or solids ($\sim 10^{-9}\ \mathrm{cm^2s^{-1}}$). Moreover, if the dynamic viscosity of liquids (between 10^{-5} Pa.s and 1 Pa.s) is so lower compared to that of solids, it is explained in terms of competition between *configurational* and *kinetic* processes. Indeed, in a solid, the displacements of atoms occur only after the breaking of the bonds that keep them in a stable configuration. At the opposite, in a gas, molecular transport is a purely kinetic process perfectly described in terms of exchanges of energy and momentum. In a liquid, the continuous rearrangement of particles and the molecular transport combine together in appropriate proportion, meaning that the liquid is an intermediate state between the gaseous and solid states.

The characterization of the three states of matter can be done in an advantageous manner by comparing the kinetic energy and potential energy as it is done in figure (1). The nature and intensity of forces acting between particles are such that the particles tend to attract each other at great distances, while they repel at the short distances. The particles are in equilibrium when the attraction and repulsion forces balance each other. In gases, the kinetic energy of particles, whose the distribution is given by the Maxwell velocity distribution, is located in the region of unbound states. The particles move freely on trajectories suddenly modified by binary collisions; thus the movement of particles in the gases is essentially an *individual movement*. In solids, the energy distribution is confined within the potential well. It follows that the particles are in tight bound states and describe harmonic motions around their equilibrium positions; therefore the movement of particles in the solids is essentially a *collective movement*. When the temperature increases, the energy distribution moves towards high energies and the particles are subjected to anharmonic movements that intensify progressively. In liquids, the energy distribution is almost entirely located in the region of bound states, and the movements of the particles are strongly anharmonic. On approaching the critical point, the energy distribution shifts towards the region of unbound states. This results in important fluctuations in concentrations, accompanied by the destruction and formation of aggregates of particles. Therefore, the movement of particles in liquids is thus the result of a combination of individual and collective movements.

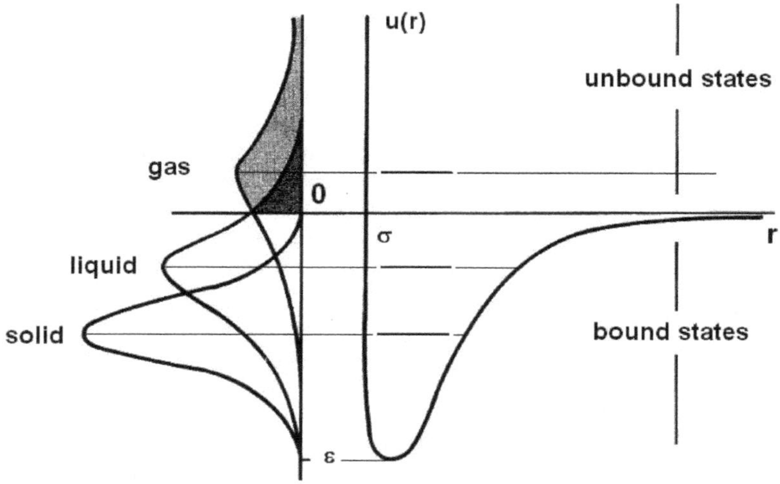

Fig. 1. Comparison of kinetic and potential energies in solids, liquids and gases.

When a crystalline solid melts, the long-range order of the crystal is destroyed, but a residual local order persits on distances greater than several molecular diameters. This local order into liquid state is described in terms of the *pair correlation function*, $g(r) = \frac{\rho(r)}{\rho_\infty}$, which is defined as the ratio of the mean molecular density $\rho(r)$, at a distance r from an arbitrary molecule, to the bulk density ρ_∞. If $g(r)$ is equal to unity everywhere, the fluid is completely disordered, like in diluted gases. The deviation of $g(r)$ from unity is a measure of the local order in the arrangement of near-neighbors. The representative curve of $g(r)$ for a liquid is formed of maxima and minima rapidly damped around unity, where the first maximum corresponds to the position of the nearest neighbors around an origin atom. It should be noted that the pair correlation function $g(r)$ is accessible by a simple Fourier transform of the experimental *structure factor* $S(q)$ (intensity of scattered radiation).

The pair correlation function is of crucial importance in the theory of liquids at equilibrium, because it depends strongly on the *pair potential* $u(r)$ between the molecules. In fact, one of the goals of the theory of liquids at equilibrium is to predict the thermodynamic properties using the pair correlation function $g(r)$ and the pair potential $u(r)$ acting in the liquids. There are a large number of potential models (hard sphere, square well, Yukawa, Gaussian, Lennard-Jones...) more or less adapted to each type of liquids. These interaction potentials have considerable theoretical interest in statistical physics, because they allow the calculation of the properties of the liquids they are supposed to represent. But many approximations for calculating the pair correlation function $g(r)$ exist too.

Note that there is a great advantage in comparing the results of the theory with those issued from the numerical simulation with the aim to test the models developed in the theory. Beside, the comparison of the theoretical results to the experimental results allows us to test the potential when the theory itself is validated. Nevertheless, comparison of simulation results with experimental results is the most efficient way to test the potential, because the simulation provides the exact solution without using a theoretical model. It is a matter of fact that simulation is generally identified to a numerical experience. Even if they are time consuming, the simulation computations currently available with thousands of interacting particles gives a role increasingly important to the simulation methods.

In the theory of simple fluids, one of the major achievements has been the recognition of the quite distinct roles played by the repulsive and attractive parts of the pair potential in determining the microscopic properties of simple fluids. In recent years, much attention has been paid in developing analytically solvable models capable to represent the thermodynamic and structural properties of real fluids. The hard-sphere (HS) model - with its diameter σ - is the natural *reference system* for describing the general characteristics of liquids, i.e. the local atomic order due to the excluded volume effects and the *solidification* process of liquids into a solid ordered structure. In contrast, the HS model is not able to predict the *condensation* of a gas into a liquid, which is only made possible by the existence of dispersion forces represented by an attractive long-ranged part in the potential.

Another reference model that has proved very useful to stabilize the local structure in liquids is the hard-core potential with an attractive Yukawa tail (HCY), by varying the hard-sphere diameter σ and screening length λ. It is an advantage of this model for modeling real systems with widely different features (1), like rare gases with a screening length $\lambda \sim 2$ or colloidal suspensions and protein solutions with a screening length $\lambda \sim 8$. An additional reason that does the HCY model appealing is that analytical solutions are available. After the search of the original solution with the mean-spherical approximation (2), valuable simplifications have been progressively brought giving simple analytical expressions for the thermodynamic properties and the pair correlation function. For this purpose, the expression for the free energy has been used under an expanded form in powers of the inverse temperature, as derived by Henderson *et al.* (3).

At this stage, it is perhaps salutary to claim that no attempt will be made, in this article, to discuss neither the respective advantages of the pair potentials nor the ability of various approximations to predict the structure, which are necessary to determine the thermodynamic properties of liquids. In other terms, nothing will be said on the theoretical aspect of correlation functions, except a brief summary of the experimental determination of the pair correlation function. In contrast, it will be useful to state some of the concepts of statistical thermodynamics providing a link between the microscopic description of liquids and classical thermodynamic functions. Then, it will be given an account of the thermodynamic perturbation theory with the analytical expressions required for calculating the thermodynamic properties. Finally, the HCY model, which is founded on the perturbation theory, will be presented in greater detail for investigating the thermodynamics of liquids. Thus, a review of the thermodynamic perturbation theory will be set up, with a special effort towards the pedagogical aspect. We hope that this paper will help readers to develop their inductive and synthetic capacities, and to enhance their scientific ability in the field of thermodynamic of liquids. It goes without saying that the intention of the present paper is just to initiate the readers to that matter, which is developed in many standard textbooks (4).

2.2 Phase stability limits versus pair potential

One success of the numerical simulation was to establish a relationship between the shape of the pair potential and the phase stability limits, thus clarifying the circumstances of the liquid-solid and liquid-vapor phase transitions. It has been shown, in particular, that the hard-sphere (HS) potential is able to correctly describe the atomic structure of liquids and predict the liquid-solid phase transition (5). By contrast, the HS potential is unable to describe the liquid-vapor phase transition, which is essentially due to the presence of attractive forces of dispersion. More specifically, the simulation results have shown that the liquid-solid phase transition depends on the steric hindrance of the atoms and that the coexistence curve of liquid-solid phases is governed by the details of the repulsive part of potential. In fact, this was already contained in the phenomenological theories of melting, like the Lindemann theory that predicts the melting of a solid when the mean displacement of atoms from their equilibrium positions on the network exceeds the atomic diameter of 10%. In other words, a substance melts when its volume exceeds the volume at 0 K of 30%.

In restricting the discussion to simple centrosymmetric interactions from the outset, it is necessary to consider a realistic pair potential adequate for testing the phase stability limits. The most natural prototype potential is the Lennard-Jones (LJ) potential given by

$$u_{LJ}(r) = 4\varepsilon_{LJ}\left[\left(\frac{\sigma_{LJ}}{r}\right)^m - \left(\frac{\sigma_{LJ}}{r}\right)^n\right], \tag{1}$$

where the parameters m and n are usually taken to be equal to 12 and 6, respectively. Such a functional form gives a reasonable representation of the interactions operating in real fluids, where the well depth ε_{LJ} and the collision diameter σ_{LJ} are independent of density and temperature. Figure (2a) displays the general shape of the Lennard-Jones potential $(m - n)$ corresponding to equation (1). Each substance has its own values of ε_{LJ} and σ_{LJ} so that, in reduced form, the LJ potentials have not only the same shape for all simple fluids, but superimpose each other rigorously. This is the condition for substances to conform to the *law of corresponding states*.

Figure (2b) represents the diagram $p(T)$ of a pure substance. We can see how the slope of the coexistence curve of solid-liquid phases varies with the repulsive part of potential: the higher the value of m, the steeper the repulsive part of the potential (Fig. 2a) and, consequently, the more the coexistence curve of solid-liquid phases is tilted (Fig. 2b).

We can also remark that the LJ potential predicts the liquid-vapor coexistence curve, which begins at the triple point T and ends at the critical point C. A detailed analysis shows that the length of the branch TC is proportional to the depth ε of the potential well. As an example, for rare gases, it is verified that $(T_C - T_T)k_B \simeq 0,55\,\varepsilon$. It follows immediately from this condition that the liquid-vapor coexistence curve disappears when the potential well is absent ($\varepsilon = 0$).

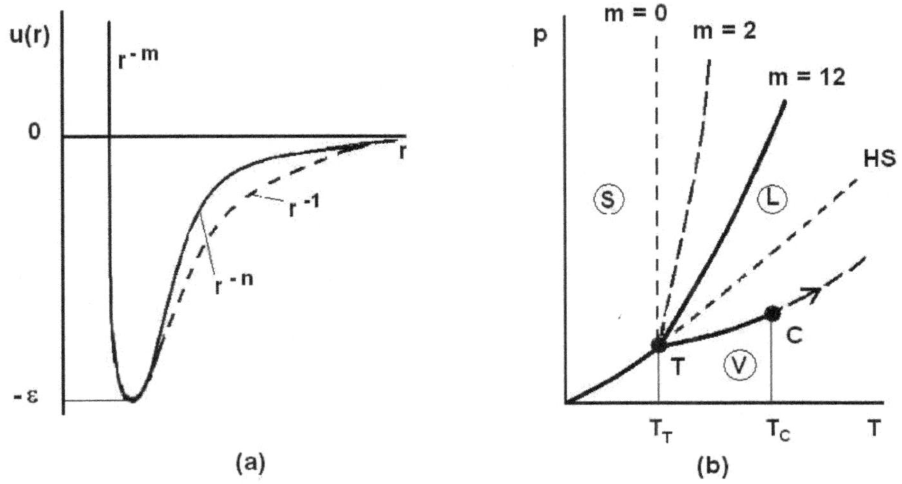

Fig. 2. Schematic representations of the Lennard-Jones potential $(m-n)$ and the diagram $p(T)$, as a function of the values of the parameters m and n.

The value of the slope of the branch TC also depends on the attractive part of the potential as shown by the Clausius-Clapeyron equation:

$$\frac{dp}{dT} = \frac{L_{vap}}{T_{vap}(V_{vap} - V_{liq})}, \tag{2}$$

where L_{vap} is the latent heat of vaporization at the corresponding temperature T_{vap} and $(V_{vap} - V_{liq})$ is the difference of specific volumes between vapor and liquid. To evaluate the slope $\frac{dp}{dT}$ of the branch TC at ambient pressure, we can estimate the ratio $\frac{L_{vap}}{T_{vap}}$ with Trouton's rule ($\frac{L_{vap}}{T_{vap}} \simeq 85$ J.K^{-1}.mol^{-1}), and the difference in volume $(V_{vap} - V_{liq})$ in terms of width of the potential well. Indeed, in noting that the quantity $(V_{vap} - V_{liq})$ is an increasing function of the width of potential well, which itself increases when n decreases, we see that, for a given well depth ε, the slope of the liquid-vapor coexistence curve decreases as n decreases.
For liquid metals, it should be mentioned that the repulsive part of the potential is softer than for liquid rare gases. Moreover, even if ε is slightly lower for metals than for rare gases, the quantity $\frac{(T_C - T_T)k_B}{\varepsilon}$ is much higher (between 2 and 4), which explains the elongation of the TC curve compared to that of rare gases. It is worth also to indicate that some *flat-bottomed* potentials (6) are likely to give a good description of the physical properties of substances that have a low value of the ratio $\frac{T_T}{T_C}$. Such a potential is obviously not suitable for liquid rare gases, whose ratio $\frac{T_T}{T_C} \simeq 0,56$, or for organic and inorganic liquids, for which $0,25 < \frac{T_T}{T_C} < 0,45$. In return, it might be useful as empirical potential for metals with low melting point such as mercury, gallium, indium, tin, etc., the ratio of which being $\frac{T_T}{T_C} < 0,1$.

3. The structure of liquids

3.1 Scattered radiation in liquids

The pair correlation function $g(r)$ can be deduced from the experimental measurement of the structure factor $S(q)$ by X-ray, neutron or electron diffractions. In condensed matter, the scatterers are essentially individual atoms, and diffraction experiments can only measure the structure of monatomic liquids such as rare gases and metals. By contrast, they provide no information on the structure of molecular liquids, unless they are composed of spherical molecules or monatomic ions, like in some molten salts.

Furthermore, each type of radiation-matter interaction has its own peculiarities. While the electrons are diffracted by all the charges in the atoms (electrons and nuclei), neutrons are diffracted by nuclei and X-rays are diffracted by the electrons localized on stable electron shells. The electron diffraction is practically used for fluids of low density, whereas the beams of neutrons and X-rays are used to study the structure of liquids, with their advantages and disadvantages. For example, the radius of the nuclei being 10, 000 times smaller than that of atoms, it is not surprising that the structure factors obtained with neutrons are not completely identical to those obtained with X-rays.

To achieve an experience of X-ray diffraction, we must irradiate the liquid sample with a monochromatic beam of X-rays having a wavelength in the range of the interatomic distance ($\lambda \sim 0{,}1$ nm). At this radiation corresponds a photon energy ($h\nu = \frac{hc}{\lambda} \sim 10^4$ eV), much larger than the mean energy of atoms that is of the order of few $k_B T$, namely about 10^{-1} eV. The large difference of the masses and energies between a photon and an atom makes that the photon-atom collision is elastic (constant energy) and that the liquid is transparent to the radiation. Naturally, the dimensions of the sample must be sufficiently large compared to the wavelength λ of the radiation, so that there are no side effects due to the walls of the enclosure - but not too much though for avoiding excessive absorption of the radiation. This would be particularly troublesome if the X-rays had to pass across metallic elements with large atomic numbers.

The incident radiation is characterized by its wavelength λ and intensity I_0, and the diffraction patterns depend on the structural properties of the liquids and on the diffusion properties of atoms. In neutron scattering, the atoms are characterized by the scattering cross section $\sigma = 4\pi b^2$, where b is a parameter approximately equal to the radius of the core ($\sim 10^{-14}$ m). Note that the parameter b does not depend on the direction of observation but may vary slightly, even for a pure element, with the isotope. By contrast, for X-ray diffraction, the property corresponding to b is the atomic scattering factor $A(q)$, which depends on the direction of observation and electron density in the isolated atom. The structure factor $S(q)$ obtained by X-ray diffraction has, in general, better accuracy at intermediate values of q. At the ends of the scale of q, it is less precise than the structure factor obtained by neutron diffraction, because the atomic scattering factor $A(q)$ is very small for high values of q and very poorly known for low values of q.

3.2 Structure factor and pair correlation function

When a photon of wave vector $\mathbf{k} = \frac{2\pi}{\lambda}\mathbf{u}$ interacts with an atom, it is deflected by an angle θ and the wave vector of the scattered photon is $\mathbf{k}' = \frac{2\pi}{\lambda}\mathbf{u}'$, where $\mathbf{u}$ and $\mathbf{u}'$ are unit vectors. If the scattering is elastic it results that $|\mathbf{k}'| = |\mathbf{k}|$, because $E \propto k^2 = cte$, and that the scattering vector (or transfer vector) $\mathbf{q}$ is defined by the Bragg law:

$$\mathbf{q} = \mathbf{k}' - \mathbf{k}, \qquad \text{and} \qquad |\mathbf{q}| = 2\,|\mathbf{k}|\sin\frac{\theta}{2} = \frac{4\pi}{\lambda}\sin\frac{\theta}{2}. \tag{3}$$

Now, if we consider an assembly of N identical atoms forming the liquid sample, the intensity scattered by the atoms in the direction θ (or $\mathbf{q}$, according to Bragg's law) is given by:

$$I(q) = A_N A_N^\star = A_0 A_0^\star \sum_{j=1}^{N} \sum_{l=1}^{N} \exp\left[i\mathbf{q}\left(\mathbf{r}_j - \mathbf{r}_l\right)\right].$$

In a crystalline solid, the arrangement of atoms is known once and for all, and the representation of the scattered intensity I is given by spots forming the Laue or Debye-Scherrer patterns. But in a liquid, the atoms are in continous motion, and the diffraction experiment gives only the mean value of successive configurations during the experiment. Given the absence of translational symmetry in liquids, this mean value provides no information on long-range order. By contrast, it is a good measure of short-range order around each atom chosen as origin. Thus, in a liquid, the scattered intensity must be expressed as a function of q by the statistical average:

$$I(q) = I_0 \left\langle \sum_{l=j=1}^{N} \exp\left[i\mathbf{q}\left(\mathbf{r}_j - \mathbf{r}_l\right)\right] \right\rangle + I_0 \left\langle \sum_{j=1}^{N} \sum_{l \neq j}^{N} \exp\left[i\mathbf{q}\left(\mathbf{r}_j - \mathbf{r}_l\right)\right] \right\rangle. \quad (4)$$

The first mean value, for $l = j$, is worth N because it represents the sum of N terms, each of them being equal to unity. To evaluate the second mean value, one should be able to calculate the sum of exponentials by considering all pairs of atoms (j, l) in all configurations counted during the experiment, then carry out the average of all configurations. However, this calculation can be achieved only by numerical simulation of a system made of a few particles. In a real system, the method adopted is to determine the mean contribution brought in by each pair of atoms (j, l), using the probability of finding the atoms j and l in the positions $\mathbf{r}'$ and $\mathbf{r}$, respectively. To this end, we rewrite the double sum using the Dirac delta function in order to calculate the statistical average in terms of the *density of probability* $P_N(\mathbf{r}^N, \mathbf{p}^N)$ of the *canonical ensemble*[1]. Therefore, the statistical average can be written by using the *distribution*

[1] It seems useful to remember that the *probability density* function in the canonical ensemble is:

$$P_N(\mathbf{r}^N, \mathbf{p}^N) = \frac{1}{N! h^{3N} Q_N(V,T)} \exp\left[-\beta H_N(\mathbf{r}^N, \mathbf{p}^N)\right],$$

where $H_N(\mathbf{r}^N, \mathbf{p}^N) = \sum \frac{p^2}{2m} + U(\mathbf{r}^N)$ is the *Hamiltonian* of the system, $\beta = \frac{1}{k_B T}$ and $Q_N(V,T)$ the *partition function* defined as:

$$Q_N(V,T) = \frac{Z_N(V,T)}{N! \Lambda^{3N}},$$

with the *thermal wavelength* Λ, which is a measure of the thermodynamic uncertainty in the localization of a particle of mass m, and the *configuration integral* $Z_N(V,T)$, which is expressed in terms of the total *potential energy* $U(\mathbf{r}^N)$. They read:

$$\Lambda = \sqrt{\frac{h^2}{2\pi m k_B T}},$$

$$\text{and} \quad Z_N(V,T) = \int \int_N \exp\left[-\beta U(\mathbf{r}^N)\right] d\mathbf{r}^N.$$

Besides, the partition function $Q_N(V,T)$ allows us to determine the free energy F according to the relation:

$$F = E - TS = -k_B T \ln Q_N(V,T).$$

The reader is advised to consult statistical-physics textbooks for further details.

function[2] $\rho_N^{(2)}(\mathbf{r},\mathbf{r}')$ in the form:

$$\left\langle \sum_{j=1}^{N}\sum_{l\neq j}^{N} \exp\left[i\mathbf{q}\left(\mathbf{r}_j-\mathbf{r}_l\right)\right]\right\rangle = \int\int_6 d\mathbf{r}d\mathbf{r}' \exp\left[i\mathbf{q}\left(\mathbf{r}'-\mathbf{r}\right)\right]\rho_N^{(2)}(\mathbf{r},\mathbf{r}').$$

If the liquid is assumed to be homogeneous and isotropic, and that all atoms have the same properties, one can make the changes of variables $\mathbf{R}=\mathbf{r}$ and $\mathbf{X}=\mathbf{r}'-\mathbf{r}$, and explicit the pair correlation function $g(|\mathbf{r}'-\mathbf{r}|)=\frac{\rho_N^{(2)}(\mathbf{r},\mathbf{r}')}{\rho^2}$ in the statistical average as[3]:

$$\left\langle \sum_{j=1}^{N}\sum_{l\neq j}^{N} \exp\left[i\mathbf{q}\left(\mathbf{r}_j-\mathbf{r}_l\right)\right]\right\rangle = 4\pi\rho^2 V\int_0^\infty \frac{\sin(qr)}{qr} g(r) r^2 dr. \tag{5}$$

One sees that the previous integral diverges because the integrand increases with r. The problem comes from the fact that the scattered intensity, for $q=0$, has no physical meaning and can not be measured. To overcome this difficulty, one rewrites the scattered intensity $I(q)$ defined by equation (4) in the equivalent form (cf. footnote 3):

$$I(q) = NI_0 + NI_0\rho\int_V \exp\left(i\mathbf{q}\mathbf{r}\right)\left[g(r)-1\right]d\mathbf{r} + NI_0\rho\int_V \exp\left(i\mathbf{q}\mathbf{r}\right)d\mathbf{r}. \tag{6}$$

To large distances, $g(r)$ tends to unity, so that $[g(r)-1]$ tends towards zero, making the first integral convergent. As for the second integral, it corresponds to the Dirac delta function[4],

---

[2] It should be stressed that the distribution function $\rho_N^{(2)}(\mathbf{r}^2)$ is expressed as:

$$\rho_N^{(2)}(\mathbf{r},\mathbf{r}') = \rho^2 g(|\mathbf{r}'-\mathbf{r}|) = \frac{N!}{(N-2)!Z_N}\int\int_{3(N-2)} \exp\left[-\beta U(\mathbf{r}^N)\right] d\mathbf{r}_3...d\mathbf{r}_N.$$

[3] To evaluate an integral of the form:

$$I = \int_V d\mathbf{r} \exp\left(i\mathbf{q}\mathbf{r}\right) g(r),$$

one must use the spherical coordinates by placing the vector $\mathbf{q}$ along the z axis, where $\theta = (\mathbf{q},\mathbf{r})$. Thus, the integral reads:

$$I = \int_0^{2\pi} d\varphi \int_0^{\pi}\int_0^{\infty} \exp\left(iqr\cos\theta\right) g(r) r^2 \sin\theta d\theta dr,$$

with $\mu=\cos\theta$ and $d\mu = -\sin\theta d\theta$. It follows that:

$$I = -2\pi\int_0^\infty \left[\int_{+1}^{-1} \exp\left(iqr\mu\right) d\mu\right] g(r) r^2 dr = 4\pi\int_0^\infty \frac{\sin(qr)}{qr} g(r) r^2 dr.$$

[4] The generalization of the Fourier transform of the Dirac delta function to three dimensions is:

$$\delta(\mathbf{r}) = \frac{1}{(2\pi)^3}\int\int\int_{-\infty}^{+\infty} \delta(\mathbf{q}) \exp\left(-i\mathbf{q}\mathbf{r}\right) d\mathbf{q} = \frac{1}{(2\pi)^3},$$

and the inverse transform is:

$$\delta(\mathbf{q}) = \int\int\int_{-\infty}^{+\infty} \delta(\mathbf{r}) \exp\left(i\mathbf{q}\mathbf{r}\right) d\mathbf{r} = \frac{1}{(2\pi)^3}\int\int\int_{-\infty}^{+\infty} \exp\left(i\mathbf{q}\mathbf{r}\right) d\mathbf{r}.$$

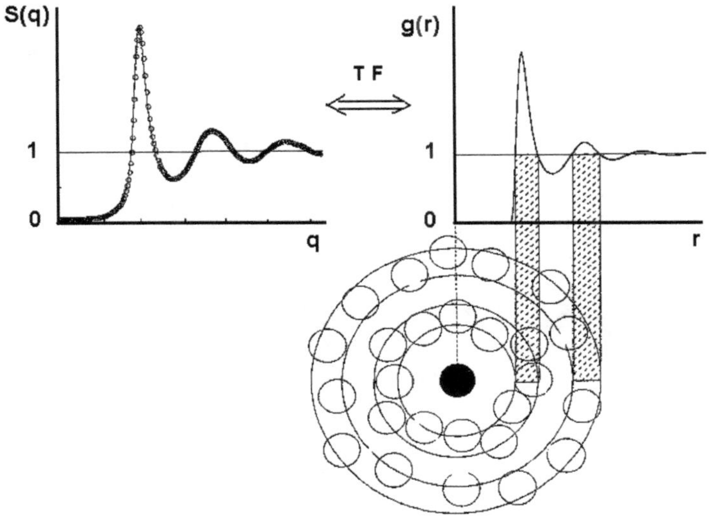

Fig. 3. Structure factor $S(q)$ and pair correlation function $g(r)$ of simple liquids.

which is zero for all values of q, except in $q = 0$ for which it is infinite. In using the delta function, the expression of the scattered intensity $I(q)$ becomes:

$$I(q) = NI_0 + NI_0\rho \int_V \exp\left(i\mathbf{qr}\right)\left[g(r) - 1\right] d\mathbf{r} + NI_0\rho(2\pi)^3\delta(\mathbf{q}).$$

From the experimental point of view, it is necessary to exclude the measurement of the scattered intensity in the direction of the incident beam ($q = 0$). Therefore, in practice, the *structure factor* $S(q)$ is defined by the following normalized function:

$$S(q) = \frac{I(q) - (2\pi)^3 NI_0\rho\delta(\mathbf{q})}{NI_0} = 1 + 4\pi\rho \int_0^\infty \frac{\sin(qr)}{qr}\left[g(r) - 1\right] r^2 dr. \tag{7}$$

Consequently, the pair correlation function $g(r)$ can be extracted from the experimental results of the structure factor $S(q)$ by performing the numerical Fourier transformation:

$$\rho\left[g(r) - 1\right] = TF\left[S(q) - 1\right].$$

The pair correlation function $g(r)$ is a dimensionless quantity, whose the graphic representation is given in figure (3). The gap around unity measures the probability of finding a particle at distance r from a particle taken in an arbitrary origin. The main peak of $g(r)$ corresponds to the position of first neighbors, and the successive peaks to the next close neighbors. The pair correlation function $g(r)$ clearly shows the existence of a short-range order that is fading rapidly beyond four or five interatomic distances. In passing, it should be mentioned that the structure factor at $q = 0$ is related to the isothermal compressibility by the exact relation $S(0) = \rho k_B T \chi_T$.

4. Thermodynamic functions of liquids

4.1 Internal energy

To express the internal energy of a liquid in terms of the pair correlation function, one must first use the following relation from statistical mechanics :

$$E = k_B T^2 \frac{\partial}{\partial T} \ln Q_N(V,T),$$

where the partition function $Q_N(V,T)$ depends on the configuration integral $Z_N(V,T)$ and on the thermal wavelength Λ, in accordance with the equations given in footnote (1). The derivative of $\ln Q_N(V,T)$ with respect to T can be written:

$$\frac{\partial}{\partial T} \ln Q_N(V,T) = \frac{\partial}{\partial T} \ln Z_N(V,T) - 3N \frac{\partial}{\partial T} \ln \Lambda,$$

with:

$$\frac{\partial}{\partial T} \ln Z_N(V,T) = \frac{1}{Z_N(V,T)} \int\int \left[\frac{1}{k_B T^2} U(\mathbf{r}^N)\right] \exp\left[-\beta U(\mathbf{r}^N)\right] d\mathbf{r}^N$$

and $$\frac{\partial}{\partial T} \ln \Lambda = \frac{1}{\Lambda}\left(-\frac{1}{2T^{3/2}}\sqrt{\frac{h^2}{2\pi m k_B}}\right) = -\frac{1}{2T}.$$

Then, the calculation is continued by admitting that the total potential energy $U(\mathbf{r}^N)$ is written as a sum of pair potentials, in the form $U(\mathbf{r}^N) = \sum_i \sum_{j>i} u(r_{ij})$. The internal energy reads:

$$E = \frac{3}{2} N k_B T + \frac{1}{Z_N(V,T)} \int\int \left[\sum_i \sum_{j>i} u(r_{ij})\right] \exp\left[-\beta U(\mathbf{r}^N)\right] d\mathbf{r}^N. \tag{8}$$

The first term on the RHS corresponds to the kinetic energy of the system; it is the ideal gas contribution. The second term represents the potential energy. Given the assumption of additivity of pair potentials, we can assume that it is composed of $N(N-1)/2$ identical terms, permitting us to write:

$$\sum_i \sum_{j>i} \frac{1}{Z_N(V,T)} \int\int u(r_{ij}) \exp\left[-\beta U(\mathbf{r}^N)\right] d\mathbf{r}^N = \frac{N(N-1)}{2} \left\langle u(r_{ij}) \right\rangle,$$

where the mean value is expressed in terms of the pair correlation function as:

$$\langle u(r_{12}) \rangle = \rho^2 \frac{(N-2)!}{N!} \int\int_6 u(r_{12}) \left[g_N^{(2}(\mathbf{r}_1,\mathbf{r}_2)\right] d\mathbf{r}_1 d\mathbf{r}_2.$$

For a homogeneous and isotropic fluid, one can perform the change of variables $\mathbf{R} = \mathbf{r}_1$ and $\mathbf{r} = \mathbf{r}_1 - \mathbf{r}_2$, where $\mathbf{R}$ and $\mathbf{r}$ describe the system volume, and write the expression of internal energy in the integral form:

$$E = \frac{3}{2} N k_B T + 2\pi\rho N \int_0^\infty u(r) g(r) r^2 dr. \tag{9}$$

Therefore, the calculation of internal energy of a liquid requires knowledge of the pair potential $u(r)$ and the pair correlation function $g(r)$. For the latter, the choice is to employ either the experimental values or values derived from the microscopic theory of liquids. Note that the integrand in equation (9) is the product of the pair potential by the pair correlation function, weighted by r^2. It should be also noted that the calculation of E can be made taking into account the three-body potential $u_3(\mathbf{r}_1, \mathbf{r}_2, \mathbf{r}_3)$ and the three-body correlation function $g^{(3)}(\mathbf{r}_1, \mathbf{r}_2, \mathbf{r}_3)$. In this case, the correlation function at three bodies must be determined only by the theory of liquids (7), since it is not accessible by experiment.

4.2 Pressure

The expression of the pressure is obtained in the same way that the internal energy, in considering the equation:

$$p = k_B T \frac{\partial}{\partial V} \ln Q_N(V,T) = k_B T \frac{\partial}{\partial V} \ln Z_N(V,T).$$

The derivation of the configuration integral with respect to volume requires using the reduced variable $\mathbf{X} = \frac{\mathbf{r}}{V}$ that allows us to find the dependence of the potential energy $U(\mathbf{r}^N)$ versus volume. Indeed, if the volume element is $d\mathbf{r} = Vd\mathbf{X}$, the scalar variable $dr = V^{1/3}dX$ leads to the derivative:

$$\frac{dr}{dV} = \frac{1}{3} V^{-2/3} X = \frac{1}{3V} r. \tag{10}$$

In view of this, the configuration integral and its derivative with respect to V are written in the following forms with reduced variables:

$$Z_N(V,T) = V^N \int \int_{3N} \exp\left[-\beta U(\mathbf{r}^N)\right] d\mathbf{X}_1 ... d\mathbf{X}_N,$$

$$\frac{\partial}{\partial V} \ln Z_N(V,T) = \frac{N}{V} + \frac{V^N}{Z_N(V,T)} \int \int_{3N} \left[-\beta \frac{\partial U(\mathbf{r}^N)}{\partial V}\right] \exp\left[-\beta U(\mathbf{r}^N)\right] d\mathbf{X}_1 ... d\mathbf{X}_N.$$

Assuming that the potential energy is decomposed into a sum of pair potentials, and with the help of equation (10), the derivation of the potential energy versus volume is performed as:

$$\frac{\partial U(\mathbf{r}^N)}{\partial V} = \frac{1}{3V} \sum_i \sum_{j>i} r_{ij} \frac{\partial u(r_{ij})}{\partial r_{ij}},$$

so that the expression of the pressure becomes:

$$p = k_B T \frac{N}{V} - \frac{1}{3V} \frac{1}{Z_N(V,T)} \sum_i \sum_{j>i} \int \int_{3N} \left[r_{ij} \frac{\partial u(r_{ij})}{\partial r_{ij}}\right] \exp\left[-\beta U(\mathbf{r}^N)\right] d\mathbf{r}_1 ... d\mathbf{r}_N. \tag{11}$$

Like for the calculation of internal energy, the additivity assumption of pair potentials permits us to write the sum of integrals of the previous equation as:

$$\sum_i \sum_{j>i} \frac{1}{Z_N(V,T)} \int \int \left[r_{ij} \frac{\partial u(r_{ij})}{\partial r_{ij}}\right] \exp\left[-\beta U(\mathbf{r}^N)\right] d\mathbf{r}^N = \frac{N(N-1)}{2} \left\langle r_{ij} \frac{\partial u(r_{ij})}{\partial r_{ij}} \right\rangle,$$

where the mean value is expressed with the pair correlation function by:

$$\left\langle r_{12}\frac{\partial u(r_{12})}{\partial r_{12}}\right\rangle = \rho^2\frac{(N-2)!}{N!}\int\int_6 r_{12}\frac{\partial u(r_{12})}{\partial r_{12}}\left[g_N^{(2}(\mathbf{r}_1,\mathbf{r}_2)\right]d\mathbf{r}_1 d\mathbf{r}_2.$$

For a homogeneous and isotropic fluid, one can perform the change of variables $\mathbf{R} = \mathbf{r}_1$ and $\mathbf{r} = \mathbf{r}_1 - \mathbf{r}_2$, and simplify the expression of pressure as:

$$p = k_B T\frac{N}{V} - \frac{2\pi}{3}\rho^2\int_0^\infty r^3\frac{\partial u(r)}{\partial r}g(r)dr. \tag{12}$$

The previous equation provides the pressure of a liquid as a function of the pair potential and the pair correlation function. It is the so-called *pressure equation of state* of liquids. It should be stressed that this equation of state is not unique, as we will see in presenting the hard-sphere reference system (§ 4. 4). As the internal energy, the pressure can be written with an additional term containing the three-body potential $u_3(\mathbf{r}_1,\mathbf{r}_2,\mathbf{r}_3)$ and the three-body correlation function $g^{(3)}(\mathbf{r}_1,\mathbf{r}_2,\mathbf{r}_3)$.

4.3 Chemical potential and entropy

We are now able to calculate the internal energy (Eq. 9) and pressure (Eq. 12) for any system, of which the potential energy is made of a sum of pair potentials $u(r)$ and the pair correlation function $g(r)$ is known. Beside this, all other thermodynamic properties can be easily derived. Traditionally, it is appropriate to derive the chemical potential μ as a function of $g(r)$ by integrating the partition function with respect to a parameter λ to be defined (8).

Firstly, the formal expression of the chemical potential is defined by the energy required to introduce a new particle in the system:

$$\mu = F(V,T,N) - F(V,T,N-1) = \left(\frac{\partial F}{\partial N}\right)_{V,T}.$$

From footnote (1), the free energy F is written:

$$F(V,T,N) = -k_B T\ln Q_N(V,T) = -k_B T\left[\ln Z_N(V,T) - \ln N! - N\ln\Lambda^3\right],$$

so that the chemical potential can be simplified as:

$$\mu = k_B T\left[-\ln\frac{Z_N(V,T)}{Z_{N-1}(V,T)} + \ln N + \ln\Lambda^3\right]. \tag{13}$$

Secondly, the procedure requires to write the potential energy as a function of the *coupling parameter* λ, under the following form, in order to assess the argument of the logarithm in the above relation:

$$U(\mathbf{r}^N,\lambda) = \lambda\sum_{j=2}^{N}u(r_{1j}) + \sum_{i}^{N}\sum_{j>i\geq 2}^{N}u(r_{ij}). \tag{14}$$

Varying from 0 to 1, the coupling parameter λ measures the degree of coupling of the particle to which it is assigned (1 in this case) with the rest of the system. In the previous relation, $\lambda = 1$

means that particle 1 is completely coupled with the other particles, while $\lambda = 0$ indicates a zero coupling, that is to say the absence of the particle 1 in the system. This allows the writing of the important relations:

$$U(\mathbf{r}^N,1) = \sum_{j=2}^{N} u(r_{1j}) + \sum_{i}^{N} \sum_{j>i\geq 2}^{N} u(r_{ij}) = \sum_{i}^{N} \sum_{j>i\geq 1}^{N} u(r_{ij}) = U(\mathbf{r}^N),$$

and

$$U(\mathbf{r}^N,0) = \sum_{i}^{N} \sum_{j>i\geq 2}^{N} u(r_{ij}) = U(\mathbf{r}^{N-1}).$$

Under these conditions, the configuration integrals for a total coupling ($\lambda = 1$) and a zero coupling ($\lambda = 0$) are respectively:

$$Z_N(V,T,\lambda = 1) = \int\int_{3N} \exp\left[-\beta U(\mathbf{r}^N)\right] d\mathbf{r}_1 d\mathbf{r}_2 ... d\mathbf{r}_N = Z_N(V,T), \tag{15}$$

$$Z_N(V,T,\lambda = 0) = \int_V d\mathbf{r}_1 \int\int_{3(N-1)} \exp\left[-\beta U(\mathbf{r}^{N-1})\right] d\mathbf{r}_2 ... d\mathbf{r}_N = V Z_{N-1}(V,T). \tag{16}$$

These expressions are then used to calculate the logarithm of the ratio of configuration integrals in equation (13):

$$\ln \frac{Z_N(V,T)}{Z_{N-1}(V,T)} = \ln \frac{Z_N(V,T,\lambda = 1)}{Z_N(V,T,\lambda = 0)} + \ln V \tag{17}$$

$$= \ln V + \int_0^1 \frac{\partial \ln Z_N}{\partial \lambda} d\lambda. \tag{18}$$

But with the configuration integral $Z_N(V,T,\lambda)$, in which potential energy is given by equation (14), we can easily evaluate the partial derivatives $\frac{\partial Z_N}{\partial \lambda}$ and $\frac{\partial \ln Z_N}{\partial \lambda}$. In particular, with the result of the footnote (1), we can write $\frac{\partial \ln Z_N}{\partial \lambda}$ as a function of the pair correlation function as:

$$\frac{\partial \ln Z_N(V,T,\lambda)}{\partial \lambda} = -\beta \rho^2 \frac{(N-1)(N-2)!}{N!} \int\int_6 u(r_{12}) \left\{ g_N^{(2)}(\mathbf{r}_1,\mathbf{r}_2,\lambda) \right\} d\mathbf{r}_1 d\mathbf{r}_2.$$

In addition, if the fluid is homogeneous and isotropic, the above relation simplifies under the following form:

$$\frac{\partial \ln Z_N(V,T,\lambda)}{\partial \lambda} = -\frac{\beta \rho^2}{N} V \int_0^\infty u(r) g(r,\lambda) 4\pi r^2 dr,$$

that remains only to be substituted in equation (18) for obtaining the logarithm of the ratio of configuration integrals. And by putting the last expression in equation (13), one ultimately arrives to the expression of the chemical potential:

$$\mu = k_B T \ln \rho \Lambda^3 + 4\pi \rho \int_0^1 \int_0^\infty u(r) g(r,\lambda) r^2 dr d\lambda. \tag{19}$$

Thus, like the internal energy (Eq. 9) and pressure (Eq. 12), the chemical potential (Eq. 19) is calculated using the pair potential and pair correlation function.
Finally, one writes the entropy S in terms of the pair potential and pair correlation function, owing to the expressions of the internal energy (Eq. 9), pressure (Eq. 12) and chemical potential (Eq. 19) (cf. footnote 1):

$$S = \frac{E-F}{T} = \frac{E}{T} - \frac{\mu N}{T} + \frac{pV}{T}. \tag{20}$$

It should be noted that the entropy can also be estimated only with the pair correlation function $g(r)$, without recourse to the pair potential $u(r)$. The reader interested by this issue should refer to the original articles (9).

4.4 Application to the hard-sphere potential

In this subsection we determine the equation of state of the hard-sphere system, of which the pair potential being:

$$u(r) = \begin{cases} \infty & \text{if } r < \sigma \\ 0 & \text{if } r > \sigma, \end{cases}$$

where σ is the hard-sphere diameter. The Boltzmann factor associated with this potential has a significant feature that enable us to express the thermodynamic properties under particularly simple forms. Indeed, the representation of the Boltzmann factor

$$\exp\left[-\beta u(r)\right] = \begin{cases} 0 & \text{if } \quad r < \sigma \\ 1 & \text{if } \quad r > \sigma, \end{cases}$$

is a step function (Fig. 4) whose derivative with respect to r is the Dirac delta function, i. e.:

$$\frac{\partial}{\partial r}\exp\left[-\beta u(r)\right] = -\beta\frac{\partial u}{\partial r}\exp\left[-\beta u(r)\right] = \delta(r-\sigma).$$

In substituting $\frac{\partial u}{\partial r}$, taken from the previous relation, in equation (12) we find the expression of the pressure:

$$p = k_B T\frac{N}{V} - \frac{2\pi}{3}\rho^2\int_0^\infty r^3\left\{-\frac{1}{\beta}\frac{\delta(r-\sigma)}{\exp\left[-\beta u(r)\right]}\right\}g(r)dr,$$

or:

$$p = k_B T\frac{N}{V} + \frac{2\pi}{3}k_B T\rho^2\sigma^3 g(\sigma)\exp\left[\beta u(\sigma)\right]. \tag{21}$$

It is important to recall that, for moderately dense gases, the pressure is usually expressed under the form of the virial expansion

$$\frac{p}{\rho k_B T} = 1 + \rho B_2(T) + \rho^2 B_3(T) + \rho^3 B_4(T) + \ldots = \frac{p^{GP}}{\rho k_B T} + \frac{p^{ex}}{\rho k_B T}.$$

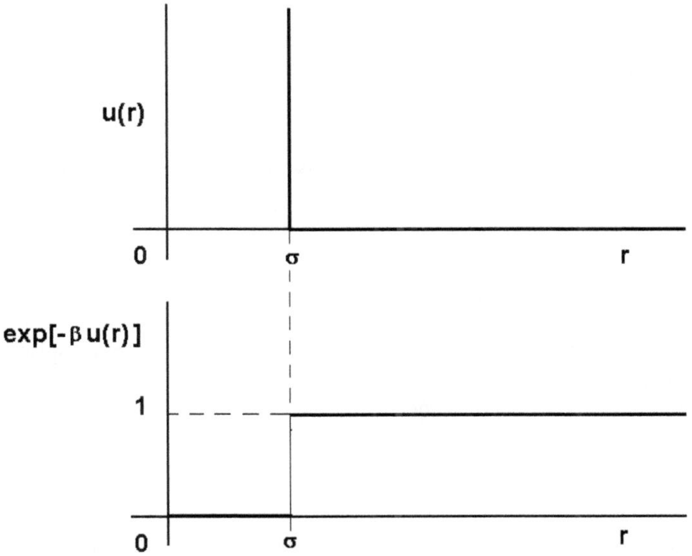

Fig. 4. Representation of the hard-sphere potential and its Boltzmann factor.

The first term of the last equality represents the contribution of the ideal gas, and the excess pressure p^{ex} comes from the interactions between particles. They are written:

$$\frac{p^{GP}}{\rho k_B T} = 1,$$

and

$$\frac{p^{ex}}{\rho k_B T} = 4\eta + \eta^2 B_3'(T) + \eta^3 B_4'(T) + ...$$

where η is the *packing fraction* defined by the ratio of the volume actually occupied by the N spherical particles on the total volume V of the system, that is to say:

$$\eta = \frac{1}{V}\frac{4\pi}{3}\left(\frac{\sigma}{2}\right)^3 N = \frac{\pi}{6}\rho\sigma^3. \tag{22}$$

Note that the first 6 coefficients of the excess pressure p^{ex} have been calculated analytically and by molecular dynamics (10), with great accuracy. In addition, Carnahan and Starling (11) have shown that the excess pressure of the hard-sphere fluid can be very well predicted by rounding the numerical values of the 6 coefficients towards the nearest integer values, according to the expansion:

$$\frac{p^{ex}}{\rho k_B T} \simeq 4\eta + 10\eta^2 + 18\eta^3 + 28\eta^4 + 40\eta^5 + 54\eta^6 ... \simeq \sum_{k=1}^{\infty}(k^2 + 3k)\eta^k. \tag{23}$$

In combining the first and second derivatives of the geometric series $\sum_{k=1}^{\infty} \eta^k$, it is found that equation (23) can be transformed into a rational fraction[5] enabling the deduction of the excess pressure in the form:

$$\frac{p^{ex}}{\rho k_B T} \simeq \sum_{k=1}^{\infty} (k^2 + 3k)\eta^k = \frac{4\eta - 2\eta^2}{(1-\eta)^3}.$$

Consequently, the equation of state of the hard-sphere fluid is written with excellent precision as:

$$\frac{p}{\rho k_B T} = \frac{1 + \eta + \eta^2 - \eta^3}{(1-\eta)^3}. \tag{24}$$

It is also possible to calculate the internal energy of the hard-sphere fluid by substituting $u(r)$ in equation (9). Given that $u(r)$ is zero when $r > \sigma$ and $g(r)$ is zero when $r < \sigma$, it follows that the integral is always zero, and that the internal energy of the hard-sphere fluid is equal to that of the ideal gas $E = \frac{3}{2} N k_B T$.

As for the free energy F, it is determined by integrating the pressure over volume with the equation:

$$p = -\left(\frac{\partial F}{\partial V}\right)_T = -\left(\frac{\partial F^{GP}}{\partial V}\right)_T - \left(\frac{\partial F^{ex}}{\partial V}\right)_T,$$

where F^{GP} is the free energy of ideal gas (cf. footnote 1, with $Z_N(V,T) = V^N$):

$$F^{GP} = N k_B T \left(\ln \rho \Lambda^3 - 1\right),$$

and F^{ex} the excess free energy, calculated by integrating equation (23) as follows:

$$F^{ex} = -\int p^{ex} dV = -\int \frac{N k_B T}{V} \left(4\eta + 10\eta^2 + 18\eta^3 + 28\eta^4 + 40\eta^5 + 54\eta^6 ...\right) \frac{dV}{d\eta} d\eta.$$

[5] To obtain the rational fraction, one must decompose the sum as:

$$\sum_{k=1}^{\infty} (k^2 + 3k)\eta^k = \sum_{k=1}^{\infty} (k^2 - k)\eta^k + \sum_{k=1}^{\infty} 4k\eta^k,$$

and combine the geometric series $\sum_{k=1}^{\infty} \eta^k$ with its first and second derivatives:

$$\sum_{k=1}^{\infty} \eta^k = \eta + \eta^2 + \eta^3 + ... = \frac{\eta}{1-\eta},$$

$$\sum_{k=1}^{\infty} k\eta^{k-1} = \frac{1}{(1-\eta)^2},$$

and

$$\sum_{k=1}^{\infty} k(k-1)\eta^{k-2} = \frac{2}{(1-\eta)^3},$$

to see appear the relation:

$$\sum_{k=1}^{\infty} (k^2 + 3k)\eta^k = \frac{2\eta^2}{(1-\eta)^3} + \frac{4\eta}{(1-\eta)^2}.$$

But, with equation (22) that gives $\frac{dV}{d\eta} = -\frac{V}{\eta}$, F^{ex} is then reduced to the series expansion:

$$\frac{F^{ex}}{Nk_BT} = 4\eta + 5\eta^2 + 6\eta^3 + 7\eta^4 + 8\eta^5 + 9\eta^6 ... = \sum_{k=1}^{\infty}(k+3)\eta^k.$$

Like the pressure, this expansion is written as a rational function by combining the geometric series $\sum_{k=1}^{\infty}\eta^k$ with its first derivative[6]. The expression of the excess free energy is:

$$\frac{F^{ex}}{Nk_BT} = \sum_{k=1}^{\infty}(k+3)\eta^k = \frac{4\eta - 3\eta^2}{(1-\eta)^2},$$

and the free energy of the hard-sphere fluid reduces to the following form:

$$\frac{F}{Nk_BT} = \frac{F^{GP}}{Nk_BT} + \frac{F^{ex}}{Nk_BT} = \ln \rho\Lambda^3 - 1 + \frac{4\eta - 3\eta^2}{(1-\eta)^2}. \tag{25}$$

Now, the entropy is obtained using the same method of calculation, by deriving the free energy with respect to temperature:

$$S = -\left(\frac{\partial F}{\partial T}\right)_V = -\left(\frac{\partial F^{GP}}{\partial T}\right)_V - \left(\frac{\partial F^{ex}}{\partial T}\right)_V,$$

where S^{GP} is the entropy of the ideal gas given by the Sackur-Tetrode equation:

$$S^{GP} = -Nk_B\left(\ln \rho\Lambda^3 - \frac{5}{2}\right),$$

and where the excess entropy S^{ex} arises from the relation:

$$S^{ex} = -\left(\frac{\partial F^{ex}}{\partial T}\right)_V = -Nk_B\frac{4\eta - 3\eta^2}{(1-\eta)^2},$$

hence the expression of the entropy of the hard-sphere fluid:

$$\frac{S}{Nk_B} = -\ln \rho\Lambda^3 + \frac{5}{2} - \frac{4\eta - 3\eta^2}{(1-\eta)^2}. \tag{26}$$

Finally, combining equations (25) and (24), with the help of equation (20), one reaches the chemical potential of the hard-sphere fluid that reads:

$$\frac{\mu}{k_BT} = \frac{F}{Nk_BT} + \frac{p}{\rho k_BT} = \ln \rho\Lambda^3 - 1 + \frac{1 + 5\eta - 6\eta^2 + 2\eta^3}{(1-\eta)^3}. \tag{27}$$

---

[6] Indeed, the identity:

$$\sum_{k=1}^{\infty}(k+3)\eta^k = \sum_{k=1}^{\infty}k\eta^k + \sum_{k=1}^{\infty}3\eta^k,$$

is yet written:

$$\sum_{k=1}^{\infty}(k+3)\eta^k = \frac{\eta}{(1-\eta)^2} + \frac{3\eta}{(1-\eta)},$$

Since they result from equation (23), the expressions of thermodynamic properties (p, F, S and μ) of the hard-sphere fluid make up a homogeneous group of relations related to the Carnahan and Starling equation of state. But other expressions of thermodynamic properties can also be determined using the pressure equation of state (Eq. 12) and the compressibility equation of state, which will not be discussed here. Unlike the Carnahan and Starling equation of state, these two equations of state require knowledge of the pair correlation function of hard spheres, $g_{HS}(r)$. The latter is not available in analytical form. The interested reader will find the Fortran program aimed at doing its calculation, in the book by McQuarrie (12), page 600. It should be mentioned that the thermodynamic properties (p, F, S and μ), obtained with the equations of state of pressure and compressibility, have analytical forms similar to those from the Carnahan and Starling equation of state, and they provide results whose differences are indistinguishable to low densities.

5. Thermodynamic perturbation theory

All theoretical and experimental studies have shown that the structure factor $S(q)$ of simple liquids resembles that of the hard-sphere fluid. For proof, just look at the experimental structure factor of liquid sodium (13) at 373 K, in comparison with the structure factor of hard-sphere fluid (14) for a value of the packing fraction η of 0.45. We can see that the agreement is not bad, although there is a slight shift of the oscillations and ratios of peak heights significantly different. Besides, numerical calculations showed that the structure factor obtained with the Lennard-Jones potential describes the structure of simple fluids (15) and looks like the structure factor of hard-sphere fluid whose diameter is chosen correctly.

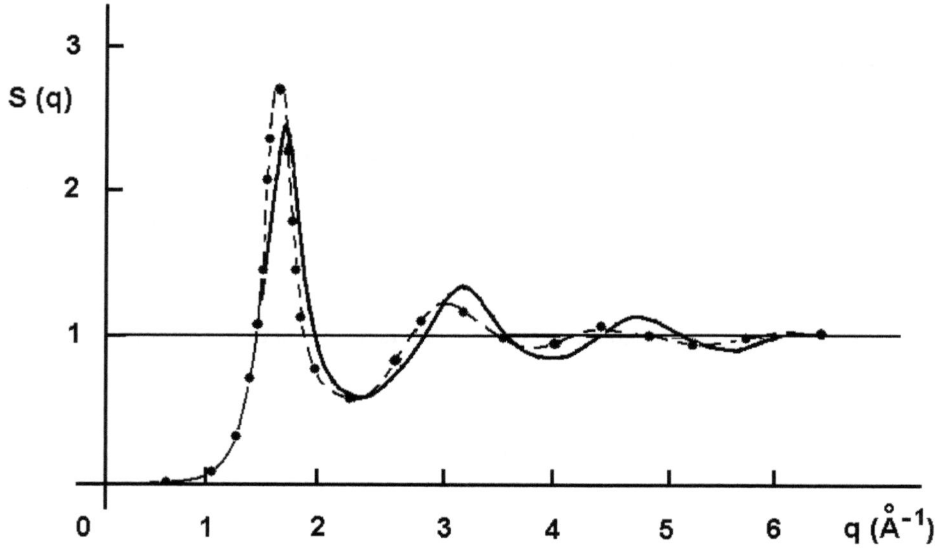

Fig. 5. Experimental structure factor of liquid sodium at 373 K (points), and hard-sphere structure factor (solid curve), with $\eta = 0,45$.

Such a qualitative success emphasizes the role played by the repulsive part of the pair potential to describe the structure factor of liquids, while the long-ranged attractive contribution has a minor role. It can be said for simplicity that the repulsive contribution of the potential determines the structure of liquids (stacking of atoms and steric effects) and the attractive contribution is responsible for their cohesion.
It is important to remember that the thermodynamic properties of the hard-sphere fluid (Eqs. 24, 25, 26, 27) and the structure factor $S_{HS}(q)$ can be calculated with great accuracy. That suggests replacing the repulsive part of potential in real systems by the hard-sphere potential that becomes the *reference system*, and precict the structural and thermodynamic properties of real systems with those of the hard-sphere fluid, after making the necessary adaptations. To perform these adaptations, the attractive contribution of potential should be treated as a perturbation to the reference system.
The rest of this subsection is devoted to a summary of thermodynamic perturbation methods[7]. It should be noted, from the outset, that the calculation of thermodynamic properties with the thermodynamic perturbation methods requires knowledge of the pair correlation function $g_{HS}(r)$ of the hard-sphere system and not that of the real system.

5.1 Zwanzig method

In perturbation theory proposed by Zwanzig (16), it is assumed that the total potential energy $U(\mathbf{r}^N)$ of the system can be divided into two parts. The first part, $U_0(\mathbf{r}^N)$, is the energy of the unperturbed system considered as reference system and the second part, $U_1(\mathbf{r}^N)$, is the energy of the perturbation which is much smaller that $U_0(\mathbf{r}^N)$. More precisely, it is posed that the potential energy depend on the coupling parameter λ by the relation:

$$U(\mathbf{r}^N) = U_0(\mathbf{r}^N) + \lambda U_1(\mathbf{r}^N)$$

in order to vary continuously the potential energy from $U_0(\mathbf{r}^N)$ to $U(\mathbf{r}^N)$, by changing λ from 0 to 1, and that the free energy F of the system is expanded in Taylor series as:

$$F = F_0 + \lambda \left(\frac{\partial F}{\partial \lambda}\right) + \frac{\lambda^2}{2}\left(\frac{\partial^2 F}{\partial \lambda^2}\right) + ... \qquad (28)$$

By replacing the potential energy $U(\mathbf{r}^N)$ in the expression of the configuration integral (cf. footnote 1), one gets:

$$Z_N(V,T) = \int\int_{3N} \exp\left[-\beta U_0(\mathbf{r}^N)\right] d\mathbf{r}^N \times \frac{\int\int_{3N}\left\{\exp\left[-\beta\lambda U_1(\mathbf{r}^N)\right]\right\}\exp\left[-\beta U_0(\mathbf{r}^N)\right] d\mathbf{r}^N}{\int\int_{3N}\exp\left[-\beta U_0(\mathbf{r}^N)\right] d\mathbf{r}^N}.$$

The first integral represents the configuration integral $Z_N^{(0)}(V,T)$ of the reference system, and the remaining term can be regarded as the average value of the quantity $\exp\left[-\beta\lambda U_1(\mathbf{r}^N)\right]$, so that the previous relation can be put under the general form:

$$Z_N(V,T) = Z_N^{(0)}(V,T)\left\langle \exp\left[-\beta\lambda U_1(\mathbf{r}^N)\right]\right\rangle_0, \qquad (29)$$

where $\langle ... \rangle_0$ refers to the statistical average in the canonical ensemble of the reference system. After the substitution of the configuration integral (Eq. 29) in the expression of the free energy

[7] The interested reader will find all useful adjuncts in the books either by J. P. Hansen and I. R. McDonald or by D. A. McQuarrie.

(cf. footnote 1), this one reads:

$$-\beta F = \ln \frac{Z_N^{(0)}(V,T)}{N!\Lambda^{3N}} + \ln \left\langle \exp\left[-\beta\lambda U_1(\mathbf{r}^N)\right]\right\rangle_0 . \tag{30}$$

The first term on the RHS stands for the free energy of the reference system, denoted $(-\beta F_0)$, and the second term represents the free energy of the perturbation:

$$-\beta F_1 = \ln \left\langle \exp\left[-\beta\lambda U_1(\mathbf{r}^N)\right]\right\rangle_0 . \tag{31}$$

Since the perturbation $U_1(\mathbf{r}^N)$ is small, $\exp(-\beta\lambda U_1)$ can be expanded in series, so that the statistical average $\langle \exp\left[-\beta\lambda U_1(\mathbf{r}^N)\right]\rangle_0$, calculated on the reference system, is expressed as:

$$\left\langle \exp\left[-\beta\lambda U_1(\mathbf{r}^N)\right]\right\rangle_0 = 1 - \beta\lambda \langle U_1\rangle_0 + \frac{1}{2!}\beta^2\lambda^2 \left\langle U_1^2\right\rangle_0 - \frac{1}{3!}\beta^3\lambda^3 \left\langle U_1^3\right\rangle_0 + \ldots \tag{32}$$

Incidentally, we may note that the coefficients of β in the preceding expansion represent statistical moments in the strict sense. Given the shape of equation (32), it is still possible to write equation (31) by expanding $\ln\langle \exp\left[-\beta\lambda U_1(\mathbf{r}^N)\right]\rangle_0$ in Taylor series. After simplifications, equation (31) reduces to:

$$\begin{aligned}\ln \left\langle \exp\left[-\beta\lambda U_1(\mathbf{r}^N)\right]\right\rangle_0 &= -\beta\lambda \langle U_1\rangle_0 + \frac{1}{2!}\beta^2\lambda^2 \left[\left\langle U_1^2\right\rangle_0 - \langle U_1\rangle_0^2\right] \\ &\quad -\beta^3\lambda^3 \left[\frac{1}{3!}\left\langle U_1^3\right\rangle_0 - \frac{1}{2}\langle U_1\rangle_0 \left\langle U_1^2\right\rangle_0 + \frac{1}{3}\langle U_1\rangle_0^3\right] + \beta^4\lambda^4 [\ldots] - \ldots\end{aligned}$$

Now if we set:

$$c_1 = \langle U_1\rangle_0 , \tag{33}$$

$$c_2 = \frac{1}{2!}\left[\left\langle U_1^2\right\rangle_0 - \langle U_1\rangle_0^2\right], \tag{34}$$

$$c_3 = \frac{1}{3!}\left[\left\langle U_1^3\right\rangle_0 - 3\langle U_1\rangle_0 \left\langle U_1^2\right\rangle_0 + 2\langle U_1\rangle_0^3\right], \quad \text{etc.} \tag{35}$$

we find that:

$$\ln \left\langle \exp\left[-\beta\lambda U_1(\mathbf{r}^N)\right]\right\rangle_0 = -\lambda\beta c_1 + \lambda^2\beta^2 c_2 - \lambda^3\beta^3 c_3 + \ldots$$

The contribution of the perturbation (Eq. 31) to the free energy is then written in the compact form:

$$-\beta F_1 = \ln \left\langle \exp\left[-\beta\lambda U_1(\mathbf{r}^N)\right]\right\rangle_0 = -\lambda\beta \sum_{n=1}^{\infty} c_n(-\lambda\beta)^{n-1}, \tag{36}$$

and the expression of the free energy F of the real system is found by substituting equation (36) into equation (30), as follows:

$$F = F_0 + F_1 = F_0 + \lambda c_1 - \lambda^2\beta c_2 + \lambda^3\beta^2 c_3 + \ldots, \tag{37}$$

where the free energy of the real system is obtained by putting $\lambda = 1$. This expression of the free energy of liquids in power series expansion of β corresponds to the *high temperature approximation*.

5.2 Van der Waals equation

As a first application of thermodynamic perturbation method, we search the phenomenological van der Waals equation of state. In view of this, consider equation (37) at zero order in β. The simplest assumption to determine c_1 is to admit that the total potential energy may be decomposed into a sum of pair potentials in the form:

$$U(\mathbf{r}^N) = U_0(\mathbf{r}^N) + U_1(\mathbf{r}^N) = \sum_i \sum_{j>i} u_0(r_{ij}) + \sum_i \sum_{j>i} u_1(r_{ij}).$$

Therefore, the free energy of the perturbation to zero order in β is given by equation (33), that is to say:

$$c_1 = \langle U_1 \rangle_0 = \frac{\int \int_{3N} \left[\sum_i \sum_{j>i} u_1(r_{ij}) \right] \exp\left[-\beta U_0(\mathbf{r}^N)\right] d\mathbf{r}^N}{\int \int_{3N} \exp\left[-\beta U_0(\mathbf{r}^N)\right] d\mathbf{r}^N}.$$

To simplify the above relation, we proceed as for calculating the internal energy of liquids (Eq. 8) by revealing the pair correlation function of the reference system in the numerator. If we assume that the sum of pair potentials is composed of equivalent terms equal to $\sum_i \sum_{j>i} u_1(r_{ij}) = \frac{N(N-1)}{2} u_1(r_{12})$, the expression of c_1 is simplified as:

$$c_1 = \frac{1}{Z_0(V,T)} \int \int \frac{N(N-1)}{2} u_1(r_{12}) \left\{ \int \int_{3(N-2)} \exp\left[-\beta U_0(\mathbf{r}^N)\right] d\mathbf{r}_3 ... d\mathbf{r}_N \right\} d\mathbf{r}_1 d\mathbf{r}_2.$$

The integral in between the braces is then expressed as a function of the pair correlation function (cf. footnote 2), and c_1 reduces to:

$$c_1 = \frac{\rho^2}{2} \int d\mathbf{R} \int u_1(r) g_0(r) d\mathbf{r}. \tag{38}$$

Yet, to find the equation of van der Waals we have to choose the hard-sphere system of diameter σ, as reference system, and suppose that the perturbation is a long-range potential, weakly attractive, the form of which is not useful to specify (Fig. 6a). Since one was unaware of the existence of the pair correlation function when the model was developed by van der Waals, it is reasonable to estimate $g_0(r)$ by a function equal to zero within the particle, and to one at the outside. According to van der Waals, suppose further that the available volume per particle[8] is $b = \frac{2}{3}\pi\sigma^3$ and the unoccupied volume is $(V - Nb)$.

With these simplifications in mind, the configuration integral and free energy of the reference system are respectively (cf. footnote 1):

$$Z_0(V,T) = \int \int_{3N} \exp\left[-\beta U_0(\mathbf{r}^N)\right] d\mathbf{r}^N = (V - Nb)^N,$$

and $$F_0 = -k_B T \ln \left[\frac{(V - Nb)^N}{N! \Lambda^{3N}} \right] = -N k_B T \left[\ln \frac{(V - Nb)}{N} - 3 \ln \Lambda + 1 \right].$$

[8] The parameter b introduced by van der Waals is the covolume. Its expression comes from the fact that if two particles are in contact, half of the excluded volume $\frac{4}{3}\pi\sigma^3$ must be assigned to each particle (Fig. 6b).

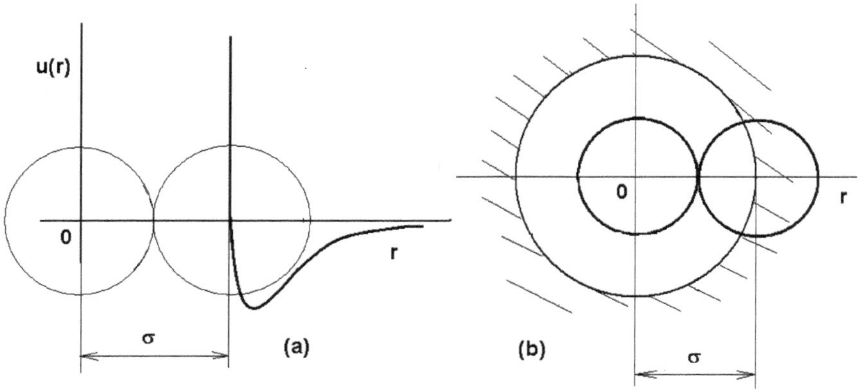

Fig. 6. Schematic representation of the pair potential by a hard-sphere potentiel plus a perturbation. (b) Definition of the covolume by the quantity $b = \frac{1}{2}\left(\frac{4}{3}\pi\sigma^3\right)$.

As for the coefficient c_1 (Eq. 38), it is simplified as:

$$c_1 = 2\pi\rho^2 V \int_\sigma^\infty u_1(r) r^2 dr = -a\rho N, \tag{39}$$

$$\text{with} \quad a = -2\pi \int_\sigma^\infty u_1(r) r^2 dr.$$

Therefore, the expression of free energy (Eq. 37) corresponding to the model of van der Waals is:

$$F = F_0 + c_1 = -Nk_BT\left[\ln\frac{(V-Nb)}{N} - 3\ln\Lambda + 1\right] - a\rho N,$$

and the van der Waals equation of state reduces to:

$$p = -\left(\frac{\partial F}{\partial V}\right)_T = \frac{Nk_BT}{V-Nb} - a\frac{N^2}{V^2}.$$

With $b = a = 0$ in the previous equation, it is obvious that one recovers the equation of state of ideal gas. In return, if one wishes to improve the quality of the van der Waals equation of state, one may use the expression of the free energy (Eq. 25) and pair correlation function $g_{HS}(r)$ of the hard-sphere system to calculate the value of the parameter a with equation (38). Another way to improve performance is to calculate the term c_2. Precisely what will be done in the next subsection.

5.3 Method of Barker and Henderson

To evaluate the mean values of the perturbation $U_1(\mathbf{r}^N)$ in equations (34) and (35), Barker and Henderson (17) suggested to discretize the domain of interatomic distances into sufficiently small intervals $(r_1, r_2), (r_2, r_3), ..., (r_i, r_{i+1}), ...,$ and assimilate the perturbating elemental potential in each interval by a constant. Assuming that the perturbating potential in the interval (r_i, r_{i+1}) is $u_1(r_i)$ and that the number of atoms subjected to this potential is N_i, the total perturbation can be written as the sum of elemental potentials:

$$U_1(\mathbf{r}^N) = \sum_i N_i u_1(r_i),$$

before substituting it in the configuration integral. The advantage of this method is to calculate the coefficients c_n and free energy (Eq. 37), not with the mean values of the perturbation, but with the fluctuation number of particles. Thus, each perturbating potential $u_1(r_i)$ is constant in the interval which it belongs, so that we can write $\langle U_1\rangle_0^2 = \sum_i \sum_j \langle N_i\rangle_0 \langle N_j\rangle_0 u_1(r_i)u_1(r_j)$. In view of this, the coefficient c_2 defined by equation (34) is:

$$c_2 = \frac{1}{2!}\left[\left\langle U_1^2\right\rangle_0 - \langle U_1\rangle_0^2\right] = \frac{1}{2!}\sum_i\sum_j \left[\left\langle N_i N_j\right\rangle_0 - \langle N_i\rangle_0 \left\langle N_j\right\rangle_0\right] u_1(r_i)u_1(r_j).$$

With the *local compressibility* approximation (LC), where ρ and g_0 depend on p, the expression of c_2 obtained by Barker and Henderson according to the method described above is written:

$$c_2(LC) = \frac{\pi N \rho}{\beta}\left(\frac{\partial \rho}{\partial p}\right)_0 \frac{\partial}{\partial \rho}\left[\int \rho u_1^2(r) g_0(r) r^2 dr\right]. \tag{40}$$

Incidentally, note the *macroscopic compressibility* approximation (MC), where only ρ is assumed to be dependent on p, has also been tested on a system made of the hard-sphere reference system and the square-well potential as perturbation. At low densities, the results of both approximations are comparable. But at intermediate densities, the results obtained with the LC approximation are in better agreement with the simulation results than the MC approximation. Note also that the coefficient c_3 has been calculated by Mansoori and Canfield (18) with the macroscopic compressibility approximation.

At this stage of the presentation of the thermodynamic perturbation theory, we are in position to calculate the first terms of the development of the free energy F (Eq. 37), using the hard-sphere system as reference system. But there is not yet a criterion for choosing the diameter d of hard spheres. This point is important because all potentials have a repulsive part that must be replaced by a hard-sphere potential of diameter properly chosen. Decisive progress has been made to solve this problem in three separate ways followed, respectively, by Barker and Henderson (19), Mansoori and Canfield (20) and Week, Chandler and Andersen (21).

Prescription of Barker and Henderson. To choose the best reference system, that is to say, the optimal diameter of hard spheres, Barker and Henderson (19) proposed to replace the potential separation $u(r) = u_0(r) + \lambda u_1(r)$, where $u_0(r)$ is the reference potential, $u_1(r)$ the perturbation potential and λ the coupling parameter, by a more complicated separation associated with a potential $v(r)$ whose the Boltzmann factor is:

$$\begin{aligned}\exp\left[-\beta v(r)\right] = &\left[1 - \Xi\left(d + \frac{r-d}{\alpha} - \sigma\right)\right]\exp\left[-\beta u(d + \frac{r-d}{\alpha})\right] \\ &+ \Xi\left(d + \frac{r-d}{\alpha} - \sigma\right) + \Xi\left(r - \sigma\right)\left\{\exp\left[-\beta\lambda u(r)\right] - 1\right\},\end{aligned} \tag{41}$$

where $\Xi(x)$ is the Heaviside function, which is zero when $x < 0$ and is worth one when $x > 0$. Note that here σ is the value of r at which the real potential $u(r)$ vanishes and d is the hard-sphere diameter of the reference potential, to be determined. Moreover, the parameters λ and α are coupling parameters that are 0 or 1. If one looks at equation (41) at the same time as figure(7a), it is seen that the function $v(r)$ reduces to the real potential $u(r)$ when $\alpha = \lambda = 1$,

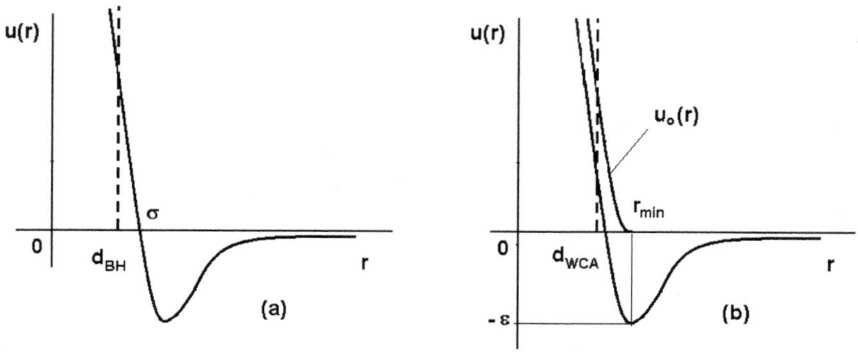

Fig. 7. Separation of the potential $u(r)$ according to (a) the method of Barker and Henderson and (b) the method of Weeks, Chandler and Andersen.

and it behaves approximately as the hard-sphere potential of diameter d when $\alpha \sim \lambda \sim 0$. The substitution of equation (41) in the configuration integral (Eq. 29), followed by the related calculations not reproduced here, enable us to express the free energy F of the real system as a series expansion in powers of α and λ, which makes the generalization of equation (28), namely:

$$F = F_{HS} + \lambda \left(\frac{\partial F}{\partial \lambda} \right) + \alpha \left(\frac{\partial F}{\partial \alpha} \right) + \frac{\lambda^2}{2} \left(\frac{\partial^2 F}{\partial \lambda^2} \right) + \frac{\alpha^2}{2} \left(\frac{\partial^2 F}{\partial \alpha^2} \right) + ... \tag{42}$$

By comparing equations (37) and (42), we see that the first derivative $\left(\frac{\partial F}{\partial \lambda} \right)$ coincides with c_1 and the second derivative $\frac{1}{2} \left(\frac{\partial^2 F}{\partial \lambda^2} \right)$ with $(-\beta c_2)$. Concerning the derivatives of F with respect to α, they are complicated functions of the pair potential and the pair correlation function of the hard-sphere system. The first derivative $\left(\frac{\partial F}{\partial \alpha} \right)$, whose the explicit form given without proof, reads:

$$\left(\frac{\partial F}{\partial \alpha} \right) = -2\pi N \rho k_B T d^2 g^{HS}(d) \left[d - \int_0^{\sigma} \{1 - \exp\left[-\beta u(r)\right]\} \, dr \right].$$

Since the Barker and Henderson prescription is based on the proposal to cancel the term $\left(\frac{\partial F}{\partial \alpha} \right)$, the criterion for choosing the hard-sphere diameter d is reduced to the following equation:

$$d = \int_0^{\sigma} \{1 - \exp\left[-\beta u(r)\right]\} \, dr. \tag{43}$$

In applying this criterion to the Lennard-Jones potential, it is seen that d depends on temperature but not on the density. Also, the calculations show that the terms of the expansion of F in α^2 and $\alpha\lambda$ are negligible compared to the term in λ^2.

Therefore, using equation (38) to evaluate c_1 and equation (40) to evaluate c_2, the expression of the free energy F of the real system (Eq. 42) is:

$$F = F_{HS} + 2\pi\rho N \int_d^{\infty} u_1(r) g_{HS}(\eta; r) r^2 dr$$
$$- \pi N\rho \left(\frac{\partial \rho}{\partial p}\right)_{HS} \frac{\partial}{\partial \rho}\left[\int_d^{\infty} \rho u_1^2(r) g_{HS}(\eta; r) r^2 dr\right], \tag{44}$$

where the first term on the RHS represents the free energy of the hard-sphere system (Eq. 25), and the partial derivative $\left(\frac{\partial \rho}{\partial p}\right)_{HS}$ can be deduced from the Carnahan and Starling equation of state (Eq. 24). Recall that the pair correlation function of hard-sphere system, $g_{HS}(\eta; r)$, must be only determined numerically. It depends on the density ρ and diameter d via the packing fraction $\eta (= \frac{\pi}{6}\rho d^3)$. Since $g_{HS}(\eta; r) = 0$ when $r < d$, either 0 or d can be used as lower limit of integration in equation (44).

5.4 Prescription of Mansoori and Canfield.

An important consequence of the high temperature approximation to first order in β is to mark out the free energy of the real system by an upper limit that can not be exceeded, because the sum of the terms beyond c_1 is always negative. The easiest way to proof this, is to consider the expression of the free energy (Eq. 30) and to write the perturbation $U_1(\mathbf{r}^N)$ around its mean value $\langle U_1 \rangle_0$ as:

$$U_1 = \langle U_1 \rangle_0 + \Delta U_1.$$

After replacing U_1 in equation (30), we obtain:

$$-\beta F = -\beta F_0 - \beta \langle U_1 \rangle_0 + \ln \langle \exp\left[-\beta \Delta U_1\right]_0 \rangle. \tag{45}$$

However, considering the series expansion of an exponential, the above relation is transformed into the so-called *Gibbs-Bogoliubov inequality*[9]:

$$F \leq F_0 + \langle U_1 \rangle_0. \tag{46}$$

A thorough study of this inequality shows that it is always valid, and it is unnecessary to consider values of n greater than zero. Equation (46), at the base of the variational method, allows us to find the value of the parameter d that makes the free energy F *minimum*. If the reference system is that of hard spheres, the value of the free energy obtained with this value of d (or $\eta = \frac{\pi}{6}\rho d^3$) is considered as the best estimate of the free energy of the real system. Its expression is:

$$F \leq F_{HS} + \frac{\rho N}{2} \int u_1(r) g_{HS}(\eta; r) d\mathbf{r}, \tag{47}$$

[9] The Mac-Laurin series of the exponential naturally leads to the inequality:

$$\exp\left(-\beta \Delta U_1\right) \geq \sum_{k=0}^{2n+1} \frac{(-\beta \Delta U_1)^k}{k!} \qquad (n = 0,\ 1,\ 2, ...).$$

For $n = 0$, the last term of equation (45) behaves as:

$$\ln \left\langle \sum_{k=0}^{1} \frac{(-\beta \Delta U_1)^k}{k!} \right\rangle_0 = \ln \langle 1 - \beta \Delta U_1 \rangle_0 \simeq -\beta \langle \Delta U_1 \rangle_0 = 0,$$

since the mean value of the deviation, $\langle \Delta U_1 \rangle_0$, is zero.

where $F_0 = F_{HS}$ is given by equation (25) and $\langle U_1 \rangle_0 = c_1$ by equation (38). Since $g_{HS}(\eta;r) = 0$ when $r < d$ and $u_1(r) = u(r) - u_{HS}(r) = u(r)$ when $r \geq d$, equation (47) can be written as:

$$F \leq F_{HS} + 2\pi\rho N \int_0^\infty u(r) g_{HS}(\eta;r) r^2 dr. \tag{48}$$

Practically, we vary the value of d (or η) until the result of the integral is minimum. And the value of F thus obtained is the best estimate of the free energy of the real system, in the sens of the variational method (20).

5.5 Prescription of Weeks, Chandler and Andersen.

At the same time that the perturbation theory was developing, Weeks, Chandler and Andersen (21) formulated another prescription for finding the hard-sphere diameter d. Its originality lies in the idea that a particle in the liquid is less sensitive to the sign of the potential $u(r)$ than to the sign of the strength, that is to say, to the derivative of the potential $\left(-\frac{\partial u}{\partial r}\right)_{\rho,T}$. That is why the authors proposed to separate the real potential into a purely repulsive contribution and a purely attractive perturbation. This separation, shown in figure (7b), is defined by the relation $u(r) = u_0(r) + u_1(r)$ with:

$$u_0(r) = \begin{cases} u(r) + \varepsilon & \text{if } r < r_{\text{min}} \\ 0 & \text{if } r \geq r_{\text{min}} \end{cases}$$

$$u_1(r) = \begin{cases} -\varepsilon & \text{if } r < r_{\text{min}} \\ u(r) & \text{if } r \geq r_{\text{min}}, \end{cases}$$

where ε is the depth of the potential well, i. e. the value of $u(r_{\text{min}}) = -\varepsilon$.
To follow the same sketch that for the Barker and Henderson prescription, define the potential $v(r)$ by the Boltzmann factor:

$$\exp\left[-\beta v(r)\right] = \exp\left[-\beta u_{HS}(d)\right] + \alpha \left\{\exp\left[-\beta\left(u_0(r) + \lambda u_1(r)\right)\right] - \exp\left[-\beta u_{HS}(d)\right]\right\}. \tag{49}$$

However, from the simultaneous observation of equation (49) and figure (7b) one remarks that the potential $v(r)$ reduces to the real potential $u(r)$ when $\alpha = \lambda = 1$, and behaves like the hard-sphere potential of diameter d when $\alpha = 0$.
The substitution of equation (49) in the configuration integral, and subsequent calculations, enables us to express the free energy F of the system as a series expansion in powers of α and λ identical to that of equation (42). The first term of this expansion, F_{HS}, is the free energy of the hard-sphere system. As for the first derivative of F with respect to α, it is written in this case:

$$\left(\frac{\partial F}{\partial \alpha}\right) = -\frac{2\pi N\rho}{\beta} \int_0^\infty \left\{\exp\left[-\beta u_0(r)\right] - \exp\left[-\beta u_{HS}(d)\right]\right\} y_{HS}(\eta;r) r^2 dr, \tag{50}$$

where $y_{HS}(\eta;r)$ is the *cavity function* of great importance in the microscopic theory of liquids[10]. It is defined by means of the pair correlation function $g_{HS}(\eta;r)$ and the Boltzmann factor $\exp\left[-\beta u_{HS}(d)\right]$ of the hard-sphere potential as:

$$y_{HS}(\eta;r) = \exp\left[\beta u_{HS}(d)\right] g_{HS}(\eta;r). \tag{51}$$

[10] The cavity function $y_{HS}(\eta;r)$ does not exist in analytical form. It must be calculated at all reduced densities (ρd^3) by solving an integro-differential equation.

It should be stressed that the cavity function depends only weakly on the potential, this is why Weeks, Chander and Anderson (WCA) suggested to use the same cavity function $y_{HS}(\eta;r)$ for all potentials, and to express the pair correlation functions of each potential as a function of $y_{HS}(\eta;r)$. As a result, the pair correlation function $g_0(r)$ related to the repulsive contribution $u_0(r)$ of the potential can be approximately written as:

$$g_0(r) \simeq \exp\left[-\beta u_0(r)\right] y_{HS}(\eta;r). \tag{52}$$

The suggestion of WCA for choosing the hard-sphere diameter d is to cancel $\left(\frac{\partial F}{\partial \alpha}\right)$, which amounts to solving the nonlinear equation given by equation (50), i. e.:

$$\int_0^\infty \{\exp\left[-\beta u_0(r)\right] - \exp\left[-\beta u_{HS}(d)\right]\} y_{HS}(\eta;r) r^2 dr = 0. \tag{53}$$

It is interesting to note that equation (53) has a precise physical meaning that appears by writing it in terms of the pair correlation functions $g_{HS}(\eta;r)$ and $g_0(r)$ drawn, respectively, from equations (51) and (52). It reads:

$$\int_0^\infty \left[g_0(r) - 1\right] r^2 dr = \int_0^\infty \left[g_{HS}(\eta;r) - 1\right] r^2 dr.$$

According to equation (7), the previous expression is equivalent to the equality $S_0(0) = S_{HS}(0)$, meaning the equality between the isothermal compressibility of the repulsive potential $u_0(r)$ and that of the hard-sphere potential $u_{HS}(d)$.
Ultimately, the expression of free energy (Eq. 28) of the real system is the sum of F_{HS}, calculated with the value of d issued from equation (53), and the term $\left(\frac{\partial F}{\partial \lambda}\right) = c_1$, calculated with equation (38) in using the pair correlation function $g_0(r) \simeq \exp\left[-\beta u_0(r)\right] y_{HS}(\eta;r)$, i. e.:

$$F = F_{HS} + 2\pi\rho N \int_0^\infty u_1(r) \{\exp\left[-\beta u_0(r)\right] y_{HS}(\eta;r)\} r^2 dr. \tag{54}$$

Note that the value of d obtained with the WCA prescription (*blip function* method) is significantly larger than that obtained with the Barker and Henderson (BH) prescription. By the fact that $y_{HS}(\eta;r)$ depends on density, the value of d calculated with equation (53) depends on temperature and density, while that calculated with equation (43) depends only on temperature.
In order to compare the relative merits of equations (44) and (54), simply note that the WCA prescription shows that not only the terms of the expansion of F in α^2 and in $\alpha\lambda$ are negligible, but also the term in λ^2. This means that the WCA treatment is a theory of first order in λ, while the BH one is a theory of second order in λ. In addition, the BH treatment is coherent as F_{HS}, c_1 and c_2 are calculated with the same reference system, whereas the WCA treatment is not coherent because it uses the hard-sphere system to calculate F_{HS} and the repulsive potential $u_0(r)$ to calculate c_1, via the pair correlation function $g_0(r)$. In contrast, the WCA treatment does not require the calculation of c_2 and improves the convergence of calculations. As an additional advantage, the WCA treatment can be used to predict the structure factor of simple liquids (22).

5.6 Application to the Yukawa attractive potential

As an application of the perturbation theory, consider the hard-core attractive Yukawa potential (HCY). This potential consists of the hard-sphere potential $u_{HS}(d)$ and the attractive perturbation $u_1(r)$. Its traditional analytical form is:

$$u(r) = \begin{cases} \infty & r < d \\ -(\varepsilon d/r)\exp[-\lambda(r/d-1)] & r \geq d, \end{cases} \tag{55}$$

where d is the diameter of hard spheres, ε the potential value at $r = d$ and λ a parameter that measures the rate of exponential decay. The shape of this potential is portrayed in figure (8a)

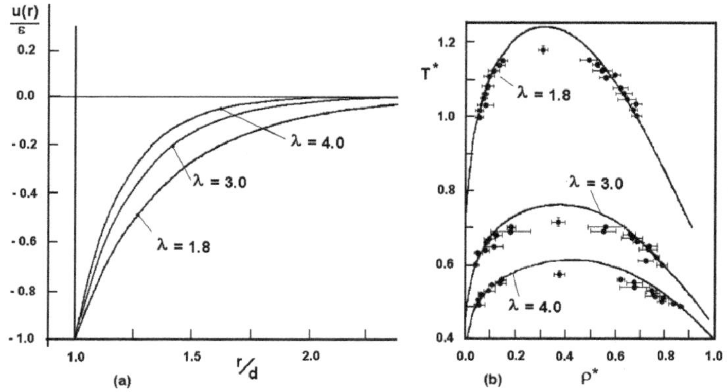

Fig. 8. (a) Representation of the HCY potential for three values of λ (1.8, 3.0, 4.0). (b) Phase diagram of the HCY potential for the same three values of λ (1.8, 3.0, 4.0).

for three different values of λ. For guidance, note that a value of $\lambda \sim 1.8$ predicts the thermodynamic properties of liquid rare gases quite so well as the Lennard-Jones potential. By contrast, a value of $\lambda \sim 8$ enables us to obtain the thermodynamic properties of colloidal suspensions or globular proteins. This wide possible range of λ values explains why the HCY potential has been used in many applications and has been the subject of numerous theoretical studies after its analytical solution was obtained by Waisman (2). Without going into details of the resolution of the problem that involves the microscopic theory of liquids, indicate that the solution reduces to determining a fundamental parameter Γ as the root of the quartic equation (23):

$$\Gamma(1+\lambda\Gamma)(1+\psi\Gamma)^2 + \beta\varepsilon w = 0, \tag{56}$$

where ε and λ are the two parameters of the HCY potential, and w and ψ two additional parameters explicitly depending on λ and η by the following relations:

$$w = \frac{6\eta}{\phi_0^2},$$

$$\psi = \lambda^2(1-\eta)^2\frac{[1-\exp(-\lambda)]}{L(\lambda)\exp(-\lambda)+S(\lambda)}$$

$$-12\eta(1-\eta)\frac{[1-\lambda/2-(1+\lambda/2)\exp(-\lambda)]}{L(\lambda)\exp(-\lambda)+S(\lambda)},$$

with:

$$\phi_0 = \frac{L(\lambda)\exp(-\lambda) + S(\lambda)}{\lambda^3(1-\eta)^2},$$
$$L(\lambda) = 12\eta[1 + 2\eta + (1 + \eta/2)\lambda],$$
$$S(\lambda) = (1-\eta)^2\lambda^3 + 6\eta(1-\eta)\lambda^2 + 18\eta^2\lambda - 12\eta(1+2\eta).$$

By expressing Γ as a function of β, Henderson *et al.* (3) managed to write the free energy F of the system according to the series expansion in powers of β:

$$F = F_{HS} - \frac{Nk_BT}{2}\sum_{n=1}^{\infty}\frac{v_n}{n}(\beta\varepsilon)^n, \tag{57}$$

where F_{HS} is the free energy of the hard-sphere system calculated, for example, with equation (25), and where the first five terms v_n have expressions derived by the authors[11].
At this stage, we see that the free energy of the HCY potential can be approximated for all values of the triplet $(\eta, \varepsilon, \lambda)$, using the analytical expressions of the five coefficients v_n given in footnote (11). From equation (57), we can also deduce the internal energy (Eq. 9), excess pressure (Eq. 12), chemical potential (Eq. 19) and entropy (Eq. 20) of the HCY system.
The HCY equation of state, which is based on the perturbation theory and expressed in terms of the relevant features of the potential, is a very handy tool for investigating the thermodynamics of systems governed by an effective hard-sphere interaction plus an attractive tail. Alternatively, the HCY system could be used vicariously to approximate any available interatomic potential for real fluids.
The reduced phase diagrams ($T^* = \frac{k_BT}{\varepsilon}$ versus $\rho^* = \rho\sigma^3 = \frac{6\eta}{\pi}$) predicted by this equation of state, for values of $\lambda = 1.8$, 3 and 4, are displayed in figure (8b), together with simulation data. A rapid glance at figure (8b) indicates that the critical temperature T_C^* predicted by the perturbation theory is noticeably greater than that predicted by the simulation. The reason is that the HCY equation of state involves a truncated series. But, also, we can speculate that the

[11] The first terms of the series expansion are:

$$v_0 = 0,$$
$$v_1 = \frac{2\alpha_0}{\phi_0},$$
$$v_2 = \frac{2w(1-\alpha_1+\alpha_0\psi)}{\lambda\phi_0},$$
$$v_3 = \frac{2w^2(1-\alpha_1+\alpha_0\psi)(1+3\lambda\psi)}{\lambda^3\phi_0},$$
$$v_4 = \frac{4w^3(1-\alpha_1+\alpha_0\psi)\left(1+4\lambda\psi+6\lambda^2\psi^2\right)}{\lambda^5\phi_0},$$
$$v_5 = \frac{10w^4(1-\alpha_1+\alpha_0\psi)\left(1+5\lambda\psi+11\lambda^2\psi^2+11\lambda^3\psi^3\right)}{\lambda^7\phi_0},$$

with:

$$\alpha_0 = \frac{L(\lambda)}{\lambda^2(1-\eta)^2} \quad ; \quad \alpha_1 = \frac{12\eta(1+\lambda/2)}{\lambda^2(1-\eta)} \quad \text{and} \quad (1-\alpha_1+\alpha_0\psi)\phi_0 = 1.$$

correlation functions, inherent in the calculations of the pressure and the chemical potential, do not represent correctly the growing correlation lengths in approaching the critical point. Looking at the evolution of the binodal lines as a function of the rate of decay λ of the Yukawa potential, we remark that higher the critical temperature is, lower λ is. At the same time, the domain below the binodal line shrinks. In contrast, when λ increases (i. e., the attraction range of the potential becomes shorter), the critical temperature decreases, and the liquid and gaseous phases become indistinguishable. For the hard-sphere potential, as a border case ($\lambda \to \infty$), there is no longer gas-liquid phase transition. On the other hand, one can see that the phase diagrams obtained with the perturbation theory agree with simulation data more favorably for the vapor branch than for the liquid branch. This is not surprising in the extent that the perturbation theory works better for low densities. Lastly, it should be mentioned that the structure and thermodynamic properties of the HCY potential have been extensively studied for the two last decades, as much by computer simulations and integral equation theory as by means of perturbation theory. Note that many other studies of the HCY potential with these various methods are available in literature (24).

6. Concluding remarks

In the foregoing, we started with a brief discussion of the liquid state, compared with the gaseous and solid states. Among other things, we mentioned that one of the most successes of the microscopic theory of liquids has been to emphasize the importance of the *pair potential* $u(r)$ to describe a wide variety of physical properties of liquids. Then, we learned a rudimentary knowledge of the structure of liquids. The latter is essentially described by the static *structure factor* $S(q)$, which can be measured directly by elastic scattering of neutrons or X-rays. But $S(q)$, or more precisely the *pair correlation function* $g(r)$, can also be determined by numerical simulations (known as virtual experiments) and with models of the microscopic theory of liquids once the nature of the pair potential $u(r)$ is known. The comparison between simulation results and those of analytical models essentially allows us to test the models, whereas the comparison of simulation results with experimental results is the ultimate test to judge the efficiency of the pair potential. Subsequently, we described in minute detail the calculations of the thermodynamic properties in terms of the pair potential and pair correlation function. In particular, with the pressure p and the chemical potential μ to apply equilibrium conditions, it has been shown that we are in a position to determine theoretical estimates for the liquid-vapor coexistence curve. Finally, we got down to basics of the thermodynamic perturbation theory, before presenting the liquid-vapor coexistence curve for the HCY fluid.

In a 1990 time warp, it became apparent that the usefulness of thermodynamic perturbation theory was on the decline compared to the more powerful simulation methods. This was primarily due to the rapid increase in the power of computers. At the same time, the integral equation theory enjoyed renewed popularity with the extensively employed concept of thermodynamic consistency (25), but this aspect goes beyond the scope of this short review. Incidentally, we recall that the thermodynamic consistency consists to adjusting the isothermal compressibility obtained by two different routes. Nevertheless, the thermodynamic perturbation theory remains undoubtedly the most tractable approach to predict the thermodynamic properties of liquids. Let us just quote few articles containing investigations of the HCY fluid with the integral equation theory (26).

As one would expect, the thermodynamic perturbation theory described previously is also important for mixtures. The thermodynamic properties established for studying pure liquids can be applied to binary mixtures, with only few modifications. Specifically, analytic expressions for the hard-sphere free energy $F_{HS}(d_1, d_2)$ and subsequent thermodynamic quantities are readily generalized for a mixture of hard spheres of different diameters d_1 and d_2. Therefore, if one estimates that a real binary mixture behaves like binary hard-sphere fluid, the variational method consists of minimizing $F_{HS}(d_1, d_2)$ versus the two hard-sphere diameters, and by taking the resulting minimum upper bound as an approximation to the free energy. The variational method is not limited neither to hard-sphere reference system nor to a specific fluid. It should be stressed that systems like pure liquid metals composed of ions embedded in an electron gas can not be treated as a binary mixture, in the strict sense of the word, but must be suitably reduced to a one-component system of pseudoions. In contrast, a liquid metal made up of two species of pseudoions forms a binary mixture, for which the approach originally developed for simple liquids can be applied. What is it that determines whether or not two metals will mix to form an alloy is a crucial issue to be answered by thermodynamic perturbation theory (27).

I would like to express my acknowledgement to Jean-Marc Bomont for its stimulating discussions.

7. References

[1] M. H. J. Hagen and D. Frenkel, J. Chem. Phys. 101, 4093 (1994).
[2] E. Waisman, Mol. Phys. 25, 45 (1973).
[3] D. Henderson, G. Stell and E. Waisman, J. Chem. Phys. 62, 4247 (1975). D. Henderson, L. Blum and J. P. Noworyta, J. Chem. Phys. 102, 4973 (1995).
[4] See, for instance, the book of J. P. Hansen and I. R. McDonald, *Theory of Simple Liquids*, Acad. Press., London (2006).
[5] W. G. Hoover and F. H. Ree, J. Chem. Phys. 47, 4873 (1967); 49, 3609 (1968).
[6] E. Chacon, M. Reinaldo-Falagan, E. Velasco and P. Tarazona, Phys. Rev. Lett. 87, 166101 (2001). D. Li and S. A. Rice, J. Phys. Chem. B. 108, 19640 (2004). J. M. Bomont and J. L. Bretonnet, J. Chem. Phys. 124, 054504 (2006).
[7] N. Jakse and J. L. Bretonnet, J. Phys.: Condens. Matter 15, S3455 (2003).
[8] J. G. Kirkwood, J. Chem. Phys. 3, 300 (1935). J. G. Kirkwood and E. Monroe, J. Chem. Phys. 9, 514 (1941). See also D. A. McQuarrie, *Statistical Mechanics*, Harper and Row, New-York (1976), p. 263.
[9] R. E. Nettleton and M. S. Green, J. Chem. Phys. 29, 1365 (1958). H. J. Raveché, J. Chem. Phys. 55, 2242 (1971).
[10] B. J. Alder and T. E. Wainwright, Phys. Rev. 127, 359 (1962); F. H. Ree and W. G. Hoover, J. Chem. Phys. 40, 939 (1964).
[11] N. F. Carnahan and K. E. Starling, J. Chem. Phys. 51, 635 (1969).
[12] D. A. McQuarrie, *Statistical Mechanics*, Harper and Row, New York (1976).
[13] A. J. Greenfield, J. Wellendorf and N. Wiser, Phys. Rev. A 4, 1607 (1971).
[14] N. W. Ashcroft and J. Lekner, Phys. Rev. 156, 83 (1966).
[15] L. Verlet, Phys. Rev. 165, 201 (1968).
[16] R. Zwanzig, J. Chem. Phys. 22, 1420 (1954).
[17] J. A. Barker and D. Henderson, J. Chem. Phys. 47, 2856 (1967).
[18] G. A. Mansoori and F. B. Canfield, J. Chem. Phys. 51, 4958 (1969).
[19] J. A. Barker and D. Henderson, Ann. Rev. Phys. Chem. 23, 439 (1972).
[20] G. A. Mansoori and F. B. Canfield, J. Chem. Phys. 51, 4967 (1969); 51 5295 (1969); 53, 1618 (1970).
[21] J. D. Weeks, D. Chandler and H. C. Andersen, J. Chem. Phys. 54, 5237 (1971); 55, 5422 (1971).

[22] J. L. Bretonnet and C. Regnaut, Phys. Rev. B 31, 5071 (1985).
[23] M. Ginoza, Mol. Phys. 71, 145 (1990).
[24] W. H. Shih and D. Stroud, J. Chem. Phys. 79, 6254 (1983). D. M. Duh and L. Mier-Y-Teran, Mol. Phys. 90, 373 (1997). S. Zhou, Phys. Rev. E 74, 031119 (2006). J. Torres-Arenas, L. A. Cervantes, A. L. Benavides, G. A. Chapela and F. del Rio, J. Chem. Phys. 132, 034501 (2010).
[25] J. M. Bomont, Advances in Chemical Physics, 139, 1 (2008).
[26] E. Lomba and N. G Almarza, J. Chem. Phys. 100, 8367 (1994). C. Caccamo, G. Pellicane, D. Costa, D. Pini and G. Stell, Phys. Rev. E 60, 5533 (1999). J. M. Caillol, F. Lo Verso, E. Scholl-Paschinger and J. J. Weis, Mol. Phys. 105, 1813 (2007). A. Reiner and J. S. Hoye, J. Chem. Phys. 128, 114507 (2008). E. B. El Mendoub, J. F. Wax and N. Jakse, J. Chem. Phys. 132, 164503 (2010).
[27] M. Shimoji, *Liquid Metals*, Acad. Press, London (1977). D. Stroud and N. W. Ashcroft, Solid State Physics, 33, 1, 1978.

9

Thermodynamics of Metal Hydrides: Tailoring Reaction Enthalpies of Hydrogen Storage Materials

Martin Dornheim
Institute of Materials Research, Department of Nanotechnology, Helmholtz-Zentrum Geesthacht
Germany

1. Introduction

Considering the increasing pollution and exploitation of fossil energy resources, the implementation of new energy concepts is essential for our future industrialized society. Renewable sources have to replace current energy technologies. This shift, however, will not be an easy task. In contrast to current nuclear or fossil power plants renewable energy sources in general do not offer a constant energy supply, resulting in a growing demand of energy storage. Furthermore, fossil fuels are both, energy source as well as energy carrier. This is of special importance for all mobile applications. Alternative energy carriers have to be found. The hydrogen technology is considered to play a crucial role in this respect. In fact it is the ideal means of energy storage for transportation and conversion of energy in a comprehensive clean-energy concept. Hydrogen can be produced from different feedstocks, ideally from water using regenerative energy sources. Water splitting can be achieved by electrolysis, solar thermo-chemical, photoelectrochemical or photobiological processes. Upon reconversion into energy, by using a fuel cell only water vapour is produced, leading to a closed energy cycle without any harmful emissions. Besides stationary applications, hydrogen is designated for mobile applications, e.g. for the zero-emission vehicle. In comparison to batteries hydrogen storage tanks offer the opportunity of fast recharging within a few minutes only and of higher storage densities by an order of magnitude. Hydrogen can be produced from renewable energies in times when feed-in into the electricity grid is not possible. It can be stored in large caverns underground and be utilized either to produce electricity and be fed into the electricity grid again or directly for mobile applications.

However, due to the very low boiling point of hydrogen (20.4 K at 1 atm) and its low density in the gaseous state (90 g/m^3) dense hydrogen storage, both for stationary and mobile applications, remains a challenging task. There are three major alternatives for hydrogen storage: compressed gas tanks, liquid hydrogen tanks as well as solid state hydrogen storage such as metal hydride hydrogen tanks. All of these three main techniques have their special advantages and disadvantages and are currently used for different applications. However, so far none of the respective tanks fulfils all the demanded technical requirements in terms of gravimetric storage density, volumetric storage density, safety, free-form, ability to store hydrogen for longer times without any hydrogen losses, cyclability as well as recyclability and costs. Further research and development is strongly required. One major advantage of hydrogen storage in metal hydrides is the ability to store hydrogen in a very energy efficient way enabling hydrogen storage at rather low pressures without further need for liquefaction or compression. Many metals and alloys are able to absorb large amounts of hydrogen. The metal-hydrogen bond offers the advantage of a very high volumetric hydrogen density under moderate pressures, which is up to 60% higher than that of liquid hydrogen (Reilly & Sandrock, 1980).

Depending on the hydrogen reaction enthalpy of the specific storage material during hydrogen uptake a huge amount of heat (equivalent to 15% or more of the energy stored in hydrogen) is generated and has to be removed in a rather short time, ideally to be recovered and used as process heat for different applications depending on quantity and temperature. On the other side, during desorption the same amount of heat has to be applied to facilitate the endothermic hydrogen desorption process - however, generally at a much longer time scale. On one side this allows an inherent safety of such a tank system. Without external heat supply hydrogen release would lead to cooling of the tank and finally hydrogen desorption necessarily stops. On the other side it implies further restrictions for the choice of suitable storage materials. Highest energy efficiencies of the whole tank to fuel combustion or fuel cell system can only be achieved if in case of desorption the energy required for hydrogen release can be supplied by the waste heat generated in case of mobile applications on-board by the hydrogen combustion process and the fuel cell respectively.

2. Basics of hydrogen storage in metal hydrides

Many metals and alloys react reversibly with hydrogen to form metal hydrides according to the reaction (1):

$$Me + x/2\, H_2 \leftrightarrow MeH_x + Q. \tag{1}$$

Here, Me is a metal, a solid solution, or an intermetallic compound, MeH_x is the respective hydride and x the ratio of hydrogen to metal, $x=c_H$ [H/Me], Q the heat of reaction. Since the entropy of the hydride is lowered in comparison to the metal and the gaseous hydrogen phase, at ambient and elevated temperatures the hydride formation is exothermic and the reverse reaction of hydrogen release accordingly endothermic. Therefore, for hydrogen release/desorption heat supply is required.

Metals can be charged with hydrogen using molecular hydrogen gas or hydrogen atoms from an electrolyte. In case of gas phase loading, several reaction stages of hydrogen with the metal in order to form the hydride need to be considered. Fig. 1 shows the process schematically.

The first attractive interaction of the hydrogen molecule approaching the metal surface is the Van der Waals force, leading to a physisorbed state. The physisorption energy is typically of the order $E_{Phys} \approx 6$ kJ/mol H_2. In this process, a gas molecule interacts with several atoms at the surface of a solid. The interaction is composed of an attractive term, which diminishes with the distance of the hydrogen molecule and the solid metal by the power of 6, and a repulsive term diminishing with distance by the power of 12. Therefore, the potential energy of the molecule shows a minimum at approximately one molecular radius. In addition to hydrogen storage in metal hydrides molecular hydrogen adsorption is a second technique to store hydrogen. The storage capacity is strongly related to the temperature and the specific surface areas of the chosen materials. Experiments reveal for carbon-based nanostructures storage capacities of less than 8 wt.% at 77 K and less than 1wt.% at RT and pressures below 100 bar (Panella et al., 2005; Schmitz et al., 2008).

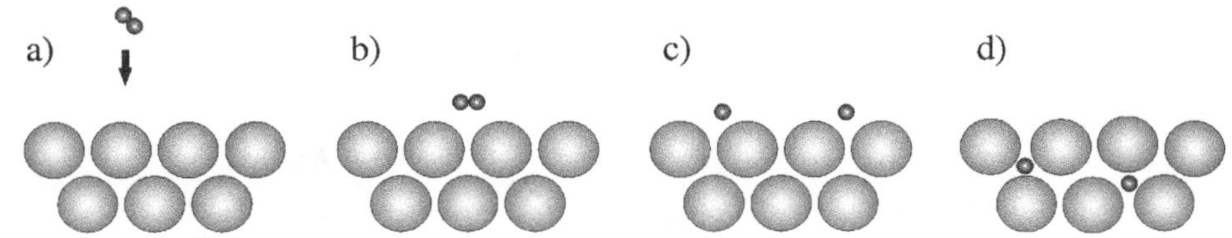

Fig. 1. Reaction of a H_2 molecule with a storage material: a) H_2 molecule approaching the metal surface. b) Interaction of the H_2 molecule by Van der Waals forces (physisorbed state). c) Chemisorbed hydrogen after dissociation. d) Occupation of subsurface sites and diffusion into bulk lattice sites.

In the next step of the hydrogen-metal interaction, the hydrogen has to overcome an activation barrier for the formation of the hydrogen metal bond and for dissociation, see Fig. 1c and 2. This process is called dissociation and chemisorption. The chemisorption energy is typically in the range of $E_{Chem} \approx 20$ - 150 kJ/mol H_2 and thus significantly higher than the respective energy for physisorption which is in the order of 4-6 kJ/mol H_2 for carbon based high surface materials (Schmitz et al., 2008).

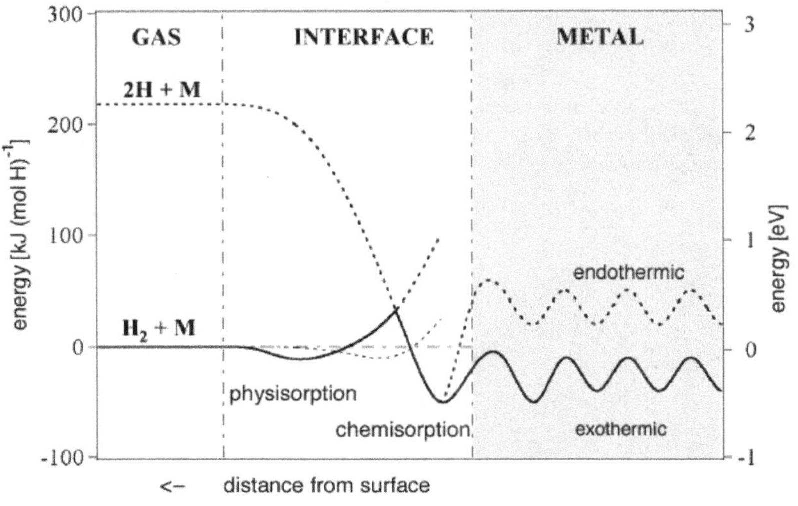

Fig. 2. Schematic of potential energy curves of hydrogen in molecular and atomic form approaching a metal. The hydrogen molecule is attracted by Van der Waals forces and forms a physisorbed state. Before diffusion into the bulk metal, the molecule has to dissociate forming a chemisorbed state at the surface of the metal (according to Züttel, 2003).

After dissociation on the metal surface, the H atoms have to diffuse into the bulk to form a M-H solid solution commonly referred to as α-phase. In conventional room temperature metals / metal hydrides, hydrogen occupies interstitial sites - tetrahedral or octahedral - in the metal host lattice. While in the first, the hydrogen atom is located inside a tetrahedron formed by four metal atoms, in the latter, the hydrogen atom is surrounded by six metal atoms forming an octahedron, see Fig. 3.

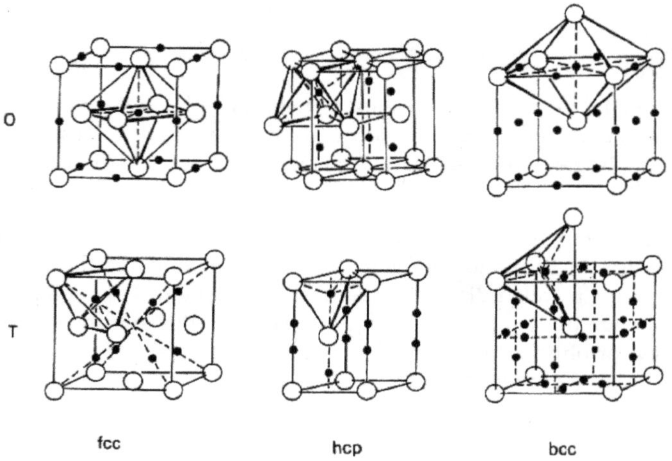

Fig. 3. Octahedral (O) and tetrahedral (T) interstitial sites in fcc-, hcp- and bcc-type metals. (Fukai, 1993).

In general, the dissolution of hydrogen atoms leads to an expansion of the host metal lattice of 2 to 3 Å³ per hydrogen atom, see Fig. 4. Exceptions of this rule are possible, e.g. several dihydride phases of the rare earth metals, which show a contraction during hydrogen loading for electronic reasons.

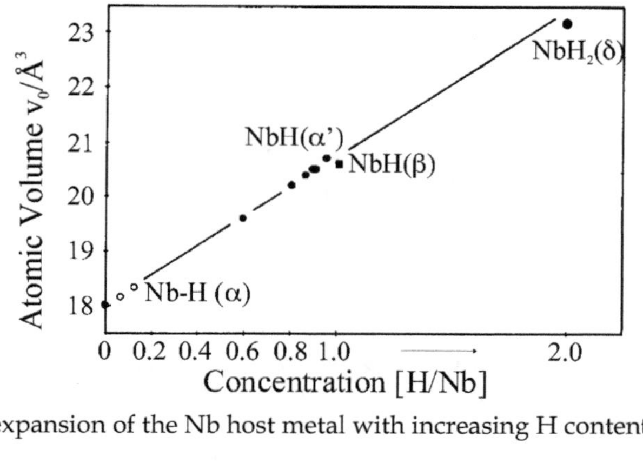

Fig. 4. Volume expansion of the Nb host metal with increasing H content. (Schober & Wenzl, 1978)

In the equilibrium the chemical potentials of the hydrogen in the gas phase and the hydrogen absorbed in the metal are the same:

$$\frac{1}{2}\mu_{gas} = \mu_{metal} \, . \tag{2}$$

Since the internal energy of a hydrogen molecule is 7/2 kT the enthalpy and entropy of a hydrogen molecule are

$$h_{gas} = \frac{7}{2} \cdot \mathrm{k} \cdot T - \mathrm{E_{Diss}} \tag{3}$$

and

$$s_{gas} = \frac{7}{2} \cdot \mathrm{k} - \mathrm{k} \cdot \ln \frac{p}{p_0(T)} \quad \text{with} \quad p_0(T) = \frac{8(\pi \mathrm{k} T)^{\frac{7}{2}} \cdot \mathrm{M_{H\text{-}H}}^{\frac{5}{2}} \cdot \mathrm{r_{H\text{-}H}}^2}{\mathrm{h}^5} \tag{4}$$

Here k is the Boltzmann constant, T the temperature, p the applied pressure, $\mathrm{E_{Diss}}$ the dissociation energy for hydrogen ($\mathrm{E_{Diss}}$ = 4.52 eV eV/H_2), $\mathrm{M_{H\text{-}H}}$ the mass of the H_2 molecule, $\mathrm{r_{H\text{-}H}}$ the interatomic distance of the two hydrogen atoms in the H_2 molecule.
Consequently the chemical potential of the hydrogen gas is given by

$$\mu_{gas} = \mathrm{k} \cdot T \cdot \ln \frac{p}{p_0(T)} - \mathrm{E_{Diss}} = \mathrm{k} \cdot \mathrm{T} \cdot \ln \frac{p}{\mathrm{p_0}} + \mu_{gas_0} \tag{5}$$

with $p_0 = 1.01325\ 10^5$ Pa.
In the solid solution (α-phase) the chemical potential is accordingly

$$\mu_\alpha = h_\alpha - T s_\alpha \quad \text{mit} \quad s_\alpha = s_{\alpha_{\mathrm{conf}}} + s_{\alpha_{\mathrm{vibr,electr}}} \, . \tag{6}$$

Here, $s_{\alpha,\mathrm{conf}}$ is the configuration entropy which is originating in the possible allocations of N_H hydrogen atoms on N_{is} different interstitial sites:

$$S_{\alpha,\mathrm{conf}} = \mathrm{k} \cdot \ln \frac{\mathrm{N_{is}}!}{\mathrm{N_H}!(\mathrm{N_{is}}\text{-}\mathrm{N_H})!} \tag{7}$$

and accordingly for small c_H using the Stirling approximation we get

$$s_{\alpha,\mathrm{conf}} = \text{-}\mathrm{k} \cdot \ln \frac{c_\mathrm{H}}{\mathrm{n_{is}}\text{-}c_\mathrm{H}} \tag{8}$$

with n_{is} being the number of interstitial sites per metal atom: $n_{is} = N_{is}/N_{Me}$ and c_H the number of hydrogen atoms per metal atom: $c_H = N_H/N_{Me}$.
Therefore the chemical potential of hydrogen in the solid solution (α-phase) is given by

$$\mu_\alpha = h_\alpha - T \cdot \left(s_{\alpha_{\mathrm{vibr,electr}}} - \mathrm{k} \cdot \ln \frac{c_\mathrm{H}}{\mathrm{n_{is}} - c_\mathrm{H}} \right) \tag{9}$$

Taking into account the equilibrium condition (2) the hydrogen concentration c_H can be determined via

$$\frac{c_H}{n_{is}-c_H}=\sqrt{\frac{p}{p_0(T)}}\cdot e^{-\frac{\Delta g_s}{k\cdot T}} \quad \text{with} \quad \Delta g_s = h_\alpha - T\cdot s_{\alpha_{vibr}} + \frac{1}{2}\mu_{g_0} \tag{10}$$

or

$$\frac{c_H}{n_{is}-c_H}=\sqrt{\frac{p}{p_0(T)}}\cdot e^{-\frac{\Delta G_s}{R\cdot T}} \quad \text{with} \quad \Delta G_s = \Delta H_s - T\Delta S\,. \tag{11}$$

Here μ_{g0} is the chemical potential of the hydrogen molecule at standard conditions and R being the molar gas constant.
For very small hydrogen concentrations $c_H|_{cH << nis}$ in the solid solution phase α the hydrogen concentration is directly proportional to the square root of the hydrogen pressure in the gas phase. This equation is also known as the **Sievert's law**, i.e.

$$c_H = \frac{1}{K_S}\sqrt{p} \tag{12}$$

with K_S being a temperature dependent constant. As the hydrogen pressure is increased, saturation occurs and the metal hydride phase $MeH_{c\beta}$ starts to form.
For higher hydrogen pressures/concentrations metal hydride formation occurs.
The conversion from the saturated solution phase to the hydride phase takes place at constant pressure p according to:

$$\text{Me-H}_{c_\alpha}|_\alpha + \frac{1}{2}\left(c_\beta - c_\alpha\right) H_2 \leftrightarrow \text{MeH}_{c_\beta}|_\beta + Q_{\alpha\to\beta}\,. \tag{13}$$

In the equilibrium the chemical potentials of the gas phase, the solid solution phase α and the hydride phase β coincide:

$$\mu_\alpha(p,T,c_\alpha)=\mu_\beta(p,T,c_\beta) = \frac{1}{2}\mu_{gas}(p,T)=\frac{1}{2}\cdot k\cdot T\cdot\ln\left(\frac{p_{eq}(T)}{p_0}\right)+\frac{1}{2}\mu_{gas_0}\,. \tag{14}$$

Following Gibb's Phase Rule $f=c-p+2$ with f being the degree of freedom, k being the number of components and p the number of different phases only one out of the four variables p, T, c_α, c_β is to be considered as independent. Therefore for a given temperature all the other variables are fixed.
Therefore the change in the chemical potential or the Gibbs free energy is just a function of one parameter, i.e. the temperature T:

$$\Delta G = \frac{1}{2}\cdot R\cdot T\cdot\ln\left(\frac{p(T)}{p_0}\right)\,. \tag{15}$$

From this equation follows the frequently-used **Van't Hoff equation** (16):

$$\frac{1}{2} \cdot \ln \frac{p}{p_0} = \frac{\Delta H}{RT} - \frac{\Delta S}{R} \tag{16}$$

The temperature dependent plateau pressure of this two phase field is the equilibrium dissociation pressure of the hydride and is a measure of the stability of the hydride, which commonly is referred to as β-phase.

After complete conversion to the hydride phase, further dissolution of hydrogen takes place as the pressure increases, see Fig. 5.

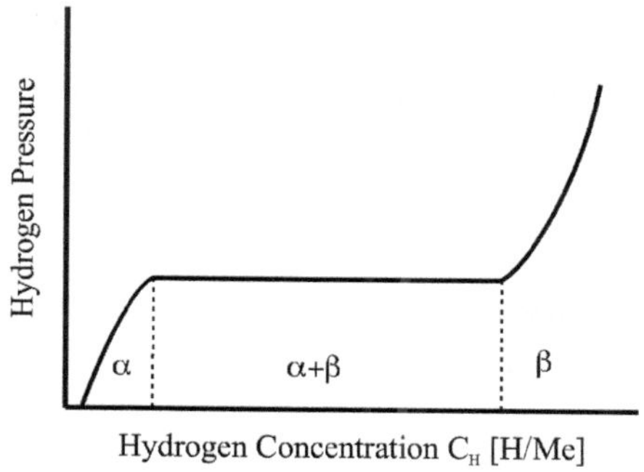

Fig. 5. Schematic Pressure/Composition Isotherm. The precipitation of the hydride phase β starts when the terminal solubility of the α-phase is reached at the plateau pressure.

Multiple plateaus are possible and frequently observed in composite materials consisting of two hydride forming metals or alloys. The equilibrium dissociation pressure is one of the most important properties of a hydride storage material.

If the logarithm of the plateau pressure is plotted vs 1/T, a straight line is obtained (van't Hoff plot) as seen in Fig. 6.

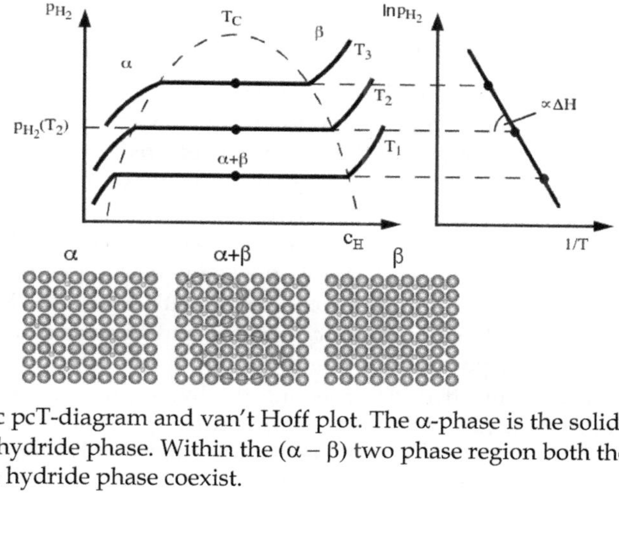

Fig. 6. Schematic pcT-diagram and van't Hoff plot. The α-phase is the solid solution phase, the β-phase the hydride phase. Within the (α – β) two phase region both the metal-hydrogen solution and the hydride phase coexist.

2.1 Conventional metal hydrides

Fig. 7 shows the Van't Hoff plots of some selected binary hydrides. The formation enthalpy of these hydrides H^0_f determines the amount of heat which is released during hydrogen absorption and consequently is to be supplied again in case of desorption. To keep the heat management system simple and to reach highest possible energy efficiencies it is necessary to store the heat of absorption or to get by the waste heat of the accompanying hydrogen utilizing process, e.g. energy conversion by fuel cell or internal combustion system. Therefore the reaction enthalpy has to be as low as possible. The enthalpy and entropy of the hydrides determine the working temperatures and the respective plateau pressures of the storage materials. For most applications, especially for mobile applications, working temperatures below 100°C or at least below 150°C are favoured. To minimize safety risks and avoid the use of high pressure composite tanks the favourable working pressures should be between 1 and 100 bar.

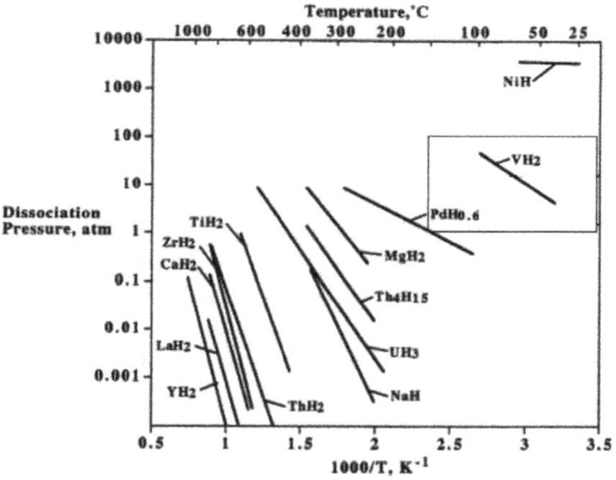

Fig. 7. Van't Hoff lines (desorption) for binary hydrides. Box indicates 1-100 atm, 0-100 °C ranges, taken from Sandrock et al. (Sandrock, 1999).

However, the Van't Hoff plots shown in Fig. 7 indicate that most binary hydrides do not have the desired thermodynamic properties. Most of them have rather high thermodynamic stabilities and thus release hydrogen at the minimum required pressure of 1 bar only at rather high temperatures (T>300°C). The values of their respective reaction enthalpies are in the range of 75 kJ/(mol H_2) (MgH_2) or even higher. Typical examples are the hydrides of alkaline metals, alkaline earth metals, rare earth metals as well as transition metals of the Sc-, Ti- and V-group. The strongly electropositive alkaline metals like LiH and NaH and CaH_2 form saline hydrides, i.e. they have ionic bonds with hydrogen. MgH_2 marks the transition between these predominantly ionic hydrides and the covalent hydrides of the other elements in the first two periods.

Examples for high temperature hydrides releasing the hydrogen at pressures of 1 bar at extremely high temperatures (T > 700°C) are ZrH_2 and LaH_2 (Dornheim & Klassen, 2009). ZrH_2 for example is characterized by a high volumetric storage density N_H. N_H values larger than 7 $\times 10^{22}$ hydrogen atoms per cubic centimetre are achievable. This value corresponds to

58 mol H_2/l or 116 g/l and has to be compared with the hydrogen density in liquid hydrogen (20 K): 4.2×10^{22} (35 mol H_2/l or 70 g/l) and in compressed hydrogen (350 bar / 700 bar): 1.3 / 2.3×10^{22} atoms/cm^3 (11 mol H_2/l or 21 g/l and 19 mol H_2/l or 38 g/l respectively) . The hydrogen density varies a lot between different hydrides. VH_2 for example has an even higher hydrogen density which amounts to 11.4×10^{22} hydrogen atoms per cubic centimetre and accordingly 95 mol H_2/l or 190 g/l. As in the case of many other transition metal hydrides Zr has a number of different hydride phases ZrH_{2-x} with a wide variation in the stoichiometry (Hägg, 1931). Their compositions extend from about $ZrH_{1.33}$ up to the saturated hydride ZrH_2. Because of the limited gravimetric storage density of only about 2 wt.% and the negligibly low plateau pressure within the temperature range of 0 – 150 °C Zr as well as Ti and Hf are not suitable at all as a reversible hydrogen storage material. Thus, they are not useful for reversible hydrogen storage if only the pure binary hydrides are considered (Dornheim & Klassen, 2009). Libowitz et al. (Libowitz et al., 1958) could achieve a breakthrough in the development of hydrogen storage materials by discovering the class of reversible intermetallic hydrides. In 1958 they discovered that the intermetallic compound ZrNi reacts reversibly with gaseous hydrogen to form the ternary hydride $ZrNiH_3$. This hydride has a thermodynamic stability which is just in between the stable high temperature hydride ZrH_2 ($\Delta_f H^0$= -169 kJ/mol H_2) and the rather unstable NiH ($\Delta_f H^0$= -8.8 kJmol$^{-1}H_2$). Thus, the intermetallic Zr-Ni bond exerts a strong destabilizing effect on the Zr-hydrogen bond so that at 300°C a plateau pressure of 1bar is achieved which has to be compared to 900°C in case of the pure binary hydride ZrH_2. This opened up a completely new research field. In the following years hundreds of new storage materials with different thermodynamic properties were discovered which generally follow the well-known semi-empirical rule of Miedema (Van Mal et al., 1974):

$$\Delta H(A_nB_mH_{x+y}) = \Delta H(A_nH_x) + \Delta H(B_mH_y) - \Delta H(A_nB_m) \tag{17}$$

Around 1970, hydrides with significantly lowered values of hydrogen reaction enthalpies, such as $LaNi_5$ and FeTi but also Mg_2Ni were discovered. While 1300 °C are necessary to reach a desorption pressure of 2 bar in case of the pure high temperature hydride LaH_2, in case of $LaNi_5H_6$ a plateau pressure of 2 bar is already reached at 20 °C only. The value of the hydrogen reaction enthalpy is lowered to $|\Delta H_{LaNi5H6}|$ = 30.9 kJmol$^{-1}H_2$. The respective values for NiH are $|\Delta H_{f,NiH}|$ = 8.8 kJmol$^{-1}H_2$ and $P_{diss,NiH,RT}$=3400 bar.

In the meantime, several hundred other intermetallic hydrides have been reported and a number of interesting compositional types identified (table 1). Generally, they consist of a high temperature hydride forming element A and a non hydride forming element B, see fig. 8.

COMPOSITION	A	B	COMPOUNDS
A_2B	Mg, Zr	Ni, Fe, Co	Mg_2Ni, Mg_2Co, Zr_2Fe
AB	Ti, Zr	Ni, Fe	TiNi, TiFe, ZrNi
AB_2	Zr, Ti, Y, La	V, Cr, Mn, Fe, Ni	$LaNi_2$, YNi_2,YMn_2, $ZrCr_2$, $ZrMn_2$,ZrV_2, $TiMn_2$
AB_3	La, Y, Mg	Ni, Co	$LaCo_3$,YNi_3,$LaMg_2Ni_9$
AB_5	Ca, La, Rare Earth	Ni, Cu, Co, Pt, Fe	$CaNi_5$, $LaNi_5$, $CeNi_5$, $LaCu_5$, $LaPt_5$, $LaFe_5$

Table 1. Examples of intermetallic hydrides, taken from Dornheim et al. (Dornheim, 2010).

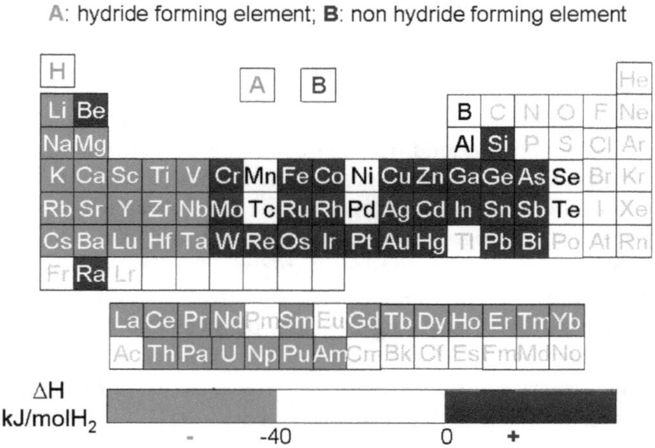

Fig. 8. Hydride and non hydride forming elements in the periodic system of elements.

Even better agreement with experimental results than by use of Miedema's rule of reversed stability is obtained by applying the semi-empirical band structure model of Griessen and Driessen (Griessen & Driessen, 1984) which was shown to be applicable to binary and ternary hydrides. They found a linear relationship of the heat of formation $\Delta H = H^0_f$ of a metal hydride and a characteristic energy ΔE of the electronic band structure of the host metal which can be applied to simple metals, noble metals, transition metals, actinides and rare earths:

$$\Delta H = \alpha \cdot \Delta E + \beta \tag{18}$$

with $\Delta E = E_F - E_S$ (E_F being the Fermi energy and E_S the center of the lowest band of the host metal, $\alpha = 59.24$ kJ (eV mol H_2)$^{-1}$ and $\beta = -270$ kJ (mol H_2)$^{-1}$ and ΔE in eV.

As described above, most materials experience an expansion during hydrogen absorption, wherefore structural effects in interstitial metal hydrides play an important role as well. This can be and is taken as another guideline to tailor the thermodynamic properties of interstitial metal hydrides. Among others Pourarian et al. (Pourarian, 1982), Fujitani et al. (Fujitani, 1991) and Yoshida & Akiba (Yoshida, 1995) report about this relationship of lattice parameter or unit cell volume and the respective plateau pressures in different material classes.

Intensive studies let to the discovery of a huge number of different multinary hydrides with a large variety of different reaction enthalpies and accordingly working temperatures. They are not only attractive for hydrogen storage but also for rechargeable metal hydride electrodes and are produced and sold in more than a billion metal hydride batteries per year. Because of the high volumetric density, intermetallic hydrides are utilized as hydrogen storage materials in advanced fuel cell driven submarines, prototype passenger ships, forklifts and hydrogen automobiles as well as auxiliary power units.

2.2 Hydrogen storage in light weight hydrides

Novel light weight hydrides show much higher gravimetric storage capacities than the conventional room temperature metal hydrides. However, currently only a very limited number of materials show satisfying sorption kinetics and cycling behaviour. The most prominent ones are magnesium hydride (MgH_2) and sodium alanate ($NaAlH_4$). In both cases a breakthrough in kinetics could be attained in the late 90s of the last century / the early 21st century.

Magnesium hydride is among the most important and most comprehensively investigated light weight hydrides. MgH_2 itself has a high reversible storage capacity, which amounts to 7.6 wt.%. Furthermore, magnesium is the eighth most frequent element on the earth and thus comparably inexpensive. Its potential usage initially was hindered because of rather sluggish sorption properties and unfavourable reaction enthalpies. The overall hydrogen sorption kinetics of magnesium-based hydrides is as in case of all hydrides mainly determined by the slowest step in the reaction chain, which can often be deduced e.g. by modelling the sorption kinetics (Barkhordarian et al, 2006; Dornheim et al., 2006). Different measures can be taken to accelerate kinetics. One important factor for the sorption kinetics is the micro- or nanostructure of the material, e.g. the grain or crystallite size. Because of the lower packing density of the atoms, diffusion along grain boundaries is usually faster than through the lattice. Furthermore, grain boundaries are favourable nucleation sites for the formation and decomposition of the hydride phase. A second important parameter is the outer dimension of the material, e.g. in case of powdered material, its particle size. The particle size (a) determines the surface area, which is proportional to the rate of the surface reaction with the hydrogen, and (b) is related to the length of the diffusion path of the hydrogen into and out of the volume of the material. A third major factor by which hydrogen sorption is improved in many hydrogen absorbing systems is the use of suitable additives or catalysts. In case of MgH_2 it was shown by Oelerich et al. (Oelerich et al., 2001; Dornheim et al., 2007) that already tiny amounts of transition metal oxides have a huge impact on the kinetics of hydrogen sorption. Using such additives Hanada et al. (Hanada et al., 2007) could show that by using such additives hydrogen uptake in Mg is possible already at room temperature within less than 1 min. The additives often do not just have one single function but multiple functions. Suitable additives can catalyze the surface reaction between solid and gas. Dispersions in the magnesium-based matrix can act as nucleation centres for the hydride or the dehydrogenated phase. Furthermore, different additives, such as liquid milling agents and hard particles like oxides, borides, etc. , can positively influence the particle size evolution during the milling process (Pranzas et al., 2006; Pranzas et al., 2007; Dornheim et al, 2007) and prevent grain i.e. crystallite growth. More detailed information about the function of such additives in MgH_2 is given in (Dornheim et al., 2007). Beyond that, a preparation technique like high-energy ball milling affects both the evolution of certain particle sizes as well as very fine crystallite sizes in the nm range and is also used to intermix the hydride and the additives/catalysts. Thus, good interfacial contact with the light metal hydride as well as a fine dispersion of the additives can be achieved.

As in case of MgH_2 dopants play also an important role in the sorption of Na-Al-hydride, the so-called Na-alanate. While hydrogen liberation is thermodynamically favorable at moderate temperatures, hydrogen uptake had not been possible until in 1997 Bogdanovic et al. demonstrated that mixing of $NaAlH_4$ with a Ti-based catalyst leads to a material, which can be reversibly charged with hydrogen (Bogdanovic, 1997). By using a tube vibration mill of Siebtechnik GmbH Eigen et al. (Eigen et al., 2007; Eigen et al., 2008) showed that upscaling of material synthesis is possible: After only 30 min milling under optimised process conditions in such a tube vibration mill in kg scale, fast absorption and desorption kinetics with charging/discharging times of less than 10 min can be obtained. The operation temperatures of this complex hydride are much lower than compared to MgH_2 and other light weight hydrides. Fast kinetics is achieved at 100 °C to 150 °C which is much less than what is required in case of MgH_2, however, still significantly higher than in case of the conventional hydrides which show only a very limited storage capacity. Such hydride working temperatures offer the possibility for combinations of metal hydride tanks based on these complex hydrides with e. g. combustion engines, high temperature PEM fuel cells or other medium to high temperature fuel cells. However, compared to MgH_2 the gravimetric hydrogen storage capacity is significantly reduced. Having a maximum theoretical storage capacity of about 5.6 wt. % $NaAlH_4$ exhibits a long term practical storage capacity of 3.5-4.5 wt. % H_2 only. Furthermore, in difference to MgH_2 $NaAlH_4$ decomposes in two reaction steps upon dehydrogenation which implies two different pressure plateaus instead of just one:

$$NaAlH_4 \leftrightarrow 1/3\ Na_3AlH_6 + 2/3\ Al + H_2(g) \leftrightarrow NaH + Al + 3/2\ H_2(g) \qquad (19)$$

The first decomposition step has an equilibrium pressure of 0.1 MPa at 30 °C, the second step at about 100 °C (Schüth et al., 2004). A maximum of 3.7 wt.% H_2 can be released during the first desorption step, 5.6 wt.% in total. The remaining hydrogen bonded to Na is technically not exploitable due to the high stability of the respective hydride.

While the reaction kinetics was optimized significantly, the desorption enthalpy of $NaAlH_4$ of 37 kJ/molH_2 and Na_3AlH_6 of 47 kJ/mol H_2 respectively remains a challenge. For many applications even this value which is much below that of MgH_2 is still too large.

3. Tailoring thermodynamics of light weight metal hydrides

While there are plenty of known hydrides with suitable thermodynamics for hydrogen uptake and release at ambient conditions (several bar equilibrium pressure at or nearby room temperature) currently no hydride is known which combines suitable thermodynamics and kinetics with such a high gravimetric storage capacity that a hydrogen storage tank based on such a material could compete with a 700 bar compressed composite vessel in regard to weight. Depending on the working temperature and pressure as well as the reversible gravimetric storage capacity of the selected hydride the achievable capacity of a metal hydride based storage tank is usually better than half of the capacity of the metal hydride bed itself (Buchner & Povel, 1982). Since modern composite pressurized gas tanks meanwhile show gravimetric hydrogen storage capacities of around 4 wt.% according to conservative extrapolations the possible choice of hydrides should be limited to those having the ability to reversibly store at least 6 wt.%H_2. All currently known high capacity hydrides, however, show either too small values of the respective reaction enthalpy and are therefore not reversible or would require several thousand bar hydrogen pressure or alternatively electrochemical loading or on the other hand are too stable and have an equilibrium pressure which around room temperature is much below the required pressures. The value of reaction enthalpy aimed at is between 20 and 30 kJ/mol H_2. Fig. 9 shows the potentially available hydrogen content of some well known hydrides plotted against their hydrogen reaction enthalpies.

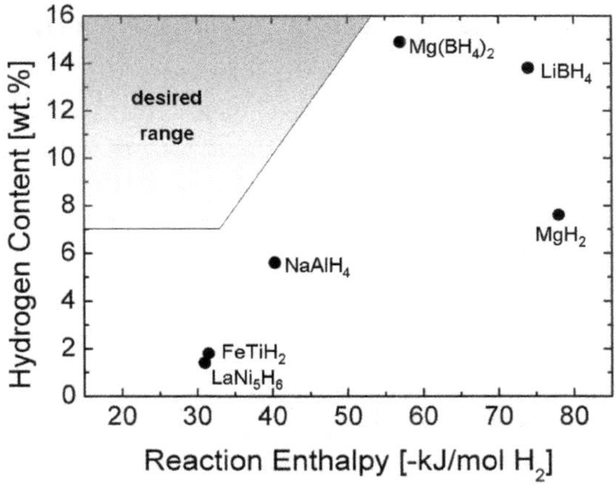

Fig. 9. Theoretically achievable reversible storage capacities and reaction enthalpies of selected hydrides. $LaNi_5H_6$ and $FeTiH_2$ are taken as examples for conventional room temperature hydrides. The reaction enthalpies and achievable hydrogen storage capacities are ΔH = -31 kJ/mol H_2, $C_{H,max}$ = 1.4 wt.% for $LaNi_5H_6$ and for the Fe-Ti system ΔH = -31.5 kJ/mol H_2, $C_{H,max}$ = 1.8 wt.%(average over two reaction steps with $\Delta H(FeTiH_2)$ = -28 kJ/mol H_2 and ΔH(FeTiH) = -35 kJ/mol H_2 respectively) (Buchner, 1982). The respective values for $NaAlH_4$ are ΔH = -40.5 kJ/mol H_2, $C_{H,max}$ = 5.6 wt.%(average over two reaction steps with $\Delta H(NaAlH_4)$ = -37 kJ/mol H_2 and $\Delta H(NaAl_3H_6)$ = -47 kJ/mol H_2 (Bogdanovic et al., 2009)), for MgH_2: ΔH = -78 kJ/mol H_2 (Oelerich, 2000) and $C_{H,max}$ = 7.6 wt.%, for $LiBH_4$: ΔH = -74 kJ/mol H_2 (Mauron, 2008) and $C_{H,max}$ = 7.6 wt.%, for $Mg(BH_4)_2$: ΔH = -57 kJ/mol H_2 (Li, 2008) and $C_{H,max}$ = 14.9 wt.%.

As shown in Fig. 9 none of the plotted hydrides, neither the conventional room temperature hydrides with their rather low gravimetric capacity nor the sophisticated novel chemical hydrides with their unsuitable reaction enthalpy, show the desired combination of properties. Therefore the tailoring of the thermodynamic properties of high capacity light weight and complex hydrides is a key issue, an imperative for future research in the area of hydrides as hydrogen storage materials.

3.1 Thermodynamic tuning of single phase light weight hydrides

The traditional way of tailoring the thermodynamic properties of metal hydrides is by formation of alloys with different stabilities as described in chapter 2.1. Thereby the value of reaction enthalpy can be reduced by stabilising the dehydrogenated state and/or destabilising of the hydride state, see Fig. 10 a. Accordingly, the total amount of reaction enthalpy is increased by destabilising the dehydrogenated state and/or stabilising the hydride, see Fig. 10 b.

This approach has been successfully applied to light weight metal hydrides also.

Mg-based hydrides

One of the first examples using this approach for tuning the thermodynamic properties of light weight metal hydrides was the discovery of the Mg-Ni –system as potential hydrogen storage system by Reilly and Wiswall (Reilly & Wiswall, 1968). Mg_2Ni has a negative heat of

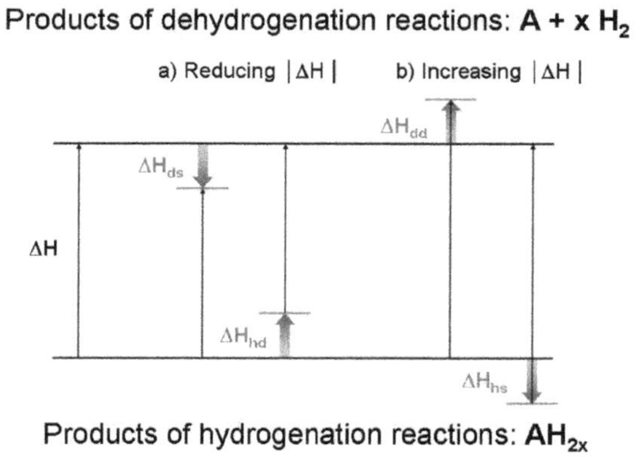

Fig. 10. Tailoring of the reaction enthalpy by altering the stability of the hydrogenated or dehydrogenated state of the metal hydrides: a) Reduction of total reaction enthalpy by stabilising the dehydrogenated phase by ΔH_{ds} or destabilising the hydride phase by ΔH_{hd}. b) Increase of total reaction enthalpy by destabilising the dehydrogenated state by ΔH_{dd} or stabilising the hydrogenated state by ΔH_{hs}.

formation of $H^0_f(Mg_2Ni)$ = -42 kJ/mol. Therefore, compared to pure Mg the dehydrogenated state is stabilised by ΔH_{ds} = -21 kJ/(mol Mg). The enthalpy of formation of Mg_2NiH_4 is $H^0_f(Mg_2NiH_4)$ = -176 kJ/mol (= -88 kJ/(mol Mg)), wherefore the hydride phase is stabilised by ΔH_{hd} = -10 kJ/(mol Mg) if compared to pure MgH_2. In total the hydrogen reaction enthalpy of Mg_2Ni

$$\left|\Delta H_{Mg_2Ni\text{-}H}\right| = \left|\Delta H_{Mg\text{-}H}\right| + \Delta H_{ds} - \Delta H_{hd} \tag{20}$$

is reduced by 11 kJ/mol H_2 to about $|\Delta H(Mg_2Ni\text{-}H)|$ = 67 kJ/mol H_2. While pure MgH_2 exhibits a hydrogen plateau pressure of 1 bar around 300 °C, in case of Mg_2NiH_4 such a plateau pressure is reached already at around 240 °C and in case of further alloying and substituting Ni by Cu at around 230°C in $Mg_2Ni_{0.5}Cu_{0.5}$ (Klassen et al., 1998). Unfortunately, the gravimetric storage capacity of Mg_2NiH_4 is reduced to 3.6 wt.% H_2 only and thus is less than half the respective value in the MgH_2 system. Darnaudery et al. (Darnaudery et al., 1983) were successful to form several quaternary hydrides by hydrogenating $Mg_2Ni_{0.75}M_{0.25}$ with different 3d elements M $\in$ {V, Cr, Fe, Co and Zn} showing stabilities very similar Mg_2NiH_4.

Increasing the amount of 3d metals Tsushio et al. (Tsushio et al., 1998) investigated the hydrogenation of $MgNi_{0.86}M_{0.03}$ with M $\in$ {Cr, Fe, Co, Mn}. Consequently, they observed a dramatic decrease in hydrogen storage capacity to 0.9 wt.% and in hydrogen reaction enthalpy which amounts to 50 kJ/(mol H_2) for $MgNi_{0.86}Cr_{0.03}$. This reaction enthalpy value is in very good agreement with the value 54 kJ/(mol H_2) given by Orimo et al. for amorphous MgNi (Orimo et al., 1998).

Lowering even more the content of Mg Terashita et al. (Terashita et al., 2001) found $(Mg_{1-x}Ca_x)Ni_2$ based alloys desorbing hydrogen at room temperature. They determined the hydride formation enthalpy and entropy of $(Mg_{0.68}Ca_{0.32})Ni_2$ to be H = -37 kJ/(mol H_2) and S = - 94 J/(mol H_2 K) respectively, which is already quite near to the envisioned target.

Unfortunately, with lowering the Mg content the hydrogen storage capacity dropped down to 1.4 wt.% only.
On the other side, as schematically shown in Fig. 10b the absolute value of reaction enthalpy can be increased by either stabilising the hydride phase or destabilising the dehydrogenated phase. In case of Mg-based hydrogen absorbing alloys this is not at all of interest for hydrogen storage itself since MgH_2 is too stable for most hydrogen storage applications , however, this is of interest for other applications like the storage of thermal energy (Dornheim & Klassen, 2009). Mg_2FeH_6 is an example of such materials with increased amount of reaction enthalpy. Furthermore, it is the one with the highest known volumetric hydrogen density which amounts to 150 kg m^{-3}. This enormously high hydrogen density is more than double the value found in case of liquid hydrogen at 20 K and moderate pressures of up to 20 bar (Klell, 2010). The gravimetric storage capacity is 5.6 wt.% and thus still rather high. Since Mg and Fe are immiscible the dehydrogenated state is destabilised compared to pure Mg: $\Delta H_{dd} > 0$ kJ/(mol H_2). Accordingly the hydride phase is more difficult to be synthesised and reversibility as well as long term stability is more difficult to be accomplished.
Nevertheless, hydrogenation is possible at hydrogen pressures of at least 90 bar and temperatures of at least 450 °C (Selvam & Yvon, 1991). Bogdanovic et al. (Bogdanovic et al., 2002) achieved very good reversibility and cycling stability with the hydrogen storage capacities remaining unchanged throughout 550-600 cycles at a level of 5-5.2 wt.% H_2. The reaction enthalpy value is reported to be in between 77 kJ/(mol H_2) and 98 kJ/(mol H_2) (Bogdanovic et al., 2002), (Konstanchuk et al, 1987), (Puszkiel et al., 2008), (Didisheim et al., 1984).
The large reaction enthalpies of MgH_2 and Mg_2FeH_6 lead to weight and volume related heat storage densities in the temperature range of 500 °C which are many times higher than that of the possible sensible or latent heat storage materials (Bogdanovic et al., 2002). The calculated and experimental heat storage densities to weight given by Bogdanovic et al. are 2814 kJ/kg and 2204 kJ/kg for the MgH_2-Mg system and 2106 and 1921 kJ/kg for the Mg_2FeH_6 - 2Mg+Fe system respectively. The corresponding calculated and experimental values for the volumetric thermal energy storage density are 3996 kJ/dm^3 and 1763 kJ/dm^3 for the MgH_2-Mg system and 5758 kJ/dm^3 and 2344 kJ/dm^3 respectively (Bogdanovic et al., 2002). These thermal energy densities ought not to be mistaken with the energy stored in the hydrogen (lower heating value) which is more than a factor of three larger.

Aluminum-based complex hydrides

As Mg_2FeH_6 decomposes during hydrogen release into 2 Mg, Fe and 3 H_2 $NaAlH_4$ decomposes during hydrogen release in 1/3 Na_3AlH_6 + 2/3 Al + H_2 and finally NaH + Al + 3/2 H_2. As written in chapter 2.2 while much lower than those of the Mg-based hydrides the reaction enthalpies of $|\Delta H| = 37$ kJ/(mol H_2) and $|\Delta H| = 47$ kJ/(mol H_2) are still two high for many applications especially for the usage in combination with low temperature PEM fuel cells. $LiAlH_4$ on the other hand is much less stable. It decomposes in two steps as is the case of the $NaAlH_4$:

$$6\,LiAlH_4 \rightarrow 2Li_3AlH_6 + 4Al + 6\,H_2 \underset{?}{\leftrightarrow} 6\,LiH + 6\,Al + 9\,H_2\,. \tag{21}$$

The first reaction step, however, the decomposition of $LiAlH_4$ is found to be exothermic with $\Delta H_{decomposition}$ = -10 kJ/(mol H_2). Since the entropy of decomposition is positive.

Rehydrogenation is not possible at all. The second reaction step, the decomposition of Li_3AlH_6 is endothermic with $\Delta H_{decomposition} = 25$ kJ/(mol H_2). The decomposition of LiH itself takes place at much higher temperatures with $\Delta H = 140$ kJ/(mol H_2) (Orimo et al., 2007). While the second reaction step, the decomposition of Li_3AlH_6 and rehydrogenation of LiH + Al shows rather suitable thermodynamic properties, sluggish kinetics prevent this system so far from being used for hydrogen storage.

To increase the storage capacity and tailor the reaction enthalpy of the $NaAlH_4$ system it is a comprehensible approach to replace some of the Na by Li. Indeed Huot et al. (Huot et al., 1999) proved the existence of Na_2LiAlH_6 and the possible formation by high energy ball-milling of NaH + LiH + $NaAlH_4$. Reversible hydrogen sorption is found to be possible in the Na-Li-Al-H system according to the following reaction:

$$2Na_2LiAlH_6 \leftrightarrow 4NaH + 2LiH + Al + 3H_2 \quad (22)$$

As in case of the pure Na-Al-H system and the Li-Al-H system kinetics can be improved by the addition of transition metal compounds like metal oxides, chlorides and fluorides, see (Ares Fernandez et al., 2007), (Ma et al., 2005) and (Martinez-Franco et al., 2010). However, due to the lack of any stable compound in the dehydrogenated state and the formation of a rather stable hydride the value of reaction enthalpy isn't decreased but increased if compared to the original single Na and Li based aluminium hydrides. Fossdal et al. (Fossdal et al., 2005) has determined the pressure-composition isotherms of TiF_3-doped Na_2LiAlH_6 in the temperature range of 170 °C – 250 °C. They determined the dissociation enthalpy and the corresponding entropy from the Van't Hoff plot: $|DH| = 56$ kJ/(mol H_2) and $S = 138$ J/(K mol H_2). Therefore, instead of a lowering the heat of reaction the opposite is observed. The heat of reaction of the hexa-hydride phase is increased by about 10 kJ/(mol H_2) if compared to the pure Na_3AlH_6 hydride phase.

In 2007 Yin et al. (Yin et al., 2007) presented DFT calculations about the doping effects of TiF_3 on Na_3AlH_6. Their calculations suggested F- substitution for the H-anion leading to a reduction of the desorption enthalpy and therefore for a favourable thermodynamic modification of the Na_3AlH_6 system which was experimentally confirmed by Brinks et al. (Brinks et al., 2008) and Eigen et al. (Eigen et al., 2009).

Borohydrides

Only a very few hydrides show a higher gravimetric storage capacity than MgH_2. For this they must be composed from very light elements. Knowing that Al-containing compounds can form reversible complex metal hydrides it is a reasonable approach to look for Boron-containing compounds as reversible hydrogen storage materials with even higher storage capacity. Borohydrides are known since 1940 when Schlesinger and Brown report about the successful synthesis of $LiBH_4$ by reaction of LiEt and diborane (Schlesinger & Brown, 1940). Despite the early patent from Goerrig in 1958 (Goerrig, 1960) direct synthesis from gaseous H_2 was not possible for long times. Until in 2004 three different groups from the USA (Vajo et al., 2005), South Korea (Cho et al., 2006) and Germany (Barkhordarian et al., 2007) independently discovered that by using MgB_2 instead of pure Boron as starting material formation of the respective borohydrides occurs at rather moderate conditions of 5 MPa H_2 pressure. Orimo et al. (Orimo et al., 2005) reports on the rehydrogenation of previously dehydrogenated $LiBH_4$ at 35 MPa H_2 pressure at 600 °C. Mauron et al. (Mauron et al., 2008) report that rehydrogenation is also possible at 15 MPa. As in case of the Mg-based alloys and the aluminum hydrides the reaction enthalpy of many borohydrides is rather unsuitable for most applications. $LiBH_4$ as one of the most investigated borohydrides with a very high gravimetric hydrogen density of 18.5 wt.% shows an endothermic desorption enthalpy of $|DH| = 74$ kJ/(mol H_2) (Mauron et al., 2008) which is almost the same as in MgH_2. Therefore the tailoring of the reaction enthalpy by substitution is a key issue for these materials as well. As in case of the aluminium hydrides there are two different possibilities for substitution in these complex hydrides: cation substitution and anion substitution. Nakamori et al. (Nakamori et al., 2006) reports about a linear relationship between the heat of formation ΔH_{boro} of $M(BH_4)_n$ determined by first principle methods and the Pauling electronegativity of the cation χ_P:

$$\Delta H_{boro} / \mathrm{kJ}(\mathrm{mol\ BH_4})^{-1} = 248.7\,\chi_P - 390.8 \tag{23}$$

Aiming to confirm their theoretical results the same group performed hydrogen desorption experiments which show that the experimentally determined desorption temperature T_d shows correlates with the Pauling electronegativity χ_P as well, see Fig. 11.

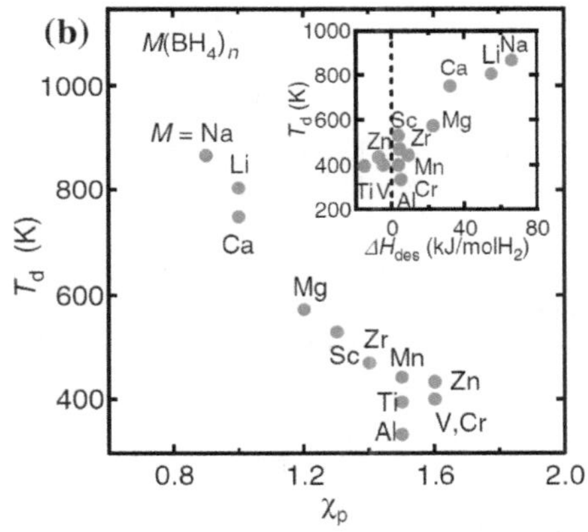

Fig. 11. The desorption temperature T_d as a function of the Pauling electronegativity χ_P and estimated desorption enthalpies ΔH_{des} (Nakamori et al., 2007).

Based on these encouraging results several research groups started to investigate the partial substitution of one cation by another studying several bialkali metal borohydrides. The decomposition temperature of the bialkali metal borohydrides like $LiK(BH_4)_2$ is approximately the average of the decomposition temperature of the mono alkali borohydrides (Rude et al., 2011). Investigations of Li et al. (Li et al., 2007) and Seballos et al. (Seballos et al., 2009) confirmed that this correlation between desorption enthalpy / observed T_d holds true for many double cation $MM'(BH_4)_n$ systems, see Fig. 12.

Several experiments are indicating that transition metal fluorides are among the best additives for borohydrides (Bonatto Minella et al., 2011). While for some cases the function of the transition metal part as additive is understood (Bösenberg et al., 2009; Bösenberg et al., 2010; Deprez et al., 2010; Deprez et al., 2011), the function of F so far remained unclear. DFT calculations performed by Yin et al. (Yin et al., 2008) suggest a favourable modification of hydrogen reaction enthalpy in the $LiBH_4$ system by substitution of the H^--ion with the F^--ion. However, no clear indicative experimental results for F^- -substitution in borohydrides are found yet. In contrast to the F the heavier and larger halides Cl, Br, I are found to readily substitute in some borohydrides for the BH_4^--ion and form solid solutions or stoichiometric compounds and are so far reported to stabilize the hydride phase leading to an increase of the desorption enthalpy $|\Delta H|$ (Rude et al., 2011).

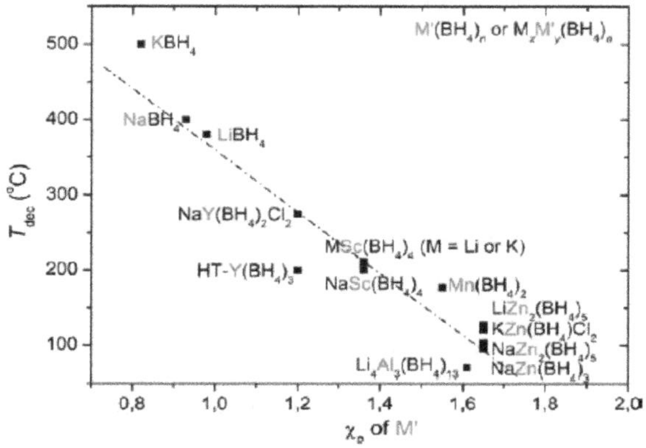

Fig. 12. Decomposition temperatures, T_{dec} for metal borohydrides plotted as a function of the electronegativity of the metal, M'. (Rude et al., 2011)

3.2 Thermodynamic tuning using multicomponent systems: reactive additives and reactive hydride composites

In 1967 Reilly and Wiswall (Reilly & Wiswall, 1967) found another promising approach to tailor reaction enthalpies of hydrides (MH_x) by mixing them with suitable reactants (A):

$$MH_x + yA \leftrightarrow MA_y + \tfrac{x}{2}H_2 \quad (24)$$

They investigated the system $MgH_2/MgCu_2$ which reversibly reacts with hydrogen according to:

$$3\,MgH_2 + MgCu_2 \leftrightarrow 2\,Mg_2Cu + 3H_2 \quad (25)$$

The formation of $MgCu_2$ from Mg_2Cu and Cu is exothermic and thus counteracts the endothermic release of hydrogen. Thereby, the total amount of hydrogen reaction enthalpy is reduced to roughly $|\Delta H| = 73$ kJ/(mol H_2) (Wiswall, 1978). The equilibrium temperature for 1 bar hydrogen pressure is reduced to about 240 °C. In spite of the lower driving force for rehydrogenation, Mg_2Cu is much more easily hydrogenated than pure Mg. A fact found in many other systems like the Reactive Hydride Composites as well.

Aluminum is another example of a reactive additive for MgH_2. The reaction occurs via two steps (Bouaricha et al., 2000):

$$17\,MgH_2 + 12\,Al \leftrightarrow 9\,MgH_2 + 4\,Mg_2Al_3 + 8\,H_2 \leftrightarrow Mg_{17}Al_{12} + 17\,H_2 \quad (26)$$

The system can reversibly store 4.4 wt.% H_2. Since the formation enthalpy ΔH_{Form} of $Mg_{17}Al_{12}$ is -102 kJ/mol the total value of reaction enthalpy of reaction (26) is reduced by ~ 6 kJ/(mol H2) if compared to pure MgH_2. An equilibrium pressure of 1 bar is reached at around 240 °C again.

To further decrease the reaction enthalpy of a Mg-based system a much more stable compound would have to be formed during dehydrogenation. A system investigated by many groups is the MgH_2-Si system. Mg_2Si has an enthalpy of formation of ΔH_{Form} = -79 kJ/mol. Due to the formation of Mg_2Si the value of reaction enthalpy of MgH_2/Si should therefore be reduced by 37 kJ/(mol H_2) to about $|\Delta H|$ = 41 kJ/(mol H_2) (Dornheim, 2010). Theoretically 5 wt.% H_2 can be stored via the reaction

$$2\,MgH_2 + Si \rightarrow Mg_2Si + 4\,H_2 \tag{27}$$

The thermodynamic data indicate a very favourable equilibrium pressure of about 1 bar at 20 °C and 50 bar at 120 °C (Vajo, 2004). While so far rehydrogenation of Mg_2Si was not shown to be possible the system LiH-Si turned out to be reversible. The enthalpy of dehydrogenation of LiH being 190 kJ/(mol H_2) an equilibrium H_2 pressure of 1 bar is reached at 910 °C (Sangster, 2000; Dornheim, 2010). LiH reversibly reacts with Si via a two step reaction with the equilibrium pressure being more than 10^4 times higher and the dehydrogenation enthalpy being reduced by 70 kJ/(mol H_2) (Vajo, 2004).

This approach has recently also been applied to borohydrides. According to Cho et al. (Cho et al., 2006) the decomposition temperature of pure $LiBH_4$ is determined by CALPHAD to 1 bar H_2 pressure at 403 °C while the corresponding equilibrium temperature for the reaction

$$2\,LiBH_4 + Al \leftrightarrow 2\,LiH + AlB_2 + 3\,H_2 \tag{28}$$

is reduced to 188 °C. Kang et al. (Kang et al., 2007) and Jin et al. (Jin et al., 2008) could show that this system indeed is reversible if suitable additives are used.

The only disadvantage of this approach is that the total reversible storage capacity per weight is reduced if something is added to the hydrogen storing material which contains no hydrogen.

The problem of reduced hydrogen capacity by using reactive additives has recently overcome by the approach of the Reactive Hydride Composites (Dornheim, 2006). Thereby, different high capacity hydrogen storage materials are combined which react exothermically with each other during decomposition, see Fig. 13.

One of the first examples of such a system is the $LiNH_2$-LiH system which was discovered by Chen et al. (Chen et al., 2002):

$$LiNH_2 + 2\,LiH \leftrightarrow Li_2NH + LiH + H_2 \leftrightarrow Li_3N + 2\,H_2 \tag{29}$$

However, the value of reaction enthalpy is $|\Delta H|$ = 80 kJ/(mol H_2) and therefore for most applications still much to high. In contrast the system

$$Mg(NH_2)_2 + 2\,LiH \leftrightarrow Li_2Mg(NH)_2 + 2H_2 \tag{30}$$

shows a much more suitable desorption enthalpy of $|\Delta H|$~40 kJ/(mol H_2) with an expected equilibrium pressure of 1 bar at approximately 90 °C (Xiong et al., 2005; Dornheim, 2010).

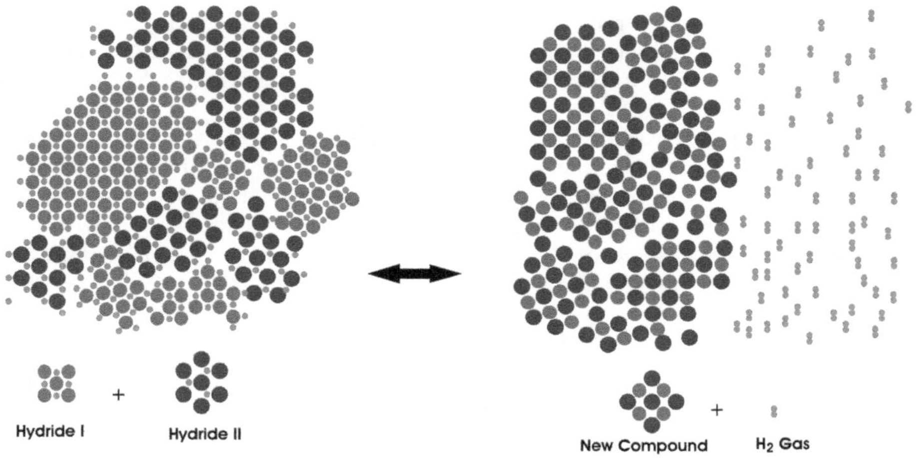

Fig. 13. Schematic of the reaction mechanism in Reactive Hydride Composite.

In 2004 Vajo et al. (Vajo et al., 2005) , Cho et al. (Cho et al., 2006) and Barkhordarian et al. (Barkordarian et al., 2007) independently discovered that the usage of borides especially MgB_2 as a starting material facilitates the formation of different borohydrides. This finding initiated the development and investigation of several new reversible systems with high storage capacities of 8 – 12 wt.% H_2 and improved thermodynamic and kinetic properties such as 2 $LiBH_4$+MgH_2 (Bösenberg et al., 2009; 2010; 2010b), 2 $NaBH_4$+MgH_2 (Garroni et al., 2010; Pistidda et al., 2010; 2011; Pottmaier et al., 2011), $Ca(BH_4)_2$+MgH_2 (Barkhordarian et al., 2008), 6 $LiBH_4$+CeH_2, 6 $LiBH_4$+CaH_2 (Jin et al., 2008b), $LiBH_4/Ca(BH_4)_2$ (Lee et al., 2009) .
One of the most intensely studied systems hereof is the 2 $LiBH_4$ + MgH_2 system. The indended reaction pathway is:

$$2\,LiBH_4 + MgH_2 \leftrightarrow 2\,LiH + MgB_2 + 4\,H_2 \tag{31}$$

However, several other reaction pathways are possible leading to products such as LiB_2, amorphous B, $Li_2B_{12}H_{12}$ or $Li_2B_{10}H_{10}$. Bösenberg et al. (Bösenberg et al., 2010b) could show that due to a higher thermodynamic driving force for the favoured reaction the competing reactions can be suppressed by applying a hydrogen back pressure and limiting the dehydrogenation temperature. Nevertheless, since long-range diffusion of metal atoms containing species is required, see Fig. 13, in bulk ball-milled samples dehydrogenation so far occurs only at temperatures higher than 350 °C, hydrogenation at temperatures higher than 250 °C.
The dehydrogenation temperatures of this Reactive Hydride Composite, however, can be significantly reduced by using nanoconfined 2 $LiBH_4$ + MgH_2 stabilised in inert nanoporous aerogel scaffold materials whereby long-range phase separation is hindered and thus the diffusion path length reduced (Gosalawit-Utke, 2011).

4. Conclusion

Metal hydrides offer a safe and compact alternative for hydrogen storage. The thermodynamic properties of them determine both their reaction heat as well as hydrogen

equilibrium pressure at given temperature and, therefore, are important parameters to be taken into account. Optimised system integration for a given application is not possible without selecting a hydride with suitable thermodynamic properties. To achieve highest possible energy efficiencies the heat of reaction and temperature of operation of the metal hydride should be adapted to the waste heat and temperature of operation of the fuel cell / fuel combustion system. It has been found that the thermodynamic properties of metal hydrides can be tailored in a wide range. Unfortunately, so far all the known conventional metal hydrides with more or less ideal reaction enthalpies and hydrogen equilibrium pressures above 5 bar at room temperature suffer from a rather limited reversible hydrogen storage capacity of less than 2.5 wt.%. With such a material it is not possible to realise a solid storage hydrogen tank with a total hydrogen storage density of more than 1.8 wt.% H_2. Such tank systems still have advantages for the storage of small quantities of hydrogen for larger quantities, however, modern high pressure composite tank shells have a clear advantage in respect of gravimetric storage density. To realise a solid storage tank for hydrogen with a comparable gravimetric storage density it is required that novel hydrogen storage materials based on light weight elements are developed. There are several promising systems with high gravimetric storage densities in the range of 8 - 12 wt.% H_2. For the applications of these novel material systems it is important to further adapt thermodynamic properties as well as the temperatures of operation towards the practical requirements of the system.
The discovery of the approach of combining different hydrides which react with each other during hydrogen release by forming a stable compound, the so-called Reactive Hydride Composites, show a great promise for the development of novel suitable hydrogen storage material systems with elevated gravimetric storage densities. However, so far, the ideal storage material with low reaction temperatures, a reaction heat in the range of $|\Delta H| = 20$-30 kJ/(mol H_2) and a on-board reversible hydrogen storage density of more than 6 wt.% H_2 has not been found.

5. References

Ares Fernandez, J.R.; Aguey-Zinsou, F.; Elsaesser, M.; Ma, X.Z.; Dornheim, M.; Klassen, T.; Bormann, R. (2007). Mechanical and thermal decomposition of LiAlH4 with metal halides. *International Journal of Hydrogen Energy*, Vol. 32, No. 8, pp. (1033-1040), ISSN: 0360-3199

Barkhordarian, G.; Klassen, T.; Bormann, R. (2006). Kinetic investigation of the effect of milling time on the hydrogen sortpion reaction of magnesium catalyzed with different Nb2O5 contents. *Journal of Alloys and Compounds*, Vol. 407, No. 1-2, pp. (249-255), ISSN: 0925-8388

Barkhordarian, G.; Klassen, T.; Dornheim, M.; Bormann, R. (2007). Unexpected kinetic effect of MgB2 in reactive hydride composites containing complex borohydrides. *Journal of Alloys and Compounds*, Vol. 440, No. 1-2, pp. (L18-L21), ISSN: 0925-8388

Barkhordarian, G.; Jensen, T.R.; Doppiu, S.; Bösenberg, U.; Borgschulte, A. ; Gremaud, R.; Cerenius, Y.; Dornheim, M., Klassen, T.; Bormann, R. (2008). Formation of Ca(BH4)2 from Hydrogenation of CaH2+MgB2 Composite. *Journal of Physical Chemistry C*, Vol. 112, No. 7, pp. (2743-2749), ISSN: 1932-7447

Bösenberg, U.; Vainio, U.; Pranzas, P.K.; Bellosta von Colbe, J.M.; Goerigk, G.; Welter, E.; Dornheim, M.; Schreyer, A.; Bormann, R. (2009). On the chemical state and distribution of Zr- and V-based additives in Reactive Hydride Composites. *Nanotechnology*, Vol. 20; No. 20, pp. (204003/1-204003/9) ISSN: 1361-6528

Bösenberg, U.; Kim, J.W.; Gosslar, D.; Eigen, N.; Jensen, T.R.; Bellosta von Colbe, J.M.; Zhou, Y.; Dahms, M.; Kim, D.H.; Guenther, R.; cho, Y.W.; Oh, K.H.; Klassen, T.; Bormann, R.; Dornheim, M. (2010). Role of additives in LiBH4-MgH2 Reactive Hydride Composites for sorption kinetics. *Acta Materialiea*, Vol, 58, No. 9; pp. (3381-3389), ISSN: 1359-6454

Bösenberg, U.; Ravnsbaek, D. B.; Hagemann, H.; D'Anna, V.; Bonatto Minella, C.; Pistidda, C.; van Beek, W.; Jensen, T.R.; Bormann, R.; Dornheim, M. (2010b). Pressure and Temperature Influence on the Desorption Pathway of the LiBH4-MgH2 Composite System. *Journal of Physical Chemistry C*, Vol. 114, No. 35, pp. (15212-15217)

Bogdanovic, B.; Schwickardi, M. (1997). Ti-doped alkali metal aluminum hydrides as potential novel reversible hydrogen storage materials. *Journal of Alloys and Compounds*, Vol. 253-254, pp. (1-9), ISSN: 0925-8388

Bogdanovic, B.; Reiser, A.; Schlichte, K.; Spliethoff, B.; Tesche, B. (2002). Thermodynamics and dynamics of the Mg-Fe-H system and its potential for thermochemical thermal energy storage. *Journal of Alloys and Compounds*, Vol. 345, No. 1-2, pp. (77-89), ISSN: 0925-8388

Bogdanovic, B.; Felderhoff, M.; Streukens, G. (2009). Hydrogen storage in complex metal hydrides. *Journal of the Serbian Chemical Society*, Vol. 74, No. 2, pp. (183-196), ISSN: 0352-5139

Bonatto Minella, Christian; Garroni, Sbastiano; Pistidda, Claudio; Gosalawit-Utke, R.; Barkhordarian, G.; Rongeat, C.; Lindeman, I.; Gutfleisch, O.; Jensen, T.R.; Cerenius, Y.; Christnsen,J.; Baro, M.D.; Bormann, R.; Klassen, T.; Dornheim, M. (2011). Effect of Transition Metal Fluorides on the Sorption Properties and Reversible Formation of $Ca(BH_4)_2$. *Journal of Physical Chemistry C*, Vol. 115, No. 5, pp (2497-2504), ISSN: 1932-7447

Bouaricha, S.; Dodelet, J.P.; Guay, D.; Huot, J. Boily, S.; Schulz, R. (2000). Hydriding behaviour of Mg-Al and leached Mg-Al compounds prepared by high-energy ball-milling. *Journal of Alloys and Compounds*, Vol. 297, pp. (282-293)

Brinks, H.; Fossdal, A.; Hauback, B. (2008). Adjustment of the stability of complex hydrides by anion substitution. *Journal of Physical Chemistry C*, Vol. 112, No. 14; pp. (5658-5661), ISSN: 1932-7447

Buchner, H. (1982). *Energiespeicherung in Metallhydriden*, Springer-Verlag, 3-211-81703-4, Wien

Buchner, H.; Povel, R. (1982). The Daimler-Benz Hydride Vehicle Project. International *Journal of Hydrogen Energy*, Vol. 7, No. 3, pp. (259-266), ISSN: 0360-3199/82/030259-08

Chen, P.; Xiong, Z.T.; Luo, J.Z.; Lin, J; Tan, K.L. (2002). Interaction of hydrogen with metal nitrides and imides. *Nature*, Vol. 420, pp. (302-304)

Cho, Y.W.; Shim, J.-H.; Lee, B.-J. (2006). Thermal destabilization of binary and complex metal hydrides by chemical reaction: A thermodynamic analysis. *CALPHAD: Computer Coupling of Phase Diagrams and Thermochemistry*, Vol. 30, No. 1, pp. (65-69), ISSN: 0364-5916

Darnaudry, J.P.; Darriet, B.; Pezat, M. (1983). The $Mg_2Ni_{0.75}M_{0.25}$ alloys (M = 3d element): their application to hydrogen storage. *International Journal of Hydrogen Energy*, Vol. 8, pp. (705-708)

Deprez, E.; Justo, A.; Rojas, T.C.; Lopez, Cartes, C.; Bonatto Minella, C.; Bösenberg, U.; Dornheim, M.; Bormann, R.; Fernandez, A. (2010) Microstructural study of the LiBH4-MgH2 Reactive Hydride Composite with and without Ti isopropoxide additive. *Acta Materialia*, Vol. 58, No. 17, pp. (5683-5694), ISSN: 1359-6454

Deprez, E.; Munoz-Marquez, M.A.; Jimenez de Haro, M.C.; Palomares, F.J.; Foria, F.; Dornheim, M.; Bormann, R.; Fernandez, A. (2011). Combined x-ray photoelectron spectroscopy and scanning electron microscopy studies of the LiBH4-MgH2 Reactive Hydride Composite with and without a Ti-based additive. *Journal of Applied Physics*, Vol. 109, No. 1, pp. (014913/1-014913/10), ISSN: 0021-8979

Didisheim, J.-J.; Zolliker, P.; Yvon, K.; Fischer, P.; Schefer, J.; Gubelmann, M.; Williams, A.F. (1984). Dimagnesium iron(II) hydride; Mg2FeH6, containing octahedral FeH64-anions. *Inorganic Chemistry*, Vol. 23, No. 13, pp. (1953-1957), ISSN: 0020-1669

Dornheim, M.; Eigen, N.; Barkhordarian, G.; Klassen, T.; Bormann, R. (2006). Tailoring Hydrogen Storage Materials Towards Application. *Advanced Engineering Materials*, Vol. 8, No. 5, pp. (377-385), ISSN: 1438-1656

Dornheim, M.; Doppiu, S.; Barkhordarian, G.; Boesenberg, U.; Klassen, T.; Gutfleisch, O.; Bormann, R. (2007) Hydrogen storage in magnesium-based hydrides and hydride composites. Viewpoint paper in: *Scripta Materialia*, Vol. 56, pp. (841-846), ISSN: 1359-6462

Dornheim, M.; Klassen, T. (2009). High Temperature Hydrides, In: *Encyclopedia of Electrochemical Power Sources*, Vol. 3, J. Garche, C. Dyer, P. Moseley, Z. Ogumi, D. Rand, B. Scrosati, pp. (459-472), Elsevier, ISBN 10: 0-444-52093-7 , Amsterdam

Dornheim, M. (2010). Tailoring Reaction Enthalpies of Hydrides, In: *Handbook of Hydrogen Storage*, Michael Hirscher, pp. (187-214), Wiley-VCH Verlag GmbH & Co, ISBN: 978-3-527-32273-2, Weinheim

Eigen, N.; Keller, C.; Dornheim, M.; Klassen, T.; Bormann, R. (2007). Industrial production of light metal hydrides for hydrogen storage. Viewpoint Set in *Scripta Materialia*, Vol. 56, No. 10, pp. (847-851), ISSN: 1359-6462

Eigen, N.; Gosch, F.; Dornheim, M.; Klassen, T.; Bormann, R. (2008). Improved hydrogen sorption of sodium alanate by optimized processing. *Journal of Alloys and Compounds*, Vol. 465, No. 1-2, pp. (310-316), ISSN: 0925-8388

Eigen, N.; Bösenberg, U.; Bellosta von Colbe, J.; Jensen, T.R.; Cerenius, Y.; Dornheim, M.; Klassen, T.; Bormann, R. (2009). Reversible hydrogen storage in NaF-Al composites. *Journal of Alloys and Compounds*. Vol. 477, No. 1-2, pp. (76-80), ISSN: 0925-8388

Fossdal, A.; Brinks, H.W.; Fonneloep, J.E.; Hauback, B.C. (2005). Pressure-composition isotherms and thermodynamic properties of TiF3-enhanced Na2LiAlH6. *Journal of Alloys and Compounds*, Vol. 397, No. 1-2, pp. (135-139), ISSN:0925-8388

Fujitani, S.; Yonezu, I.; Saito, T.; Furukawa, N.; Akiba, E.; Hayakawa, H.; Ono, S. (1991). Relation between equilibrium hydrogen pressure and lattice parameters in pseudobinary Zr—Mn alloy systems. *Journal of the Less Common Metals*, vol. 172-174, No. 1, pp. (220-230)

Fukai, Y. (1993). The Metal-Hydrogen System. *Springer Series in Materials Science*, Vol. 21, Springer, Berlin

Garroni, S.; Milanese, C.; Girella, A.; Marini, A.; Mulas, G.; Menendez, E.; Pistidda, C.; Dornheim, M.; Surinach, S.; Baro, M. D. (2010). Sorption properties of NaBH4/MH2 (M = Mg, Ti) powder systems. *International Journal of Hydrogen Energy*, Vol. 35, No. 11, pp. (5434-5441)

Goerrig, D. (1960). Borohydrides of alkali and alkaline earth metals. German Patent 1077644, F27373 IVa/12i, Application No. DE 1958-F27373

Gosalawit-Utke, R.; Nielsen, T.K.; Saldan, I.; Laipple, D.; Cerenius, Y.; Jensen, T.R.; Klassen, T.; Dornheim, M. (2011). Nanoconfined 2LiBH4-MgH2 prepared by direct melt infiltration into nanoporous materials. *Journal of Physical Chemistry C*, Vol. 115, No. 21, pp. (10903-10910)

Griessen, R.; Driessen, A.(1984). Heat of formation and band structure of binary and ternary metal hydrides. *Phys. Rev. B*, Vol. 30, No. 8, pp. (4372-81), ISSN: 0163-1829

Hägg, G. (1931). Röntgen investigations on the hydrides of titanium zirconium, vanadium and tantalum. *Zeitschrift für physikalische Chemie* B, Vol. 11, pp. (433-445)

Hanada, N.; Ichikawa, T.; Fujii, H. (2007). Hydrogen absorption kinetics of the catalyzed MgH2 by niobium oxide. *Journal of Alloys and Compounds*, Vol. 446-447, pp. (67-71), ISSN: 0925-8388

Huot, J.; Boily, S.; Güther, V.; Schulz, R. (1999). Synthesis of Na3AlH6 and Na2LiAlH6 by mechanical alloying. *Journal of Alloys and Compounds*, Vol. 283, No. 1-2, pp. (304-306), ISSN: 0925-8388

Jin, S.-A.; Shim, J.-H.; Cho, Y.W.; Yi, K.-W.; Zabara, O.; Fichtner, M. (2008). Reversible hydrogen storage in LiBH4-Al-LiH composite powder. *Scripta Materialia*, Vol. 58, No. 11, pp. (963-965), ISSN: 1359-6462

Jin, S.-A.; Lee, Y.-S.; Shim, J.-H.; Cho, Y.W. (2008b). Reversible Hydrogen Storage in LiBH4-MH2 (M = Ce, Ca) Composites. *Journal of Physical Chemistry C*, Vol. 112, No. 25, pp. (9520-9524), ISSN: 1932-7447

Kang, X.-D.; Wang, P.; Ma, L.-P.; Cheng, H.-M. (2007). Reversible hydrogen storage in LiBH4 destabilized by milling with Al. *Applied Physics A: Materials Science & Processing*, Vol. 89, No. 4, pp. (963-966)

Klassen, T.; Oelerich, W.; Zeng, K.; Bormann, R.; Huot, J. (1998). Nanocrystalline Mg-based alloys for hydrogen storage, In: *Magnesium Alloys and their Applications*, B.L. Mordike and K. U. Kainer, pp. (308-311), Werkstoff-Informationsgesellschaft mbH Frankfurt, CODEN: 68TSA9, Wolfsburg, Germany

Klell, M. (2010). Storage of Hydrogen in the Pure Form, In: *Handbook of Hydrogen Storage*, Michael Hirscher, pp. (187-214), Wiley-VCH Verlag GmbH & Co, ISBN: 978-3-527-32273-2, Weinheim

Konstanchuk, I.G.; Ivanov, E.Y.; Pezat, M.; Darriet, B.; Bodyrev, V.V.; Hagenmüller, P. (1987). The hydrideng properties of a mechanical alloy with composition Mg-25% Fe. *J. Less-Common Met.*, Vol. 131, pp. (181-189)

Lee, J.Y.; Ravnsbaek, D.; Lee, Y.-S.; Kim, Y.; Cerenius, Y., Shim, J.-H.; Jensen, T.R.; Hur, N.H.; Cho, Y.W. (2009). Decomposition Reactions and Reversibility of the LiBH4-Ca(BH4)2 Composite. *Journal of Physical Chemistry C*, Vol. 113, No. 33, pp. (15080-15086), ISSN: 1932-7447

Libowitz, G.G.; Hayes, H.F.; Gibb, T.R.G.Jr. (1958). The system zirconium-nickel and hydrogen. *Journal of Physical Chemistry*, Vol. 62 pp. (76-79), ISSN: 0022-3654

Li, H.-W.; Orimo, S.; Nakamori, Y.; Miwa, K.; Ohba, N.; Towata, S.; Züttel, A. (2007). Materials designing of metal borohydrides: Viewpoints from thermodynamical stabilities. *Journal of Alloys and Compounds*, Vol. 446-447, pp. (315-318)

Li, H.-W.; Kikuchi, K.; Nakamori, Y.; Ohba, N.; Miwa, K.; Towata, S.; Orimo, S. (2008) Dehydriding and Rehydriding Processes of Well-Crystallized Mg(BH4)2 Accompanying with Formation of Intermediate Compounds. *Acta Materialia*, Vol. 56, pp. (1342–1347)

Ma, X.Z.; Martinez-Franco, E.; Dornheim, M.; Klassen, T.; Bormann, R. (2005). Catalyzed Na2LiAlH6 for hydrogen storage. *Journal of Alloys and Compounds*, Vol. 404-406, pp. (771-774), ISSN: 0925-8388

Martinez-Franco, E.; Klassen, T.; Dornheim, M.; Bormann, R.; Jaramillo-Vigueras, D. (2010). Hydrogen sorption properties of Ti-oxide/chloride catalyzed Na_2LiAlH_6. *Ceramic Transactions*, Vol. 209, No. 13-20, pp. (13-20), ISSN: 1042-1122

Mauron, P., Buchter, F., Friedrichs, O., Remhof, A., Bielmann, M., Zwichy, C.N., Züttel, A. (2008) Stability and Reversibility of LiBH4. *J. Phys. Chem. B*, Vol. 112, pp. (906–910)

Nakamori, Y.; Miwa, K.; Ninomiya, A.; Li, H.; Ohba, N.; Towata, S.; Züttel, A.; Orimo, S. (2006). Correlation between thermodynamical stabilities of metal borohydrides and cation electronegativites: First-principles calculations and experiments. *Physical Review B*, Vol. 74, pp. (045126 - 1-9)

Nakamori, Y.; Miwa, K.; Li, H.; Ohba, N.; Towata, S.; Orimo, S. (2007). Tailoring of Metal Borohydrides for Hydrogen Storage Applications. *Mater. Res. Soc. Symp. Proc.* Vol. 971, 0971-Z02-01

Oelerich, W. (2000). *Sorptionseigenschaften von nanokristallinen Metallhydriden für die Wasserstoffspeicherung*, Helmholtz-Zentrum Geesthacht Zentrum für Material- und Küstenforschung GmbH (formerly GKSS-Forschungszentrum Geesthacht GmbH), ISSN 0344-9626, Geesthacht, Germany

Oelerich, W. (2001). Metal oxides as catalysts for improved hydrogen sorption in nanocrystalline Mg-based materials. *Journal of Alloys and Compounds*, Vol.315, No. 1-2, pp. (237-242), ISSN: 0925-8388

Orimo, S.; Ikeda, K.; Fujii, H.; Saruki, S.; Fukunaga, T.; Züttel, A.; Schlapbach, L. (1998). Structural and hydriding properties of (Mg1-xAlx)Ni-H(D) with amorphous or CsCl-type cubic structure (x=0-0.5), *Acta Materialia*, Vol. 46, No. 13, pp. (4519-4525), ISSN: 1359-6454

Orimo, S.; Nakamori, Y.; Kitahara, G.; Miwa, K.; Ohba, N.; Towata, S. Zuettel, A. (2005). Dehydriding and rehydriding reactions of LiBH4. *Journal of Alloys and Compounds*, Vol. 404-430, pp. (0925-8388), ISSN: 0925-8388

Orimo, S.; Nakamori, Y.; Eliseo, J.R.; Züttel, A.; Jensen, C.M. (2007). Complex Hydrides for Hydrogen Storage. *Chemical Reviews*, Vol. 107, No. 10, pp. (4111-4132). ISSN:0009-2665

Panella, B.; Hirscher, M.; Roth, S. (2005). Hydrogen adsorption in different carbon nanostructures. *Carbon*, Vol. 43, pp. (2209-2214)

Pistidda, C.; Garroni, S.; Bonatto Minella, C.; Coci, F.; Jensen, T.R.; Nolis, P.; Bösenberg, U.; Cerenius, Y.; Lohstroh, W.; Fichtner, M.; Baro, M.D.; Bormann, R.; Dornheim, M.

(2010). Pressure Effect on the 2NaH+MgB2 Hydrogen Absorption Reaction. *Journal of Physical Chemistry C*, Vol. 114, No. 49, pp. (21816-21823)

Pistidda, C. Barkhordarian, G.; Rzeszutek, A.; Garroni, S.; Minella, C. Bonatto; Baro, M. D.; Nolis, P.; Bormann, Ruediger; Klassen, T.; Dornheim, M. (2011). Activation of the reactive hydride composite $2NaBH_4$ + MgH_2. *Scripta Materialia*, Vol. 64, No. 11, pp. (1035-1038)

Pottmaier, D.; Pistidda, C.; Groppo, E.; Bordiga, S.; Spoto, G.; Dornheim, M.; Baricco, M. (2011). Dehydrogenation reactions of $2NaBH_4$ + MgH_2 system. International *Journal of Hydrogen Energy*. Vol. 36, No. 13, pp. (7891-7896)

Pourarian, F.; Shinha, V.K.; Wallace, W.E; Smith, H.K. (1982) Kinetics and thermodynamics of $ZrMn_2$-based hydrides. *Journal of the Less Common Metals*, Vol. 88, No. 2, pp. (451-458)

Pranzas, P.K.; Dornheim, M.; Bellmann, D.; Aguey-Zinsou, K.-F., Klassen, T.; Schreyer, A. (2006). SANS/USANS investigations of nanocrystalline MgH2 for reversible storage of hydrogen. *Physica B: Condensed Matter*, Vol. 385-386, No. 1, pp. (630-632), ISSN: 0921-4526

Pranzas, P.K.; Dornheim, M.; Boesenberg, U.; Ares Fernandez, J.R.; Goerigk, G.; Roth, S.V.; Gehrke, R.; Schreyer, A. (2007). Small-angle scattering investigations of magnesium hydride used as a hydrogen storage material. *Journal of Applied Crystallography*, Vol 40, No. S1, pp. (383-387), ISSN: 0021-8898

Puszkiel, J.A.; Larochette, P.A.; Gennari, F.C. (2008). Thermodynamic-kinetic characterization of the synthesized Mg2FeH6-MgH2 hydrides mixture. *International Journal of Hydrogen Energy*, Vol. 33, No. 13, pp. (3555-3560), ISSN: 0360-3199

Reilly, J.J.; Wiswall, R. H. (1967). Reaction of hydrogen with alloys of magnesium and copper. *Inorganic Chemistry*, Vol. 6, pp. (2220-2223), ISSN: 0020-1669

Reilly, J.J.; Wiswall, R.H. (1968). Reaction of hydrogen with alloys of magnesium and nickel and the formation of Mg_2NiH_4. *Inorganic Chemistry*, Vol. 7, pp. (2254-2256), ISSN: 0020-1669

Reilly, J.J., Sandrock, G.D. (1980). Hydrogen storage in metal hydrides. *Scientific American*, Vol. 242, No2, pp. (5118- 5129), ISSN: 00368733

Rude, L.H.; Nielsen, T. K.; Ravnsbaek, D.B.; Bösenberg, U.; Ley, M.B.; Richter, B.; Arnbjerg, L.M.; Dornheim, M.; Filinchuk, Y.; Besenbacher, F.; Jensen, T.R. (2011). Tailoring properties of borohydrides for hydrogen storage: A review. *Phys. Status Solidi A*, DOI: 10.1002/pssa.201001214

Sandrock, G. (1999).A panoramic overview of hydrogen storage alloys from a gas reaction point of view. *Journal of Alloys and Compounds*, Vol. 293-295, pp. (877-888)

Sangster, J.J.; Pelteon, A. D. (2000). In: *Phase Diagrams of Binary Hydrogen Alloys*; Manchester, F.D. ; ASM International: Materials Park, OH, pp. 74

Schlesinger, H.I.; Brown, H.C. (1940). Metallo borohydrides. III. Lithium borohydride. *Journal of the American Chemical Society*, Vol. 62, pp. (3429-3435), ISSN: 0002-7863

Schmitz, B.; Mueller, U.; Trukhan, N.; Schubert, M.; Ferey, G.; Hirscher, M. (2008). Heat of adsorption in microporous high-surface-area materials. *Chem Phys Chem*, Vol. 9, pp. (2181-2184)

Schober, T.; Wenzl, H. (1978). The systems niobium hydride (deuteride), tantalum hydride (deuteride), vanadium hydride (deuteride): structures, phase diagrams, morphologies, methods of preparation, In: *Topics in Applied Physics Vol. 29, Hydrogen in Metals, Vol. 2,* G. Alefeld and J. Völkl, pp. (11-71), Springer, ISSN: 0303-4216, Berlin

Schüth, F.; Bogdanovic, B., Felderhoff, M. (2004). Light metal hydrides and complex hydrides for hydrogen storage. *Chemical Communications,* No. 20, pp. (2249-2258), ISSN: 1359-7345

Seballos, L.; Zhang,J.Z.; Rönnebro, E.; Herberg,J.L.; Majzoub, E.H. (2009). Metastability and crystal structure of the bialkali complex metal borohydrides $NaK(BH_4)_2$. *Journal of Alloys and Compounds,* Vol. 476, pp. (446-450)

Selvam, P.; Yvon, K. (1991). Synthesis of magnesium iron hydride (Mg2FeH6), magnesium cobalt hydride (Mg2CoH5) and magnesium nickel hydride (Mg2NiH4) by high-pressure sintering of the elements , *International Journal of Hydrogen Energy,* Vol. 16, No. 9, pp. (615-617), ISSN: 0360-3199

Terashita, N.; Kobayashi, K.; Sasai, T.; Akiba, E. (2001). Structural and hydriding properties of (Mg1-xCax)Ni2 Laves phase alloys. *Journal of Alloys and Compounds,* Vol. 327, No. 1-2, pp. (275-280), ISSN: 0925-8388

Tsushio, Y.; Enoki, H.; Akiba, E. (1998). Hydrogenation properties of MgNi0.86M10.03 (M1=Cr, Fe, Co, Mn) alloys. *Journal of Alloys and Compounds,* Vol. 281, pp. (301-305).

Vajo, J.J.; Mertens, F.; Ahn, C.C.; Bowman, R.C.Jr.; Fultz, B. (2004). Altering Hydrogen Storage Properties by Hydride Destabilization through Alloy Formation: LiH and MgH2 Destabilized with Si. *Journal of Physical Chemistry B,* Vol. 108, No. 37, pp. (13977-13983), ISSN: 1520-6106

Vajo, J.J.; Skeith, S.L.; Mertens, F. (2005). Reversible Storage of Hydrogen in Destabilized LiBH4. *Journal of Physical Chemistry B,* Vol. 109, No. 9, pp. (3719-3722), ISSN: 1520-6106

Van Mal, H.H.; Buschow, K.H.H.; Miedema, A.R. (1974). Hydrogen absorption in lanthanium-nickel (LaNi5) and related compounds. Experimental observations and their explanation. *Jounal of the Less-Common Metals,* Vol. 35, No. 1, pp. (65-76), ISSN: 0022-5088

Wiswall, R. (1978). Hydrogen Storage in Metals. In: *Topics of Appl. Phys. , Vol. 29 – Hydrogen in Metals II,* G. Alefeld and J. Völkl, pp. (201-242), Springer-Verlag, ISS: 0303-4216, Berlin, Heidelberg, New York

Xiong, Z.T.; Hu, J.J.; Wu, G.T.; Chen, P.; Luo, W.; Gross, K.; Wang, J. (2005). Thermodynamic and kinetic investigations of the hydrogen storage in the Li-Mg-N-H system. *Journal of Alloys and Compounds,* Vol. 398, No. 1-2, pp. (235-239), ISSN: 0925-8388

Yin, L.-C.; Wang, P.; Kang, X.-D.; Sun, C.-H.; Cheng, H.-M. (2007). Functional anion concept: Effect of fluorine anion on hydrogen storage of sodium alanate. *Physical Chemistry Chemical Physics,* Vol. 9, No. 12, pp. (1499-1502), ISSN: 1463-9076

Yin, L.; Wang, P.; Fang, Z.; Cheng, H. (2008). Thermodynamically tuning LiBH4 by fluorine anion doping for hydrogen storage: A density functional study. *Chemical Physics Letters*, Vol. 450, No. 4-6, pp. (318-321), ISSN: 0009-2614

Yoshida, M. ; Akiba, E. (1995). Hydrogen absorbing-desorbing properties and crystal structure of the Zr-Ti-Ni-Mn-V AB2 Laves phase alloys. *Journal of Alloys and Compds.*, Vol. 224, pp. (121-126)

Züttel, A. (2003). Materials for Hydrogen Storage. *Materials Today*, Vol. 6, pp. (24-33), ISSN : 1369 7021

10

Probing Solution Thermodynamics by Microcalorimetry

Gregory M. K. Poon
Department of Pharmaceutical Sciences, Washington State University
U.S.A.

1. Introduction

Solution microcalorimetry has entrenched itself as a major technique in laboratories concerned with studying the thermodynamics of chemical systems. Recent developments in the calorimeter marketplace will undoubtedly continue to popularize microcalorimeters as mainstream instruments. The technology of microcalorimetry has in turn benefited from this trend in terms of enhanced sensitivity, signal stability, physical footprint and user-friendliness. As the popularity of solution microcalorimeters has grown, so has an impressive body of literature on various aspects of microcalorimetry, particularly with respect to biophysical characterizations. The focus of this chapter is on experimental and analytical aspects of solution microcalorimetry that are novel or represent potential pitfalls. It is hoped that this information will aid bench scientists in the formulation and numerical analysis of models that describe their particular experimental systems. This is a valuable skill, since frustrations often arise from uninformed reliance on turnkey software that accompany contemporary instruments. This chapter will cover both differential scanning calorimetry (DSC) and isothermal titration calorimetry (ITC). It targets physical chemists, biochemists, and chemical engineers who have some experience in calorimetric techniques as well as nonlinear regression (least-square analysis), and are interested in quantitative thermodynamic characterizations of noncovalent interactions in solution.

2. Differential scanning calorimetry

DSC measures the heat capacity (C_P) of a sample as the instrument "scans" up or down in temperature. For reversible systems, direct interpretation of the data in terms of thermodynamic parameters requires that chemical equilibrium be re-established much more rapidly than the scan rate. This can be verified by comparing data obtained at different scan rates. For transitions involving a change in molecularity (*e.g.*, self-association/dissociation, ligand binding/unbinding), reversibility can also be confirmed by the lack of hysteresis between heating and cooling experiments. The optimal scan rate is ultimately a compromise between the requirement for reversibility and the desire for reasonable throughput; typically this falls between 0.5 to 1.0 °C/min for most systems in dilute aqueous solutions.

2.1 Experimental conditions for DSC

The observed or apparent thermodynamics of solution systems generally include linked contributions from other solutes in addition to the species of interest. They include buffers, salts, neutral cosolutes, and cosolvents. Of these, the choice of buffer, or any ionizable species in general, must take into account the change in pK_a with respect to temperature *i.e.*, the enthalpy of ionization (ΔH_{ion}). Unless a buffer's ΔH_{ion} is negligibly small, its pK_a will exhibit a temperature dependence, leading to a change in pH of the solution upon heating or cooling. Failure to take this fact into account may introduce significant artifacts into the observed melting behavior. Such changes in pH represent a different issue from any coupled ionization enthalpy arising from the release or uptake of protons associated with the transition of interest.

The direction and extent of the temperature of pH for a buffered solution depends on the sign and magnitude of ΔH_{ion} as well as the concentration of the buffering species. Consider the ionization of a buffer A^Z in the direction of deprotonation to produce its conjugate base B^{Z-1}:

$$A^z \rightleftharpoons B^{z-1} + H^+ \tag{1}$$

According to the van't Hoff equation,

$$\frac{dK_{ion}}{d(1/T)} = \frac{d10^{-pKa}}{d(1/T)} = -\frac{dH_{ion}}{R}, \tag{2}$$

where R is the gas constant, T is absolute temperature, and

$$K_{ion} = \frac{[H^+][B^{z-1}]}{[A^z]}. \tag{3}$$

Thus, for a buffer with a positive (endothermic) ΔH_{ion}, its ionization equilibrium shifts towards deprotonation as temperature increases, leading to a drop in pH. Conversely, the pH of a solution buffered by an exothermic buffer rises with increasing pH.

Table 1 lists several common buffers for aqueous solutions (King 1969; Disteche 1972; Lo Surdo et al. 1979; Kitamura and Itoh 1987; Goldberg et al. 2002). As a group, substituted ammonium compounds exhibit substantial positive values of ΔH_{ion}, making them poor choices for DSC experiments. These compounds include the so-called "Good buffers" (Good et al. 1966) that are prevalent in biochemistry. Among these, Tris, is a particular offender: the pH of a 25 mM solution initially buffered at pH 9.0 drops by more than one pH unit from 0 to 37°C (Poon et al. 2002). In contrast, the ionization of carboxylic acids and their analogues is far less sensitive to temperature. Generally, buffers based on acetate, cacodylate, and phosphate, for example, are preferred choices for DSC experiments.

Another important note relates to polyprotic species such as phosphates, citrates, and borates, whose pK_a also depends markedly on ionic strength. This relationship is quantitatively given by the volume changes of ionization ($\Delta V°$) and interpreted in terms of electrostriction of solvent water molecules. Thus, the addition of salts such as NaCl or guanidinium salts (the latter commonly used to denature proteins) will systematically reduce the pH of a solution buffered by polyprotic acids. The pH of a 0.1 M phosphate buffer at pH 7.2, for example, can fall by 0.5 pH unit upon addition of 0.5 M of NaCl. Molar concentrations of guanidinium hydrochloride will produce an even greater drop. On the other hand, inorganic cosolvents have the opposite effect by affecting the solution dielectric. Of course, once the pH of these buffers is adjusted to a value that is compatible with the apparent pK_a, it will be stable with respect to temperature. As seen in Table 1, it is generally the case that a buffer is either sensitive to temperature or ionic strength in aqueous solution.

Buffer	pK$_a$	$\Delta H°$, kJ mol^{-1}	$\Delta C_p°$, J K^{-1} mol^{-1}	$\Delta V°$, mL mol^{-1} [a]
Acetate	4.756	-0.41	-142	-10.6
Bicine	8.334	26.34	0	-2.0
Bis-tris	6.484	28.4	27	3.1
Cacodylate	6.28	-3	-86	-13.3
	3.128	4.07	-131	-10.7
Citrate	4.761	2.23	-178	-12.3
	6.396	-3.38	-254	-22.3
Glycine	2.351	4	-139	-6.8
	9.780	44.2	-57	
HEPES	7.564	20.4	47	4.8
Imidazole	6.993	36.64	-9	1.8
MES	6.27	14.8	5	3.9
MOPS	7.184	21.1	25	4.7
	2.148	-8	-141	-16.3
Phosphate	7.198	3.6	-230	-25.9
	12.35	16	-242	-36.0
Succinate	4.207	3.0	-121	
	5.636	-0.5	-217	
Tris	8.072	47.45	-142	4.3

[a] Ionization volume at atmospheric pressure at infinite dilution.

Table 1. Thermodynamic properties of ionization by common aqueous buffers

2.2 Analysis of DSC data

A complete DSC experiment consists of matched scans of a sample and a sample-free reference solution. Blank-subtracted data can be empirically analyzed to obtain model-independent thermodynamic parameters. The difference between pre- and post-transition baselines gives the change in heat capacity, ΔC_p. After subtracting a suitable baseline across the transition range, the arithmetic integration of the C_p vs. T trace yields the so-called calorimetric enthalpy:

$$\Delta H_{\text{cal}} = \int_{Ti}^{Tf} C_p dT \,. \tag{4}$$

ΔH_{cal} is the value of the transition enthalpy at the transition temperature, T_m. The entropy at T_m is

$$\Delta S(T_m) = \frac{\Delta H_{\text{cal}}}{T_m} \,. \tag{5}$$

Thus, a single DSC experiment yields the complete thermodynamics of a transition. Model-free determination of thermodynamics, including the direct measurement of ΔC_p, is a unique feature of DSC not possible with optical techniques (such as absorption and fluorescence spectroscopy) which access ΔH via the van't Hoff equation. However, model fitting by least-square analysis can extract considerably more useful information and facilitate quantitative hypothesis testing.

In general, the reference-subtracted DSC data represent the heat capacity of the initial state 0 ($C_{p,0}$) as well as the excess heat capacity function, $\langle \Delta C_p(T) \rangle$:

$$C_p(T) = C_{p,0} + \langle \Delta C_p(T) \rangle . \tag{6}$$

Consider a general model in which the sample undergoes a transition from initial state 0 through intermediates 1, 2, ... , i to the final state n. (One can readily envisage extensions of this model in which a heterotypic complex dissociates into subunits which then go on to further, independent transitions.) The excess heat capacity function is (Privalov and Potekhin 1986)

$$\begin{aligned} \langle \Delta C_p(T) \rangle &= \frac{d\langle \Delta H(T) \rangle}{dT} = \frac{d}{dT}\left[\sum_{i=1}^{n} \Delta H_i(T)\alpha_i(T) \right] \\ &= \sum_{i=1}^{n} \Delta H_i \frac{d\alpha_i(T)}{dT} + \sum_{i=1}^{n} \alpha_i(T)\Delta C_{p,i} \quad , \\ &= \sum_{i=1}^{n} \langle \delta C_{p,i}^{tr}(T) \rangle + \sum_{i=1}^{n} \langle \delta C_{p,i}^{int}(T) \rangle \end{aligned} \tag{7}$$

where ΔH_i and $\Delta C_{p,i}$ are the enthalpy and heat capacity change, respectively, from state 0 to state i, and $\alpha_i(T)$ is the fractional conversion at state i. $\langle \delta C_{p,i}^{tr}(T) \rangle$ and $\langle \delta C_{p,i}^{int}(T) \rangle$ are called "transition" and "intrinsic" heat capacities, respectively. The intrinsic heat capacity, which represents the summed heat capacities of the various species present at T, is the baseline function of the observed DSC trace (Figure 1). Some analytical protocols invite the user to

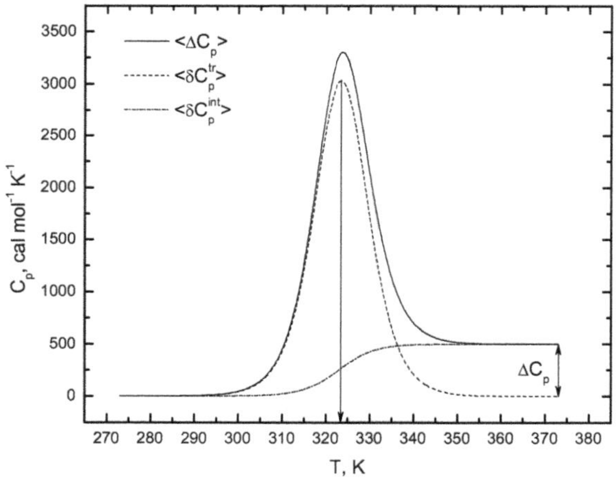

Fig. 1. An excess heat capacity function and its constituent transition and intrinsic heat capacities. Integration of the transition heat capacity, $\langle \delta C_{p,i}^{tr}(T) \rangle$, gives ΔH_{cal} at the transition temperature (50°C here).

perform manual baseline subtraction before fitting a excess heat capacity function. This is intended to eliminate $\langle \delta C_{p,i}^{int}(T) \rangle$ from the fitting function. In the transition region, manual baseline subtraction requires either heuristic or semi-empirical criteria to connect the pre- and post-transitional states. This is both unnecessary and questionable practice, since manual editing may (and probably do) bias the data. The most appropriate approach is to fit both the excess and intrinsic heat capacities directly according to Eq (7). Since both terms are functions of $\alpha_i(T)$, the fitted baseline will objectively track the progress of each transition. Note that $C_{p,0}$ and $\Delta C_{p,i}$ are taken to be constants in Eqs (6) and (7) since their temperature dependence is generally weak over the experimental range. They can be formulated, if desired, as polynomials to define nonlinear baselines. Care must be taken, however, to ensure that such curvature is not masking some low-enthalpy transition such as conformation changes of proteins in the native state (Privalov and Dragan 2007).

2.2.1 Formulation of DSC models

The principal task in formulating DSC models is deriving expressions for $\alpha_i(T)$ from the relevant equilibrium expressions and equations of state. Implicit in this task is the computation of K_i from its corresponding thermodynamic parameters. This in turn requires the choice of a reference temperature, the most convenient of which is the characteristic temperature $T°$ at which $\Delta G(T°) = 0$:

$$\Delta G(T) = \Delta H(T°)\left(1 - \frac{T}{T°}\right) + \Delta C_p\left(T - T° + T\ln\frac{T°}{T}\right). \tag{8}$$

Again ΔC_p is taken to be independent of temperature in the experimental range. The simplest DSC model involves the isomeric conversion of a species in a strictly two-state manner (*i.e.*, no intermediate state is populated at equilibrium). The denaturation of many single-domain proteins exemplifies this model. This excess heat capacity function is

$$\langle \Delta C_p(T) \rangle = \frac{K}{(K+1)^2}\frac{\Delta H^2}{RT^2} + \frac{K}{K+1}\Delta C_p\,. \tag{9}$$

The two terms on the right side represent $\langle \delta C_{p,i}^{tr}(T) \rangle$ and $\langle \delta C_{p,i}^{int}(T) \rangle$, respectively. The DSC traces in Figure 1 are simulated using Eq (9) with ΔH = 50 kcal/mol, ΔC_p = 500 cal mol^{-1} K^{-1}, and $T°$ = 50°C (1 cal ≡ 4.184 J). In this model, $T°$ is the midpoint of the transition (*i.e.*, $K = 1$ and $\alpha = \frac{K}{K+1} = 0.5$) and marks the maximum of the $\langle \delta C_{p,i}^{int}(T) \rangle$ function.

For transitions involving changes in molecularity, $\alpha_i(T)$ includes total sample concentration, c_t in addition to equilibrium constants. While the mechanics of formulating such models is not different, a potential source of inconsistency arises from the choice of unit in thermodynamic parameters. Specifically, every intensive thermodynamic parameter can be defined either per unit of monomer or oligomer. Either choice is correct, of course, but the resultant differences may lead to some confusion. Take for example a two-state homo-oligomeric transition (Privalov and Potekhin 1986; Freire 1989):

$$X_n \xrightleftharpoons{K} nX\,, \tag{10}$$

where K is the equilibrium dissociation constant. Table 2 shows the subtle differences in accounting arising out of the two definitions.

Variable/parameter	Per unit monomer	Per unit oligomer
c_t	$[X] + n[X_n]$	$[X]/n + [X_n]$
α	$\frac{[X]}{c_t}$	$\frac{[X]}{nc_t}$
$\frac{d\alpha}{dT}$	$\frac{n\alpha(1-\alpha)}{n-\alpha(n-1)}\frac{\Delta H^2}{RT^2}$	$\frac{\alpha(1-\alpha)}{n-\alpha(n-1)}\frac{\Delta H^2}{RT^2}$
K	$nc_t^{n-1}\frac{\alpha^n}{1-\alpha}$	$n^n c_t^{n-1}\frac{\alpha^n}{1-\alpha}$
ΔG	$-nRT\ln K$	$-RT\ln K$

Table 2. Equivalent formulations of a two-state homo-oligomeric transition

In the author's experience (Poon et al. 2007), the choice of per unit monomer is more convenient, particularly when oligomers of different molecularities are compared. In addition, it can be seen that K is a polynomial in α of order n. Even in cases where α can be solved explicitly in terms of K and c_t ($n \leq 4$), it is advisable to use numerical procedures such as Newton's method instead to minimize potential algebraic errors and avoid a loss of significance in the fitting procedure.

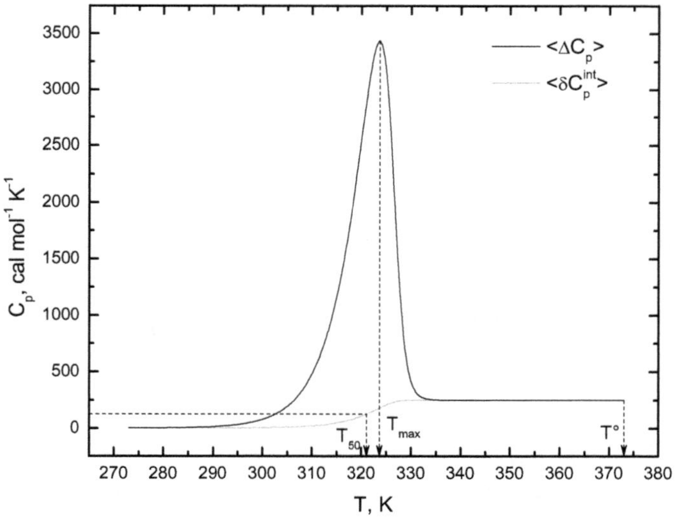

Fig. 2. Two-state dissociation of an homo-oligomer. The traces are simulated for a pentamer (n = 5) where ΔH = 50 kcal mol^{-1}, ΔC_p = 250 cal mol^{-1} K^{-1} and $T°$ = 100°C. All thermodynamics parameters are per unit *monomer*. Note the asymmetry in both heat capacity functions.

An additional consideration for transitions involving changes in molecularity concerns the choice and interpretation of the reference temperature. In contrast with isomeric transitions, $T°$ is neither the midpoint of a transition (in the context of concentrations) nor does it mark the maximum of the transition heat capacity function. Both of the latter temperatures are lower than $T°$. In addition, the midpoint of the transition, T_{50}, is below the temperature of the transition heat capacity maximum, T_m. These relationships are illustrated for the two-state dissociation model in Figure 2. The non-equivalence of T_m and T_{50} also introduces a systematic difference between the calorimetric and van't Hoff enthalpies (ΔH_{vH}, reported at T_m) (Freire 1989; Freire 1995). Moreover, the actual values of T_{50} and T_m are concentration-dependent, and this serves as a diagnostic for a change in molecularity in the transition. For data fitting purposes, $T°$ remains the most efficient choice because it is independent of concentration. After data fitting, estimates of T_m and T_{50} can also be easily obtained from the fitted curve.

Extension of the foregoing discussion applies readily to multi-state transitions. However, an explicit, statistical thermodynamic approach is generally used to derive the required equations for each state as a function of the partition function (Freire and Biltonen 1978). Details in deriving these models have been discussed extensively by Privalov's and Freire's groups (Privalov and Potekhin 1986; Freire 1994). From the standpoint of numerical analysis, it is worth noting that the excess enthalpy is the summed contributions from each state:

$$\langle \Delta C_p \rangle = \frac{d\langle \Delta H(T) \rangle}{dT} = \frac{d}{dT}\left[\sum_{i=1}^{n} \alpha_i(T)\Delta H_i(T) \right] \tag{11}$$

Depending on the number of states considered, the expansion of the derivative on the right side of Eq (11) can be formidable. Commercial programs such as Mathematica (Wolfram Research, Champaign, IL, USA) are thus recommended for symbolic manipulation for all but the most trivial derivations. Less preferably, one can numerically integrate the raw C_p vs. T data and fit $\langle \Delta H(T) \rangle$ directly. There are generally enough data points (at 0.1°C resolution) in an experiment that any loss of resolution should be negligibly small.

3. Isothermal titration calorimetry

As its name indicates, ITC measures the heat change accompanying the injection of a titrant into titrate at constant temperature. In contemporary instruments, this is accomplished by compensating for any temperature difference between the sample and reference cells (the latter lacking titrate, usually just water). The raw ITC signal is therefore power P, a time-dependent variable. Integration with respect to time therefore yields heat q which is the primary dependent variable that tracks the progress of the titration of interest:

$$q(\Delta t) = \int_0^{\Delta t} P\,dt\,. \tag{12}$$

Typically, ITC is operated in incremental mode in which the titrant is injected in preset aliquots after successive re-equilibration periods. A feature of ITC that distinguishes it from most titration techniques is that the measured heat does not accumulate from one injection to the next, but dissipates as the instrument measures the heat signal by returning the sample and reference cells to isothermal conditions. ITC is therefore a differential technique with respect to the concentration of the titrant (X) *i.e.*, the derivative of q with respect to the total titrant concentration, $[X]_t$:

$$\frac{dq}{d[X]_t} = V\sum_{i=1}^{n} \Delta H_i \frac{d[X_i]_b}{d[X]_t}. \tag{13}$$

where $[X_i]_b$ is the concentration of X in the i-th bound state. This contrasts with most other physical binding signals (*e.g.*, absorbance, fluorescence intensity or anisotropy, pH) which are integrative in nature.

3.1 Experimental conditions for ITC

As a thermodynamic tool for studying molecular interactions, the singular strength of ITC is the direct measurement of binding enthalpies. Model-based analysis of ITC data, the subject of Section 3.2, allows the extraction of binding affinity and additional parameters in complex systems. As a label-free technique, ITC compares favorably with other titration techniques such as fluorescence and radioactivity. However, despite much-improved sensitivity (minimum detectable thermal energy <0.1 μJ), baseline stability, and titrant control found in contemporary instruments, sensitivity of ITC is relatively limited. The actual limit of detection depends primarily on the intrinsic enthalpy of the binding system at the temperature of interest, and to a lesser extent, the physical configuration of the instrument. Roughly speaking, a typical ITC experiment requires at least 10^{-6} M of titrate in a 1-mL volume and 10^{-4} M of titrant in a 100-μL syringe. (Recently, so-called "low volume" instruments equipped with 200-μL cells and 50-μL syringes have become available.)

The sensitivity of ITC is helped considerably by the differential nature of its signal (which is proportional to $\frac{dq}{d[X]_t}$) with respect to titrant concentration (Figure 3). In practice, the requirement for sufficient concentration in the sample cell to produce sufficient heat signals poses a direct limit on the tightest binding that may be reliably determined. Specifically, depletion of titrant dominates at titrate concentrations that are high relative to the dissociation constant K. Under this condition, the titration approaches a discontinuity in the first derivative at the equivalence point (Poon 2010). In the case of simple 1:1 binding, an empirical relationship that the product $1 < c_{\text{titrate}}/K < 1000$, where c_{titrate} is the titrate concentration, is optimal for reliable estimation of K (Wiseman et al. 1989). One way of getting around this problem for very tight binding is to lower the apparent value of K by including a suitable, fixed concentration of competitor in the cell. Another possibility is continuous ITC, which will be discussed in Section 3.1.2. It should be noted that even under conditions where K cannot be determined, ΔH can still be determined by integration of the measured heat:

$$\Delta H = \frac{1}{c_{\text{titrate}}} \int_0^V \frac{q(v)}{v} dv, \tag{14}$$

where V is the volume of the (fixed) sample cell. (A volume correction is generally necessary; see Section 3.2.1.3.) A requirement for Eq (14) is, of course, that the titration is complete. Whichever the case, repeating the titration at different temperatures provides an estimate of the change in heat capacity by Kirchoff's relation:

$$\Delta C_p = \frac{d\Delta H(T)}{dT} \quad (15)$$

If the interaction under investigation occurs in a buffered solution, the earlier discussion on the temperature dependence of pK_a would again be relevant. More generally, binding that involves coupled uptake or release of protons will contain the buffer's enthalpy of ionization in the apparent binding enthalpy (Fisher and Singh 1995).

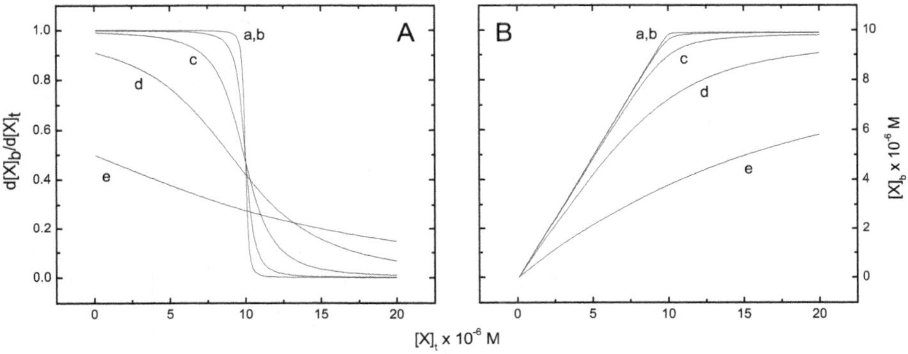

Fig. 3. The differential nature of ITC data. The volume and enthalpy are intentionally omitted to illustrate the differential nature of ITC data with respect to total titrant concentration. Simulated 1:1 binding to 10 μM titrate is shown in Panel *A*, with the corresponding integrated binding curves in Panel *B*. The values of *K* are as follows: a, 10^{-9} M; b, 10^{-8} M; c, 10^{-7} M; d, 10^{-6} M; and e, 10^{-5} M. The values of $c_{titrate}/K$ range from 10^4 to 1.

Finally, ITC has been used as an "analog" of DSC for studying the stability of complexes. A concentrated solution of complex in the syringe is titrated into pure water or buffer in the cell. The resulting dilution drives complex dissociation and the attendant enthalpy is measured. This method has been used to characterize complexes through the spectrum of stoichiometries, from dimers (Burrows et al. 1994; Lovatt et al. 1996) to higher oligomers (Lassalle et al. 1998; Luke et al. 2005) to polymeric species (Stoesser and Gill 1967; Arnaud and Bouteiller 2004). Again, given the sensitivity of ITC, relatively high concentrations are required, so this technique is limited to the measurement of relatively weak complexes. It has been shown (Poon 2010) that the ITC data can be used to diagnose a dimeric or higher-order complex based on the presence of an inflection point in the latter.

3.1.1 Baseline signals

Two types of baselines are operative in an ITC experiment. One is instrument noise. Drifts on the order of 0.02 μW/h are routinely achievable in contemporary instruments. Another source of baseline arises from the injection of titrant. At the very least, viscous mixing makes a measurable if small exothermic contribution to the observed heat. This effect can be observed in a blank-to-blank injection (Figure 4), and serves as a casual useful indicator of the cleanliness of the cell and syringe between sample runs. Moreover, any mismatch in the matrices of the titrant and titrate will be manifest as a dilution enthalpy with each injection.

For small molecules, solids or lyophilized samples are usually dissolved in water or buffer. To complicate matters, hydrophobic solutes often require a cosolvent such as DMSO or DMF to achieve initial solubility before addition of the aqueous solvent; dilution of the cosolvent will therefore contribute to the observed heat at each injection. In other cases, ionizable solutes can perturb solution pH due to their substantial (>10^{-3} M) concentration in the syringe. Unless the solvent is adequately buffered, the pH in the cell and syringe will differ significantly and neutralization heats will contribute to the observed signal.

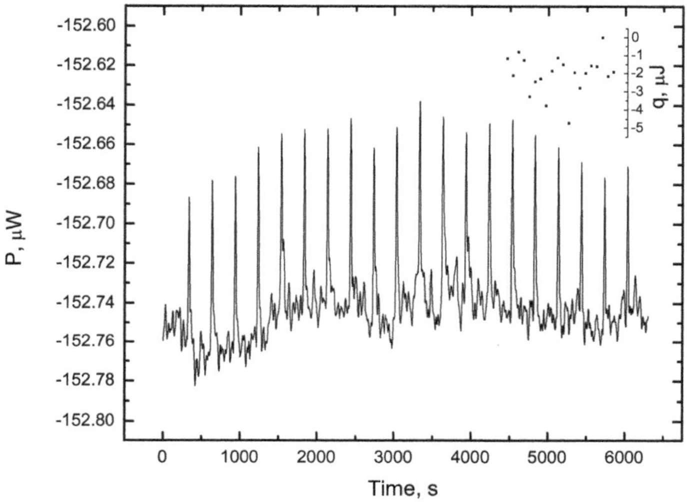

Fig. 4. Typical water-water control "titration." *Inset,* integrated areas of the water peaks. Given the low signal-to-noise ratios, the values must be considered very crude. However, the scatter also attests to the cleanliness of the cell and syringe.

In the case of macromolecular titrant and titrates, the solution matrices can be closely matched by extensive co-dialysis in the same solution. Care must be taken, however, with ionic polymers such as nucleic acids. These solutes can alter the distribution of ions in their compartment during dialysis due to the Donnan effect. Specifically, a non-diffusible polyionic solute excludes diffusible ions of the same charge from their compartment and therefore induces an asymmetric distribution of diffusible ions across the semipermeable membrane at equilibrium. To illustrate, for a simple system consisting only of the non-diffusible polyion M and a monovalent salt AB, the ratio concentrations of A^+ or B^- across the membrane is (Cantor and Schimmel 1980):

$$r_D = \frac{z[M]}{2[AB]_t} + \sqrt{1 + \left(\frac{z[M]}{2[AB]_t}\right)^2} \equiv \Psi + \sqrt{1+\Psi^2}\ , \tag{16}$$

where z is the charge on the polyion (shown simply as M for brevity). At the concentrations required for the syringe, z[M] can be substantial. For example, a duplex oligonucleotide consisting of 20 base-pairs represents at 100 μM contributes 4 mM in total anions (phosphates). The so-called Donnan ratio, r_D, is significantly above unity even at low polyion:salt ratios (Ψ). Eq (16) shows that a ten-fold excess of salt concentrations (Ψ=0.1) leads to a 10% exclusion of A or B from the compartment occupied by polyion. If AB is a buffer salt, the result is also a change of pH. Such asymmetry will modify the heat observed by ITC. As is well known, and illustrated by Eq (16), Donnan effects can be suppressed by ensuring a sufficient excess of diffusible salt in the dialysate. Unfortunately, the required salt concentrations may interfere with the investigation of interactions mediated substantially by electrostatic effects (as is usually the case for polyions). If possible, therefore, it is advisable to arrange for the polyionic species to be in the cell, where concentrations are lower.

In practical terms, none of these baselines effects are significant if the heat generated by the interaction of interest is sufficiently strong. This is not always achievable, however, for several reasons. Availability or solubility of the sample, particularly biological samples, may be limiting. It may also be desirable, for example, to perform titrations at a range of cell concentrations for binding to oligomerizing systems. Characterization of binding to a polyion at low salt concentration may require a reduced concentration. Thus, strategies for handling relatively low signal-to-noise situations are helpful in many situations. The most basic of these involve inspection and, where necessary, manual editing of the power baseline to mitigate the occasion excursion due to instrumental noise. To this end, an increase in the time between injection may be indicated to unambiguously identify the restoration of baseline. In addition, the residual heats (which may be up to 10 μJ) after the equivalence point are unlikely to be negligible. Subtraction with data from a titrant-to-blank run would likely introduce more noise into the data and be no more helpful than taking a simple average of the final post-equivalence heats. If the data is to be fitted to models, a more appropriate solution is to add a constant parameter B to the fitting equation:

$$\frac{dq_{\mathrm{obsd}}}{d[\mathrm{X}]_{\mathrm{t}}} = V\sum_{i=1}^{n}\Delta H_i \frac{d[\mathrm{X}_i]_{\mathrm{b}}}{d[\mathrm{X}]_{\mathrm{t}}} + B\,. \tag{17}$$

3.1.2 Continuous ITC titrations

While incremental titrations most commonly performed in ITC, an alternative mode of operation is a continuous titration (cITC) (Markova and Hallén 2004). In cITC, the titrant is continuously into injected the cell at a low rate (~0.1 μL/s). The primary advantage of cITC is throughput. An incremental ITC experiment typically requiring 20 injections of 5 μL at intervals of 300 s takes (neglecting time for baseline stabilization) 100 min; at 0.1 μL/s, cITC would require approximately 17 min. Another potential motivation for cITC is increased resolution in terms of model-dependent analysis. In incremental ITC, peak-by-peak integration of thermal power is performed to obtain $\frac{dq}{d[\mathrm{X}]_{\mathrm{t}}}$. This step reduces the number of collected data points (typically 1 s^{-1} over 10^3 s, or about one to two hours) for nonlinear regression to the number of injections (usually 10 to 30). By using the thermal power data directly for fitting, cITC can in principle discriminate the curvature required to define tight binding.

Maintenance of quasi-equilibrium conditions throughout the titration is essential to correct interpretation of derived thermodynamic parameters and is a major concern for cITC. To this end, the stirring rate in cITC must be considerably increased (up to 700 rpm) relative to incremental ITC to facilitate mixing of titrant into the titrate solution. Additionally, the instrument must be able to provide sufficient thermal compensation during the titration to maintain isothermal conditions. Finally, the kinetics of the interaction of interest must be fast relative to the injection rate. Generally speaking, these criteria are most likely met by relatively high-affinity interactions with moderate binding enthalpies.

3.2 Analysis of ITC data

The direct measurement of ΔH is considered a significant advantage over non-calorimetric binding methods since the latter access the binding thermodynamics indirectly in terms of the equilibrium constant K via the van't Hoff equation:

$$\frac{d\ln K}{d(1/T)} = -\frac{\Delta H_{\mathrm{vH}}}{R}. \tag{18}$$

Estimation of ΔC_{p} by non-calorimetric methods, therefore, involves taking a second derivative of the measured data. Another potential source of difficulty is the interpretation of K, which is model-dependent. As in DSC, however, ITC becomes a considerably more useful analytical technique when model fitting is used for parameter estimation and hypothesis testing. Moreover, agreement between the directly-fitted, calorimetric ΔH and the van't Hoff ΔH_{vH} obtained from Eq (18) is a strong indication of the physical correctness of the model at hand.

The most fundamental concept in the analysis of ITC data is the differential nature of the heat signal with respect to titrant concentration (Figure 3). Recalling Eq (13),

$$\frac{dq}{d[\mathrm{X}]_{\mathrm{t}}} = V\sum_{i=1}^{n}\Delta H_i \frac{d[\mathrm{X}_i]_{\mathrm{b}}}{d[\mathrm{X}]_{\mathrm{t}}}.$$

Integration of this fundamental equation gives the is the functional form of conventional binding models:

$$q([\mathrm{X}_i]_{\mathrm{b}}) = V\sum_{i=1}^{n}\Delta H_i [\mathrm{X}_i]_{\mathrm{b}}([\mathrm{X}_i]_{\mathrm{t}}) \tag{19}$$

The ideal approach to fitting ITC data is to directly fit Eq (13) (Poon 2010). For simple models, it is possible to write analytical expressions for $\frac{d[\mathrm{X}_i]_{\mathrm{b}}}{d[\mathrm{X}]_{\mathrm{t}}}$ explicitly in terms of $[\mathrm{X}]_{\mathrm{t}}$ i.e., the functional form of Eq (19). Take, for example, simple 1:1 binding of $\mathrm{X} + \mathrm{Y} \rightleftharpoons \mathrm{XY}$:

$$[\mathrm{X}]_{\mathrm{t}}[\mathrm{Y}]_{\mathrm{t}} - ([\mathrm{X}]_{\mathrm{t}} + [\mathrm{Y}]_{\mathrm{t}} + K)[\mathrm{X}]_{\mathrm{b}} + [\mathrm{X}]_{\mathrm{b}}^2 = 0$$

$$[\mathrm{X}]_{\mathrm{b}} = \frac{[\mathrm{X}]_{\mathrm{t}} + [\mathrm{Y}]_{\mathrm{t}} + K - \sqrt{([\mathrm{X}]_{\mathrm{t}} + [\mathrm{Y}]_{\mathrm{t}} + K)^2 - 4[\mathrm{X}]_{\mathrm{t}}[\mathrm{Y}]_{\mathrm{t}}}}{2}, \tag{20}$$

where $K = \frac{[\mathrm{X}][\mathrm{Y}]}{[\mathrm{XY}]} = \frac{[\mathrm{X}][\mathrm{Y}]}{[\mathrm{X}]_{\mathrm{b}}}$ is the equilibrium dissociation constant. Its derivative is

$$\frac{d[\mathrm{X}]_{\mathrm{b}}}{d[\mathrm{X}]_{\mathrm{t}}} = \frac{1}{2} + \frac{1 - \Phi - r}{2\sqrt{(\Phi + 1 + r)^2 - 4\Phi}}, \tag{21}$$

where $r = K/[X]_t$ and $\Phi = [X]_t/[Y]_t$. Eq (21) is sometimes referred to as the Wiseman isotherm (Wiseman et al. 1989). As complexity of the model increases, however, the algebra involved rapidly becomes prohibitive. In any case, numerical methods provide the means for generating solutions for Eq (13).

3.2.1 Numerical aspects of ITC data analysis

Given the usual practice of formulating models in terms of total or unbound concentrations, rather than their derivatives, conventional ITC data analysis has handled the differential nature of calorimetric data by fitting to a finite-difference version of Eq (13). Thus, for the *j*-th injection:

$$\Delta q_j = q_j - q_{j-1} - \delta q_j, \tag{22}$$

where δq_j is a volume-correction factor. This approach has the apparent advantage that models which have been formulated in terms of $[X]_b$ vs. $[X]_t$ can be used directly. However, Eq (22) represents poor practice in nonlinear regression. Since each computation of Δq_j requires evaluation of q corresponding to two consecutive injections, j and j-1, the data points (and their errors, albeit small) are not independent. Independence of observations constitutes a major assumption of nonlinear regression, one which the recursive form of Eq (22) clearly violates. Specifically, the residual in Δq_j during fitting becomes increasing correlated with increasing j. In addition, the resultant propagation of error likely violates the assumption of homoscedasticity (uniform variance) as well. Unless specialized regression techniques are employed (such as correlated least-squares), violations of these assumptions potentially calls the errors of the parameters, and possibly the parameters themselves, into question.

From the perspective of numerical analysis, Eq (22) is also entirely unnecessary. As stated previously, the most appropriate approach to handling ITC data is to fit Eq (13) directly. Posed in the form of Eq (13), ITC models (more specifically, the solution of $\frac{d[X]_b}{d[X]_t}$) represent classic initial value problems (IVPs). IVPs are first-order ordinary differential equations (ODEs) with a specific initial condition ($[X]_b = 0$ at $[X]_t = 0$) for which numerical methods for their solution are well-established. In addition avoiding the statistical pitfalls of Eq (22), formulating ITC models as differential equations simplify the algebra considerably. This is because implicit differentiation offers a welcome shortcut that obviates the need for an explicit solution for $[X]_b$. This is illustrated for $X + Y \rightleftharpoons XY$ with Eq (20):

$$\begin{gathered}
[Y]_t - ([Y]_t + [X]_t + K)\frac{d[X]_b}{d[X]_t} - [X]_b + 2[X]_b\frac{d[X]_b}{d[X]_t} = 0 \\
\frac{d[X]_b}{d[X]_t}(2[X]_b - [Y]_t - [X]_t - K) = [X]_b - [Y]_t \\
\frac{d[X]_b}{d[X]_t} = \frac{[X]_b - [Y]_t}{2[X]_b - [Y]_t - [X]_t - K}
\end{gathered} \tag{23}$$

Recognizing that $[X]_t$, $[Y]_t$ are constants (and K being the parameter to be estimated), Eq (23) takes on a noticeable simpler (but equivalent) form compared to the Wiseman isotherm, Eq (21). More importantly, implicit differentiation always yields an explicit ODE even when no analytical expression for $[X]_b$ exists *e.g.*, polynomials of order >4. Direct substitution of Eq (23) into Eq (13) directly yields the ODE needed to generate values for nonlinear regression. This approach of formulating explicit titration models for ITC has been demonstrated for many empirical models in common use, including the multi-site model, homotropic cooperativity, and two-state self-association (Poon 2010). To further illustrate the this approach, a competitive binding model will be examined. This is a method of measuring tight binding by ITC by adding a competitive species in the cell to reduce the apparent affinity (Sigurskjold 2000). The mechanism is an example of the general multi-site model involving two ligands competing for a single site (Wells 1992). The equilibrium distribution of bound titrate is a function of the affinities of the titrant X_1 and competitor X_2 for the titrate Y as well as the total concentrations of all three species. For the titrant-titrate complex, X_1Y:

$$a_0 + a_1[X_1Y] + a_2[X_1Y]^2 + a_3[X_1Y]^3 = 0\,, \tag{24}$$

where

$$\begin{aligned} a_0 &= -K_2[X_1]_t^2[Y]_t \\ a_1 &= [X_1]_t\{K_1(K_2 + [X_2]_t - [Y]_t) + K_2([X_1]_t + 2[Y]_t)\} \\ a_2 &= K_1^2 - K_2(2[X_1]_t + [Y]_t) - K_1(K_2 - [X_1]_t + [X_2]_t - [Y]_t) \\ a_3 &= K_2 - K_1 \end{aligned}$$

and K_1 and K_2 are the equilibrium dissociation constants of X_1 and X_2 for the titrate, respectively. The corresponding expression for the competitor-titrate complex, X_2Y, can be obtained from symmetry arguments:

$$b_0 + b_1[X_2Y] + b_2[X_2Y]^2 + b_3[X_2Y]^3 = 0\,, \tag{25}$$

where

$$\begin{aligned} b_0 &= -K_1[X_2]_t^2[Y]_t \\ b_1 &= [X_2]_t\{K_2([X_1]_t - [Y]_t) + K_1(K_2 + [X_2]_t + 2[Y]_t)\} \\ b_2 &= K_2(K_2 - [X_1]_t + [X_2]_t + [Y]_t) - K_1(K_2 + 2[X_2]_t + [Y]_t) \\ b_3 &= K_1 - K_2 \end{aligned}.$$

It is possible to solve the cubic equation in Eqs (24) and (25) explicitly, followed by differentiation of the solutions to obtain $\frac{d[X_1Y]}{d[X_1]_t}$ and $\frac{d[X_2Y]}{d[X_1]_t}$ as was done with the Wiseman isotherm. All of this is avoided, however, by implicit differentiation with respect to $[X_1]_t$ which directly yields the required derivatives:

$$\begin{aligned} \frac{d[X_1Y]}{d[X_1]_t} &= \frac{-\left(\frac{da_2}{d[X_1]_t}[X_1Y]^2 + \frac{da_1}{d[X_1]_t}[X_1Y] + \frac{da_0}{d[X_1]_t}\right)}{3a_3[X_1Y]^2 + 2a_2[X_1Y] + a_1} \\ \frac{d[X_2Y]}{d[X_1]_t} &= \frac{-\left(\frac{db_2}{d[X_1]_t}[X_2Y]^2 + \frac{db_1}{d[X_1]_t}[X_2Y] + \frac{db_0}{d[X_1]_t}\right)}{3b_3[X_2Y]^2 + 2b_2[X_2Y] + b_1} \end{aligned}. \tag{26}$$

The observed heat is now the sum of the enthalpy of unbinding of X_2 from Y and the binding of X_1 to Y:

$$\frac{dq}{d[\mathrm{X}]_\mathrm{t}} = V\left(\Delta H_{\mathrm{X1Y}}\frac{d[\mathrm{X_1Y}]}{d[\mathrm{X_1}]_\mathrm{t}} + \Delta H_{\mathrm{X2Y}}\frac{d[\mathrm{X_2Y}]}{d[\mathrm{X_1}]_\mathrm{t}}\right). \quad (27)$$

As shown in Figure 5, a judicious choice of competitor reduces the apparent affinity of the titrant to a range more amenable for regression. In this case, the initial condition for $[X_2Y]$ is *not* zero because the titrate is essentially pre-equilibrated with competitor before any titrant has been injected. The initial value for numerical integration is supplied by Eq (20):

$$[\mathrm{X_2Y}] = \frac{[\mathrm{X_2}]_\mathrm{t} + [\mathrm{Y}]_\mathrm{t} + K_2 - \sqrt{([\mathrm{X_2}]_\mathrm{t} + [\mathrm{Y}]_\mathrm{t} + K_2)^2 - 4[\mathrm{X_2}]_\mathrm{t}[\mathrm{Y}]_\mathrm{t}}}{2}.$$

Thus, the competitive model requires prior knowledge of both the concentration and affinity of the competitor. Since the competitor would therefore require its own characterization, a weak competitor is preferred, which means it would need to be present at significant concentrations (i.e., $[Y]_t \geq K$). Since binding of the titrate involves the unbinding of the competitor, the enthalpy of competitor unbinding may be substantially convoluted in the observed heat as indicated by Eq (27).

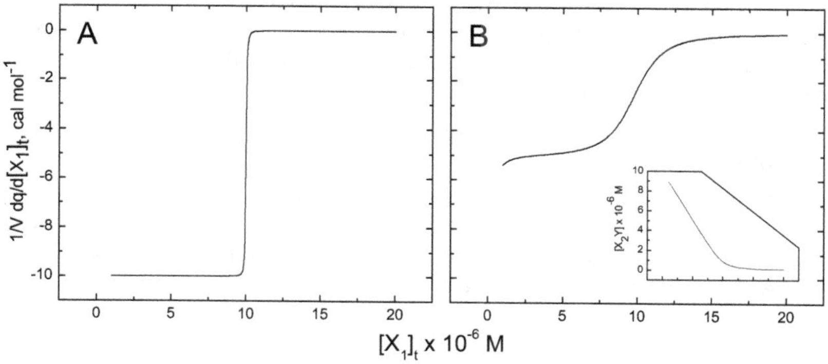

Fig. 5. Titrant-titrate binding in the presence of a competitor. *A*, simulated 1:1 titrant-titrate binding in the absence of competitor. The parameters are $K_1 = 10^{-10}$ M, $\Delta H_1 = -10$ kcal mol^{-1}, and $[Y]_t = 10$ μM. Data representing this level of affinity would be unfit for model-fitting; $K_1/[Y]_t = 10^4$. *B*, in the presence a competitor, X_2, at 10 μM ($K_2 = 10^{-8}$ M, $\Delta H_2 = -5$ kcal mol^{-1}). *Inset*, titration of the competitor.

Explicit titration models are also amenable to formulating models for continuous ITC. In the case of cITC, thermal power P is directly used in model fitting, so Eq (19) needs to be differentiated with respect to *time*:

$$P = \frac{dq(t)}{dt} = \frac{d}{dt}\left\{V\sum_{i=1}^{n}\Delta H_i[\mathrm{X}_i]_\mathrm{b}(t)\right\} = V\left[\sum_{i=1}^{n}\Delta H_i\frac{d[\mathrm{X}_i]_\mathrm{b}(t)}{dt}\right]. \quad (28)$$

Applying the chain rule and the relation $[\mathrm{X}_i]_{\mathrm{tot}} = \frac{c_{\mathrm{syr}} v_{\mathrm{inj}}}{V}$, where c_{syr} is the concentration of titrant in the syringe and v_{inj} is the injection volume,

$$\frac{d[\mathrm{X}_i]_{\mathrm{b}}}{dt} = \frac{d[\mathrm{X}_i]_{\mathrm{b}}}{d[\mathrm{X}_i]_{\mathrm{tot}}} \frac{d[\mathrm{X}_i]_{\mathrm{tot}}}{dt} = \frac{d[\mathrm{X}_i]_{\mathrm{b}}}{d[\mathrm{X}_i]_{\mathrm{tot}}} \left(\frac{c_{\mathrm{syr}}}{V} \frac{dv_{\mathrm{inj}}}{dt} \right). \quad (29)$$

(The subscript "tot" has been used to denote total concentration to avoid ambiguity with the variable t for time.) Note that $\frac{dv_{\mathrm{inj}}}{dt}$ is the (constant) injection rate. Substituting into Eq (28) gives

$$P = c_{\mathrm{syr}} \frac{dv_{\mathrm{inj}}}{dt} \left[\sum_{i=1}^{n} \Delta H_i \frac{d[\mathrm{X}_i]_{\mathrm{b}}}{d[\mathrm{X}_i]_{\mathrm{tot}}} \right]. \quad (30)$$

Thus explicit expressions of $\frac{d[\mathrm{X}_i]_{\mathrm{b}}}{d[\mathrm{X}_i]_{\mathrm{tot}}}$ can be directly used as in incremental ITC. At a sufficiently low injection rate, Eq (30) has the potential of "flattening out" the titration by transforming it in the time domain. This feature, in addition to the higher density of data points available for regression, may allow cITC to characterize much tighter binding than is practicable with incremental ITC (Markova and Hallén 2004).

In the foregoing discussion, the need to correct for various displacement and dilution effects due to the injection process has not been considered. In the author's experience, this is best handled during preliminary data reduction, before nonlinear regression. This aspect will be discussed in Section 3.2.1.3.

3.2.1.1 Choice of dependent variable

Another benefit of formulating ITC models as explicit ODEs is the flexibility in the choice of dependent variable for implicit differentiation, as long as it is a function in $[\mathrm{X}]_{\mathrm{t}}$. For the 1:1 binding model, formulation in terms of the unbound titrant X gives the (and simple) functional form of the familiar Langmuir isotherm:

$$[\mathrm{X}]_{\mathrm{b}} = \frac{[\mathrm{X}][\mathrm{Y}]_{\mathrm{t}}}{[\mathrm{X}] + K}. \quad (31)$$

Applying the chain rule of calculus,

$$\frac{d[\mathrm{X}]_{\mathrm{b}}}{d[\mathrm{X}]_{\mathrm{t}}} = \frac{d[\mathrm{X}]_{\mathrm{b}}}{d[\mathrm{X}]} \frac{d[\mathrm{X}]}{d[\mathrm{X}]_{\mathrm{t}}} = \frac{[\mathrm{X}]K}{([\mathrm{X}] + K)^2} \frac{d[\mathrm{X}]}{d[\mathrm{X}]_{\mathrm{t}}}. \quad (32)$$

At the same time, using the equation of state $[\mathrm{X}] = [\mathrm{X}]_{\mathrm{t}} - [\mathrm{X}]$,

$$\frac{d[\mathrm{X}]}{d[\mathrm{X}]_{\mathrm{t}}} = \frac{d([\mathrm{X}]_{\mathrm{t}} - [\mathrm{X}]_{\mathrm{b}})}{d[\mathrm{X}]_{\mathrm{t}}} = 1 - \frac{d[\mathrm{X}]_{\mathrm{b}}}{d[\mathrm{X}]_{\mathrm{t}}}. \quad (33)$$

Substituting the results from Eq (23) into Eq (33),

$$\frac{d[X]}{d[X]_t} = 1 - \frac{[X]_b - [X]_t}{2[X]_b - [X]_t - [Y]_t - K} = \frac{[X] + K}{2[X] - [X]_t + [Y]_t + K} \tag{34}$$

The simultaneous equations (32) and (34) represent another formulation of the same model, except now [X] is the explicit dependent variable instead of $[X]_b$. (The initial condition is [X] = 0 at $[X]_t$ = 0.) Of course, we have previously derived Eq (23) directly, so this approach is regressive for this simple model. However, The flexibility to use any dependent variable of $[X]_t$ is useful, for example, for models that are formulated in terms of the binding polynomial which is based on [L] (Schellman 1975; Freire et al. 2009).

3.2.1.2 Practical considerations in implementing explicit ITC models

Successful implementation of Eq (13) requires numerical procedures for solving IVPs. The explicit, closed-form ODEs encountered in most models are typically ratios of polynomials. These functions are generally amenable any of the standard Runge-Kutta methods, which are widely available. A fast CPU is helpful, but not required. To this end, the tolerance for iteration should not be unnecessarily stringent in relation to the concentrations used and the value of *K* expected. Generally, a value of 10^{-6} will suffice. Even average single-core CPUs will handle ODEs of practical complexity with reasonable dispatch. Any significant delay is almost always related to input/output issues and can be alleviated by suppressing intermediate output. Numerical failures, if they occur, usually does so when there are two or more very different scales of the independent variable on which the dependent variable is changing *e.g.*, extremely tight binding to one site in the multi-site model. One such example is competitive binding in the presence of a very strong competitor in Eq (27). These pathological scenarios are associated with "stiff" differential equations that require more specialized algorithms for numerical solution. Fortunately, these situations are unlikely to be encountered as they are usually incompatible with experimental data in the first place.

Several technical software suites, such as Mathematica (Wolfram Research, Champaign, IL, USA), MATLAB (the MathWorks, Natick, MA USA), Maple (Maplesoft, Waterloo, Ontario, Canada), and IgorPro (WaveMetrics, OR, USA) which have built-in numerical ODE and least-square minimization capabilities, represent full-featured, integrated solutions. Alternatively, pre-compiled libraries containing optimized algorithms for numerical ODEs and least-square minimization are available commercially (the Numerical Algorithm Group Library; Numerical Algorithms Group [NAG], Oxford, UK) or free (GNU Scientific Library [GSL]) for most computing platforms. Functions from these libraries can be called under standard programming environments (*e.g.*, C/C++, FORTRAN) to perform the required procedures. Some commercial data analysis software such as Origin (Northampton, MA, USA) can interface with external libraries such as the NAG Library or GSL to perform numerical ODEs within their least-squares routines. If access to external libraries is not available, an adequate alternative is to code a numerical ODE algorithm (such as Runge-Kutta-Fehlberg) as part of the target function within the data fitting routines of the analysis program. "Cookbook" recipes for a variety of ODE solvers are straightforward and can be found in most texts of numerical analysis (Press 2007).

3.2.1.3 Volume correction

The sample and reference cells are typically overfilled for both DSC and ITC. Overfilling maximizes heat transfer between the solution and the wall of the cell as air is a poor thermal conductor. In the case of ITC, overfilling also minimizes stray signal arising from mechanical agitation of the solution-air-cell interface caused by the stirring paddle.

However, the introduction of titrant into an overfilled ITC sample cell leads to displacement effects that need to be taken into account. Specifically, each injected volume of titrant simultaneously displaces an equal volume of titrate and any previously injected titrant out of the sample cell (into the access tube). The accounting for these displaced volumes and their effect on titrant and titrate concentration is made on the assumption that the displaced material is immediately and completely excluded from the titration. This implies that no mixing occurs between the injected and displaced materials at the time injection. The concentrations of the titrant X after the i-th injection in the cell is therefore

$$[\mathrm{X}]_{\mathrm{t},i} = \frac{v_{\mathrm{inj}} c_{\mathrm{syr}} + [\mathrm{X}]_{\mathrm{t},i\text{-}1}(V - v_{\mathrm{inj}})}{V}, \quad [\mathrm{X}]_{\mathrm{t},0} = 0, \tag{35}$$

where V is the cell volume and v_{inj} is the injection volume. The corresponding concentrations of the titrate Y is

$$[\mathrm{Y}]_{\mathrm{t},i} = \frac{[\mathrm{Y}]_{\mathrm{t},i\text{-}1}(V - v_{\mathrm{inj}})}{V}. \tag{36}$$

There are two ways to handle volume corrections. One is to incorporate Eqs (35) and (36) as additional terms in the fitting equation. In the author's experience (Poon 2010), it is more efficient instead to perform the volume corrections on the dataset at the outset, and simply treat $[\mathrm{Y}]_{\mathrm{t},i}$ as an additional dependent variable in the least-square procedure.

3.2.1.4 Error analysis in ITC

Compared to other titrations, particularly in the biochemical laboratory, that requires extensive manual manipulation *e.g.*, electrophoretic mobility shift, filter binding, ITC instrumentation offers a greatly reduced level of statistical error in the measured data. Nonetheless, detailed theoretical and experimental studies of the nature and magnitude of statistical errors in ITC have offered insight into how instrumental errors can be minimized in fitted parameters. Although such studies have so far only focused on 1:1 binding, it has become clear that at least two régimes of instrumental errors exist (Tellinghuisen 2003; 2005b). Specifically, for titrations associated with relatively large heats (>300 μcal or "high-q"), proportional errors in the injected volumes dominate, and the inclusion of statistical weights is indicated to optimize fitted parameters. For "low-q" titrations, which typify low-concentration titrations needed for tight binding interactions, constant errors in thermal detection and compensation dominate, and unweighted fitting does not adversely affect parameter optimization. In either case, the optimal number of injections is considerably lower than the norm of 20 to 25 injections, especially if variable volume procedures are employed (Tellinghuisen 2005a). This approach could substantially reduce titration time and increase experimental throughput. It may be pointed out, however, that in some applications, sample-to-sample variation may be greater than any statistical error inherent in the analytical technique. In these cases, standard errors in parameters estimated from replicate experiments will be higher but more representative indicators of experimental uncertainty than the fitting error extracted from the variance-covariance matrix.

4. Conclusion

Commercial development of microcalorimetry has greatly increased the accessibility of this technique for the thermodynamic characterization of chemical systems in solution.

Unfortunately, the "black-box" nature of commercial software has engendered unwarranted reliance by many users on the turnkey software accompanying their instruments, and an attendant tendency to fit data to models of questionable relevance to the actual chemistry. This chapter discusses several novel aspects and potential pitfalls in the experimental practice and analysis of both DSC and ITC. This information should enable users to tailor their experiments and model-dependent analysis to the particular requirements.

5. Acknowledgement

Financial support by the College of Pharmacy at Washington State University is acknowledged.

6. References

Arnaud, A., and Bouteiller, L. (2004). Isothermal titration calorimetry of supramolecular polymers. *Langmuir* 20: 6858-6863

Burrows, S.D., Doyle, M.L., Murphy, K.P., Franklin, S.G., White, J.R., Brooks, I., McNulty, D.E., Scott, M.O., Knutson, J.R., and Porter, D. (1994). Determination of the monomer-dimer equilibrium of interleukin-8 reveals it is a monomer at physiological concentrations. *Biochemistry* 33: 12741-12745

Cantor, C.R., and Schimmel, P.R. (1980). *Biophysical Chemistry: the behavior of biological macromolecules*. W. H. Freeman, 0716711915, San Francisco, USA

Disteche, A. (1972). Effects of pressure on the dissociation of weak acids. *Symp Soc Exp Biol* 26: 27-60

Fisher, H.F., and Singh, N. (1995). Calorimetric methods for interpreting protein-ligand interactions. *Methods Enzymol* 259: 194-221

Freire, E. (1989). Statistical thermodynamic analysis of the heat capacity function associated with protein folding-unfolding transitions. *Comments Mol Cell Biophys* 6: 123-140

Freire, E. (1994). Statistical thermodynamic analysis of differential scanning calorimetry data: Structural deconvolution of heat capacity function of proteins. *Methods Enzymol* 240: 502-530

Freire, E. (1995). Thermal denaturation methods in the study of protein folding. *Methods Enzymol* 259: 144-168

Freire, E., and Biltonen, R.L. (1978). Statistical mechanical deconvolution of thermal transitions in macromolecules. I. Theory and application to homogeneous systems. *Biopolymers* 17: 463-479

Freire, E., Schön, A., and Velazquez-Campoy, A. (2009). Isothermal Titration Calorimetry: General Formalism Using Binding Polynomials. *Methods Enzymol* 455: 127-155

Goldberg, R.N., Kishore, N., and Lennen, R.M. (2002). Thermodynamic Quantities for the Ionization Reactions of Buffers. *J Phys Chem Ref Data* 31: 231-370

Good, N.E., Winget, G.D., Winter, W., Connolly, T.N., Izawa, S., and Singh, R.M.M. (1966). Hydrogen Ion Buffers for Biological Research. *Biochemistry* 5: 467-477

King, E.J. (1969). Volume changes for ionization of formic, acetic, and butyric acids and the glycinium ion in aqueous solution at 25°C. *J Phys Chem* 73: 1220-1232

Kitamura, Y., and Itoh, T. (1987). Reaction volume of protonic ionization for buffering agents. Prediction of pressure dependence of pH and pOH. *J Solution Chem* 16: 715-725

Lassalle, M.W., Hinz, H.J., Wenzel, H., Vlassi, M., Kokkinidis, M., and Cesareni, G. (1998). Dimer-to-tetramer transformation: loop excision dramatically alters structure and stability of the ROP four alpha-helix bundle protein. *J Mol Biol* 279: 987-1000

Lo Surdo, A., Bernstrom, K., Jonsson, C.A., and Millero, F.J. (1979). Molal volume and adiabatic compressibility of aqueous phosphate solutions at 25.degree.C. *J Phys Chem* 83: 1255-1262

Lovatt, M., Cooper, A., and Camilleri, P. (1996). Energetics of cyclodextrin-induced dissociation of insulin. *Eur Biophys J* 24: 354-357

Luke, K., Apiyo, D., and Wittung-Stafshede, P. (2005). Dissecting homo-heptamer thermodynamics by isothermal titration calorimetry: entropy-driven assembly of co-chaperonin protein 10. *Biophys J* 89: 3332-3336

Markova, N., and Hallén, D. (2004). The development of a continuous isothermal titration calorimetric method for equilibrium studies. *Anal Biochem* 331: 77-88

Poon, G.M. (2010). Explicit formulation of titration models for isothermal titration calorimetry. *Anal Biochem* 400: 229-236

Poon, G.M., Brokx, R.D., Sung, M., and Gariépy, J. (2007). Tandem Dimerization of the Human p53 Tetramerization Domain Stabilizes a Primary Dimer Intermediate and Dramatically Enhances its Oligomeric Stability. *J Mol Biol* 365: 1217-1231

Poon, G.M., Gross, P., and Macgregor, R.B., Jr. (2002). The sequence-specific association of the ETS domain of murine PU.1 with DNA exhibits unusual energetics. *Biochemistry* 41: 2361-2371

Press, W.H. (2007). *Numerical recipes : the art of scientific computing,* 3rd ed. Cambridge University Press, 0521880688, Cambridge, UK ; New York, USA

Privalov, P.L., and Dragan, A.I. (2007). Microcalorimetry of biological macromolecules. *Biophys Chem* 126: 16-24

Privalov, P.L., and Potekhin, S.A. (1986). Scanning microcalorimetry in studying temperature-induced changes in proteins. *Methods Enzymol* 131: 4-51

Schellman, J.A. (1975). Macromolecular binding. *Biopolymers* 14: 999-1018

Sigurskjold, B.W. (2000). Exact analysis of competition ligand binding by displacement isothermal titration calorimetry. *Anal Biochem* 277: 260-266

Stoesser, P.R., and Gill, S.J. (1967). Calorimetric study of self-association of 6-methylpurine in water. *J Phys Chem* 71: 564-567

Tellinghuisen, J. (2003). A study of statistical error in isothermal titration calorimetry. *Anal Biochem* 321: 79-88

Tellinghuisen, J. (2005a). Optimizing experimental parameters in isothermal titration calorimetry. *J Phys Chem B* 109: 20027-20035

Tellinghuisen, J. (2005b). Statistical error in isothermal titration calorimetry: variance function estimation from generalized least squares. *Anal Biochem* 343: 106-115

Wells, J.W. (1992). Analysis and interpretation of binding at equilibrium. In: *Receptor-Ligand Interactions: a Practical Approach.* E.C. Hulme(Ed., pp. 289-395. IRL Press at Oxford University Press, 0199630909, Oxford, England; New York, USA

Wiseman, T., Williston, S., Brandts, J.F., and Lin, L.N. (1989). Rapid measurement of binding constants and heats of binding using a new titration calorimeter. *Anal Biochem* 179: 131-137

Permissions

The contributors of this book come from diverse backgrounds, making this book a truly international effort. This book will bring forth new frontiers with its revolutionizing research information and detailed analysis of the nascent developments around the world.

We would like to thank Juan Carlos Moreno-Piraján, for lending his expertise to make the book truly unique. He has played a crucial role in the development of this book. Without his invaluable contribution this book wouldn't have been possible. He has made vital efforts to compile up to date information on the varied aspects of this subject to make this book a valuable addition to the collection of many professionals and students.

This book was conceptualized with the vision of imparting up-to-date information and advanced data in this field. To ensure the same, a matchless editorial board was set up. Every individual on the board went through rigorous rounds of assessment to prove their worth. After which they invested a large part of their time researching and compiling the most relevant data for our readers. Conferences and sessions were held from time to time between the editorial board and the contributing authors to present the data in the most comprehensible form. The editorial team has worked tirelessly to provide valuable and valid information to help people across the globe.

Every chapter published in this book has been scrutinized by our experts. Their significance has been extensively debated. The topics covered herein carry significant findings which will fuel the growth of the discipline. They may even be implemented as practical applications or may be referred to as a beginning point for another development. Chapters in this book were first published by InTech; hereby published with permission under the Creative Commons Attribution License or equivalent.

The editorial board has been involved in producing this book since its inception. They have spent rigorous hours researching and exploring the diverse topics which have resulted in the successful publishing of this book. They have passed on their knowledge of decades through this book. To expedite this challenging task, the publisher supported the team at every step. A small team of assistant editors was also appointed to further simplify the editing procedure and attain best results for the readers.

Our editorial team has been hand-picked from every corner of the world. Their multi-ethnicity adds dynamic inputs to the discussions which result in innovative outcomes. These outcomes are then further discussed with the researchers and contributors who give their valuable feedback and opinion regarding the same. The feedback is then collaborated with the researches and they are edited in a comprehensive manner to aid the understanding of the subject.

Apart from the editorial board, the designing team has also invested a significant amount of their time in understanding the subject and creating the most relevant covers. They scrutinized every image to scout for the most suitable representation of the subject and create an appropriate cover for the book.

The publishing team has been involved in this book since its early stages. They were actively engaged in every process, be it collecting the data, connecting with the contributors or procuring relevant information. The team has been an ardent support to the editorial, designing and production team. Their endless efforts to recruit the best for this project, has resulted in the accomplishment of this book. They are a veteran in the field of academics and their pool of knowledge is as vast as their experience in printing. Their expertise and guidance has proved useful at every step. Their uncompromising quality standards have made this book an exceptional effort. Their encouragement from time to time has been an inspiration for everyone.

The publisher and the editorial board hope that this book will prove to be a valuable piece of knowledge for researchers, students, practitioners and scholars across the globe.